"This excellent volume enriches the study of bioethics by critically assessing the principles of bioethics and by articulating African insights and contributions, which are relational and holistic, foster social justice and solidarity, and are environmentally concerned. Readers will appreciate the engagement with philosophical and theological scholarship, in Africa and beyond. The book's culturally situated approach to bioethics will contribute to addressing some of the ongoing bioethical challenges that plague our world today."

—ANDREA VICINI, SJ, professor of bioethics, Boston College

"Jude Buyondo's voice bursts onto the scene with energy and a gift for synthesis. The text cuts through vapid appeals to the 'global' by demonstrating the power of Black African voices for complementing bioethics poised to alter clinical action. The bibliography alone is worth the price of entry for Anglo-American audiences, but even more delightful is the playful intellectual parrying in the venerable tradition of critical scholarship—highly recommended."

—CYRUS P. OLSEN III, associate professor of theology/religious studies, University of Scranton

"Jude Buyondo searches for a more comprehensive and inclusive approach to bioethics in a unique modus that responds to different lacunae presented both in Western principlism and the African approach to bioethics. His systematic way of thought does not take any approach as superior but rather brings the Western and African approaches to bioethics into a possible integral link. Buyondo opens our eyes to pursue this vital connection if we are to conceive a more inclusive and comprehensive approach to bioethics."

—RICHARD RWIZA, faculty of theology, Catholic University of Eastern Africa

"This is an extraordinary contribution to intercultural dialogue on bioethics. Jude Buyondo refrains from separating African ethics from Western ethics by claiming its 'Africanness.' The author argues that bioethical principles are common ground but, in order to be fruitfully applied in Africa, need to be interpreted in the light of the African view of the connectedness of all life. What is more, such an interpretation allows African ontologically founded solidarity to complement and permeate Western bioethical principles."

—SIGRID MÜLLER, faculty of Catholic theology, University of Vienna

The Critique *of* Bioethical
Principlism *in* Contrast *to a*
Black African Approach *to* Bioethics

The Critique *of* Bioethical
Principlism *in* Contrast *to a*
Black African Approach *to* Bioethics

JUDE THADDAEUS BUYONDO
Foreword by Sigrid Müller

WIPF & STOCK · Eugene, Oregon

THE CRITIQUE OF BIOETHICAL PRINCIPLISM IN CONTRAST TO A
BLACK AFRICAN APPROACH TO BIOETHICS

Wipf & Stock
An Imprint of Wipf and Stock Publishers
199 W. 8th Ave., Suite 3
Eugene, OR 97401

www.wipfandstock.com

PAPERBACK ISBN: 979-8-3852-1744-1
HARDCOVER ISBN: 979-8-3852-1745-8
EBOOK ISBN: 979-8-3852-1746-5

08/14/24

To my little Jude, my Caroline, and my refined twin Maria, who endlessly breathed life into my work, day, and night, and in whom I constantly found that missing piece of myself. To my darling mum, Rose and dad, Gerald, thank you for the finest chance and such an indelible seed of knowledge you planted in me, and may our memorable dad savor our anticipated ancestral immortality.

"Oluganda nkovu terugwa kumubiri."

Kinship or solidarity is like a scar, you can never loose.

BAGANDA WISDOM

Contents

Foreword by Sigrid Müller   ix

Preface   xi

Acknowledgments   xiii

List of Abbreviations and Acronyms   xv

1   In Search of a More Comprehensive and Inclusive Approach to Bioethics   1

PART I | CRITIQUE OF BIOETHICAL PRINCIPLISM IN AN AFRICAN CONTEXT

2   General Critique of Bioethical Principlism   19

3   Critique of Bioethical Principlism in an African Context   49

PART II | AFRICAN MORAL THOUGHT: AN AFRICAN INTERPRETATION OF BIOETHICS

4   An African Interpretation of Bioethics   87

5   The Criterion of the Normative Structure of the Black African Ethical System   153

6   Conclusion: Résumé of Part II and Conclusive Indications of a Comprehensive and Inclusive Approach to Bioethics   235

Bibliography   247

Foreword

In promoting a more comprehensive and richer functional bioethics, Dr. Buyondo proposes a complementary contribution to Western principlism from a Black African approach to bioethics. For him all life is connected and can be understood outside the hospital setting—and in view of the current global challenges to the climate and the environment, global relationships, and intercultural bioethical interventions, his insight is now also becoming increasingly imperative in Western countries. However, while the interconnections between the continents are made visible by exploitive relationships in bioethical interventions and broad threatening events, in African culture, knowledge of the social interactions traditionally has a positive contextual benefit. It testifies to a lived attitude of integration and concern toward fellow human beings and their relationships, animals, plants, and forces of nature as a whole. But how can a Western culture, more developed in technology, and an African culture, with a strong metaphysical attitude and background of interconnectedness, be brought into a dynamic, fruitful conversation on an equal footing? How can there be a respectful and inspiring exchange between seemingly very different ways of thinking? What can we do to contribute to a better exchange of ideas between the two cultural worlds? How can hegemonic asymmetrical relationships convert into symmetrical genuine respectful partnerships? Listening is certainly a first step toward a sincere encounter. I therefore hope that this book will inspire many readers to engage in this genuine dialogue.

Prof. Chair Dr. Sigrid Müller
Chair, Department of Systematic Theology and Ethics
Faculty of Catholic Theology, University of Vienna
Vienna, April 25, 2024

Preface

This study investigates the problem of bioethical principlism (as majorly proposed by Thomas L. Beauchamp and James F. Childress) within African contexts south of the Sahara (Bantu people), citing how principlistic interpretations of bioethics don't sit well in diverse Sub-Saharan African contexts. Taking various criticisms (raised within the African scholarship of Godfrey B. Tangwa, Bénézet Bujo, Richard Rwiza, etc.) and weaknesses of the African moral thought itself seriously, the book shows what is missing for better-functioning global bioethical relationships. The study claims, on the one hand, that although there is the universality of fundamental principles, appropriated relevance and applicability ought to conform to concrete local realities since principles cannot be valid everywhere without contextual concretization. On the other hand, the study claims that local treasuries of knowledge ought to equally contribute and inform the universalized approach, i.e., bioethical principlism. The investigation makes use of systematic and hermeneutical methods to re-appraise scientific scholarships not only in order to bridge gaps between the universalized approach of *bioethical principlism* and *local approaches to bioethics*, but also suggests a Black African possible contribution so that a global approach to bioethics can be gradually unfolded to *bring together, collectively enrich each other*, while *amply complementing each other*, rather than *taking apart* the African approach to bioethics from a universally oriented principlism.

My curiosity to bridge this gap in knowledge originates from what has struck me since my childhood. I continuously used to see Bantu people making use of every entity, force, and their integral social reality as part of their broader medicinal healing palaver, which to me seems different from the general practice of medicine I have experienced in the Western world, where bioethics seems more limited to hospital settings. In Bantu concrete realities of life, one comes to appreciate how there seems to be no creation and life force that is lifeless in their stronger practice of medicine within their lived social realities. And indeed, it is here where Black African moral thought partially points to some possible missing approach to bioethics.

So, the task before me is to systematically translate *daily lived life* experiences and numerous observations into something more credibly understandable not only to African moral thought but also to Western moral thought along our continuous search for a more inclusive and functional approach to bioethics. What can never be underestimated is the magnitude of both moral thoughts: Western and African. And I must say that attaining good knowledge of both approaches is an extremely huge, unending project for us all, and given their wide-ranging basics, it has been an immense challenge throughout my rigorous investigation.

Mag. Dr. Jude Thaddaeus Buyondo
Department of Systematic Theology and Ethics
Faculty of Catholic Theology, University of Vienna
Vienna April 24, 2024

Acknowledgments

In this academic and publishing journey, I am forever grateful to my mentor and current boss, Prof. Mag. Dr. Sigrid Müller, who has never abandoned us (me and this project) right from our infancy till our maturity, from the mornings till the evenings of this project. Without her, I could hardly have realized its creation, progress, in-depth research, funding, and eventual completion. I am not only thankful to her for devoting time to go through the manuscript and write the foreword, but I am also thankful for her interest, listening, patience, provision, and organization of weekly discussions and international conferences in theological bioethics across Europe. It is here where I richly interacted with knowledgeable colleagues and likewise benefitted from varied experienced international figures in scientific research.

I am thankful to Prof. Dr. Andrea Vicini, Prof. Dr. Cyrus P Olsen III, and Prof. Dr. Richard N. Rwiza for their generous mentorship and devoting some time toward the eventual realization of this manuscript.

I am extremely thankful to Gerfried-Werner-Hunold Stiftung for making it possible for me to conclude the final necessary requirements of putting together this book and generously funding the realization of its publication.

I am eternally thankful to Dr. Med. Thomas Schenk and Dr. Ingeborg Schipp for their generous presence, adorable love, and support throughout my academic career.

From my heart, I am profoundly indebted to you all.

Mag. Dr. Jude Thaddaeus Buyondo

List of Abbreviations and Acronyms

AIDS	Acquired Immunodeficiency Syndrome
ART	artificial reproductive technologies/antiretroviral therapy
CD4	cluster of differentiation 4
CIOMS	Council for International Organizations of Medical Sciences
DNA	Deoxyribonucleic acid (the biological uniqueness of each species)
ECMO	extracorporeal membrane oxygenation
ed.	edition
ed(s)./ Hrsg.	editor(s)/Herausgeber or edited/herausgegeben
ERC	ethical review committee
ET	embryo transfer
EVD	Ebola Virus Disease
DoH	Declaration of Helsinki
FAO	Food and Agriculture Organization (of the United Nations)
FGM	female genital mutilation/cutting
FIA	fair-innings argument
GMA	genetically modified animal
H3Africa	Human Heredity and Health in Africa
HIC	high-income country
HIV	human immunodeficiency virus
HRE	health research ethics
IAB	International Association of Bioethics
IBC	International Bioethics Committee (of UNESCO)
ILO	International Labor Organization
IUCN	International Union for Conservation of Nature
LMICs	low- and middle-income countries
MTCT	mother-to-child transmission

NABRE	New American Bible (Revised Edition)
NIH	National Institutes of Health
PBE	*Principles of Biomedical Ethics*, by Tom L. Beauchamp and James F. Childress, 8th ed. (New York: Oxford University Press, 2019)
PhD diss.	PhD dissertation
REC	research ethics committee
SMC	safe male circumcision
SOP	standard operating procedure
SSA	Sub-Saharan Africa
STD/STI	sexually transmitted disease/infection
TRC	Truth and Reconciliation Commission
UDBHR	Universal Declaration of Bioethics and Human Rights
UNAIDS	Joint United Nations Programme on HIV/AIDS
UNEP	United Nations Environment Programme
UNESCO	United Nations Educational, Scientific and Cultural Organization
IVF	in vitro fertilization
WHO	World Health Organization
WMA	World Medical Association
WWF	World Wildlife Fund

1

In Search of a More Comprehensive and Inclusive Approach to Bioethics

POINT OF DEPARTURE

There has been remarkable progress in the domain of medicine and bio-medical research ethics since the last half of the twentieth century. It is here that we drastically notice the rise of principlism, with its emphasis on "common morality" as a collection of very general norms to which every seriously committed moral person subscribes, together with its four ethical principles—respect for autonomy, beneficence, non-maleficence, and justice. The four principles were directly proposed by Thomas L. Beauchamp (1939—present) and James F. Childress (1940–present), in their now magnum opus book *Principles of Biomedical Ethics* (*PBE*), to cater for both local (cultural) and global bioethical/biomedical (universalized) challenges.[1] One notices that the universalized principlism has on the one side connected the world, but on the other side, awakened latent knowledge of local approaches to bioethics as well as inequalities, since the universalized bioethics has not conformed to numerous relevant accurate local realities. Some inequalities are widely spotted by Africanists

---

1 *PBE* by Beauchamp and Childress remains a classic in the field of medical ethics (understood in this investigation as bioethics or biomedical ethics) worldwide. "The first edition was published in 1979 and 'unleashed' the four principles of respect for autonomy, non-maleficence, beneficence, and justice." Holm, *Principles of Biomedical Ethics*, 332.

and African scholars. The foremost example is Tangwa B. Godfrey,[2] one of the world-celebrated bioethicists in the twenty-first century of African origin. His critique of principlism (broadly understood in this study as Western thought in contrast to African thought) and his defense of local approaches to bioethics could be of help in our eventual search for a comprehensive approach to bioethics. Tangwa recognizes that fundamental ethical principles are important in every human community—though, even if they are, they should have concrete real-life applications and particular relevance to the local realities. For him, under peculiar human conditions, principlism ought not be regarded as universally valid (absolute).[3] This is why numerous African scholarships are of the view that the absence of relevant functioning infrastructure in bioethical interventions in Africa,[4] or rather a deficiency of a transparent bioethical

2. Godfrey Bayuy Tangwa is an emeritus Cameroonian professor, researcher, and former head of the Department of Philosophy at the University of Yaounde. Though he specialized in the areas of epistemology and metaphysics, his teaching and research interests stretch from practical matters in medical ethics/bioethics, global bioethics, research ethics, clinical ethics, research regulation and governance, international ethics guidelines, to ethics committee work and capacity building in bioethics. He is one of the leading contemporary bioethicists of Sub-Saharan Africa who has gained international recognition. He has been a member, vice-president, and member of the board of directors for the International Association of Bioethics (IAB). He was elected Fellow of the Cameroon Academy of Sciences (CAS) in 2007 and was proposed as a Fellow of the African Academy of Sciences (AAS), the chairperson and founder of the Cameroon Bioethics Initiative (CAMBIN), an executive committee member of the Pan-African Bioethics Initiative (PABIN), and the chairperson of the Cultural, Anthropological, Social and Economic (CASE) working group of the Global Emerging Pathogens Treatment Consortium (GET). Some of his research grants and projects include the following: Exploring Perspectives on Genomics and Sickle Cell Public Health Interventions, funded by the National Institutes of Health (NIH); documenting facilities and needs of ethics committees and implementing a training intervention to strengthen ethical review capacity in Central Africa, funded by the European and Developing Countries Clinical Trial Partnership; Mapping of Francophone African Research Ethics Committees, funded by the Council on Health Research for Development (COHRED); Strengthening Ethical Review Capacity in Africa, funded by the African Malaria Network Trust (AMANET). Tangwa, "Leaders in Ethics Education," 91–92.

3. See Tangwa, "Ethical Principles," s2, s6; Tangwa, "Universalism and Relativism," 63.

4. What is meant by "Africa" ("African")? The focus is on Black Bantu of Sub-Saharan African, whose cultures and mores are similar. See Tangwa et al., "Global Health Inequalities," 245; Metz, "African and Western," 49–58. Likewise, scholarship like that of Tangwa is in dialogue with my position thus: "I am also not unaware of some of the shortcomings and weaknesses of African culture, traditional and modern. I could equally catalogue these in a whole book, but this is not the appropriate occasion. I use the word 'African' throughout in the same way that I use 'Western,' without any necessary implication that some differences or exceptions may not be found within what is thus bracketed. I insist on this point because, quite often, people who themselves use

infrastructure that conforms to African local realities, makes it not just difficult,[5] but quite impossible to contextualize bioethical principlism in different cultural settings, at least those with shared commonalities.[6] I am fully aware that generalizing African commonalities, for instance, in African moral thought—employing the definitive adjective "African"—might raise some ethical questions at some moment in the course of this study, since I am doing a systematic investigation. But regardless of the fallacious argument that may arise based on what the majority thinks (*argumentum ad populum*), which is sometimes erected towards common beliefs in *African* thought, in this investigation, however, I refer strictly to Africa south of the Sahara. The prepositional phrase "in Africa" connotes not only the geographical area of my contributions—Africa south of the Sahara, labelled as *Afrique Noire*—but also refers to cultural constructs (academic and non-academic, latent and forensic) that I employ to differentiate between the imported Western categories of thought in contrast to what is truly African, without necessarily imposing my arguments as implying any unchallengeable permissibility. As common knowledge may similarly dictate, Africa as a diverse continent with multiple cultures remains with uncontroversial observable commonalities, to which it would be uncharitable to override them, just simply because one foreign tourist or researcher saw some part of Africa where the common description does not entirely apply. The reader should nonetheless note that, even though there are other ethnic groups in Sub-Saharan Africa, my focus remains largely on Black African Bantu.

The study bilaterally aims at finding out more about how the universalized bioethical principles (Western moral thought) could be more inclusively adapted through local approaches (African moral thought), on the one side, while on the other side the universalized approach to bioethics could also be more informed by local values, so that bioethics

---

such terms as 'the Western world,' 'Western civilization,' 'Western liberalism,' 'Western democracy,' 'Western technology,' 'Western philosophy,' 'Western ethical values,' etc., have objected to my description of something as 'African' on the ground that they happen to know some part of Africa where the description does not apply." Tangwa, "Biomedical and Environmental Ethics," 388–89.

5. Tangwa et al., "Ebola Vaccine Trials," 57.

6. See Tangwa and Henk, "Colony of Genes," 1041–42; In other studies, Tangwa believes there "is great variety and diversity between the different African ethnicities, but they are all united by commonalities that give them a remarkable family resemblance analogous to the family resemblance of groupings . . ." Tangwa, "Voice to African Thought," 104.

would be more inclusively and comprehensively appreciated throughout the academic, social, and global communities.

THE PROBLEMATIC

Critique of Bioethical Principlism from an African Context

At the origin of bioethics (*bios* implying life and *ethike* ethics), it was intended to sort the global bioethical problems of that time. Van Rensselaer Potter, as the first proponent in the 1970s, understood it as a new discipline which would combine biological knowledge with knowledge of human value systems.[7] As from 1990s, bioethics became a multidisciplinary mode of inquiry, which answered ethical questions about the value and meaning of life[8] as concretized in various applications of biomedical ethics. Narrowing down to Beauchamp and Childress's *Principle of Biomedical Ethics*, in this masterpiece, they claim that their four-principle-approach (I refer to it as "bioethical principlism"[9])—"(1) respect for autonomy (a norm of respecting and supporting autonomous decisions), (2) nonmaleficence (a norm of avoiding the causation of harm), (3) beneficence (a group of norms pertaining to relieving, lessening, or preventing harm and providing benefits and balancing benefits against risks and costs), and (4) justice (a cluster of norms for fairly distributing benefits, risks, and costs)"[10]—is a globally applicable framework of broad moral principles since it descends from the *common morality*.

African scholarship in bioethics, citing for instance that of Tangwa as a main protagonist in bioethics, and more so a foremost leading scholar in global bioethics of African descent, understands this multidisciplinary mode somewhat differently. For him, in his book *Elements of African Bioethics in a Western Frame*, bioethics is "a broad and rather untidy general framework of beliefs, attitudes, dispositions, outlooks and practices given which, any of the myriad specific ecological or bio-medical problems of

---

7. Gillon, "Bioethics, Overview," 2.

8. Harris, *Bioethics*, 4, 15; Gillon, "Bioethics, Overview," 4.

9. Over the years Beauchamp and Childress have developed this approach and vigorously defended it (till the 8th ed of *PBE* in 2019) against various criticisms. They have maintained that their four-principles approach, otherwise known as "principlism," is a globally applicable framework for biomedical ethics. Herissone-Kelly, "Common Morality's Norms," 584, 587.

10. *PBE*, 13, see also 25.

the contemporary world might be very differently perceived, appreciated and evaluated."[11] From this understanding, we get an impression that an African understanding of bioethics goes beyond the typically known hospital setting, suggesting we may have to understand bioethical principlism and its related critiques rather differently. But before proceeding with some critiques, we should note the broad sense in which the term "bioethical principlism" is used in numerous African criticisms, where it relates not only to the principlism of Beauchamp and Childress, but also to other "Western theories and types of bioethics." This further reiterates my earlier accent that we will throughout this investigation understand bioethical principlism as Western moral thought with its antithesis as African moral thought.

Let me briefly state some four critiques arising from various African studies done and my interpretation of African moral thought. *Firstly*, for African people, bioethics stems from a *lived life* within the existential cultural realities, values, traditions, customs, and practices.[12] We note, in contrast to principlism, that African moral thought is not only humanistic—concentrating on human fraternal solidarity—but it is further shaped by life outside the hospital setting like its *ecological, biological,* and *cultural diversity*. That is to say, the *distinction amongst them all forms a single ethical continuum* that Tangwa labels "eco-bio-ethics."[13] Put differently, he understands bioethics beyond a clinic setting as resting broadly on the eco-bio-communitarianism,[14] "implying recognition and acceptance of *interdependence and peaceful coexistence between earths, plants, animals and humans,* by contrast with the Western outlook which could be described as anthropocentric and individualistic."[15] Therefore, on the one hand, my reading of Tangwa understands bioethics to build on a shared life that aims at bringing "rationality, moral sensibility and sensitivity to human relationship and interaction with fellow living beings, including plants, animals and other humans, *at all levels and in all dimensions*."[16] On the other hand, as we try to establish the second critique, my reading picks up an additional bioethical novelty of African thought—something

11. Tangwa, *African Bioethics in a Western Frame*, 52.

12. Tangwa, *Contemporary Bioethics Problems*, 16–17; Tangwa, "Leaders in Ethics Education," 91–105.

13. Tangwa, "Biomedical and Environmental Ethics," 388. Emphasis added.

14. Tangwa, *Contemporary Bioethics Problems*, 12–14, 25–26, 36, 45–46, 60.

15. Tangwa, *African Bioethics in a Western Frame*, 57. Emphasis added.

16. Tangwa, "Leaders in Ethics Education," 104. Emphasis added.

that Tangwa slightly subscribes to—whereby the attribution of life (*bios*) is not only confined to the living things of the visible world, but also to the invisible world, inanimate things and all forces of life. This is how, for instance, Africans generally express their own understanding of life in their own concrete existence. In his study, "Some African Reflections on Biomedical and Environmental Ethics," Tangwa writes:

> Within the African traditional outlook, human beings tend to be *more cosmically humble and therefore not only more respectful of other people but also more cautious in their attitude to plants, animals, and inanimate things, and to the various invisible forces of the world.* One might say, in short, that they are more disposed toward an attitude of *live and let live* . . . Within that world-view the distinction between plants, animals, and inanimate things, between the sacred and the profane, matter and spirit, the communal and the individual, is a slim and flexible one . . . very significant implications for the way nature is approached. In like manner, metaphysical conceptions, ethics, customs, laws, and taboos form *a single unbroken chain* within this African outlook.[17]

This indicates that an African approach to bioethics emphasizes a life simply cosmically lived out as a single unbroken rhythm, inclusive of the life of the living and the nonliving members, who are part of an uninterrupted life, a life of the three-dimensional African community of the living, the dead, and those not yet born (alternatively known as a "triad community" or a "three-dimensional community" or a "tripartite community"). It is this understanding that highlights the *second critique* of principlism.

The *third critique* of principlism is the overemphasis of the common morality as the sole solution to local understandings and

17. Tangwa, "Biomedical and Environmental Ethics," 389. Emphasis added, except "*live and let live.*" See also Tangwa, *African Bioethics in a Western Frame*, 41. Tangwa's perspective is no different from other African scholars like Molefe. Molefe rejects humanism as a basis for the African moral thought just because it fails to capture some of the prevalent thoughts and intuitions that Africans typically have about their intimate duties towards the natural environment—the substantive belief envisaged here suggests that Africans are one substance with nature. In Molefe's broader studies, humanism is not a plausible interpretation of African moral thought (ethics) since for African relational metaphysics, some non-human components matter for their own sakes. He argues to the effect that it is implausible to interpret African moral theories to be anthropocentric, as do other African scholars like Gyekye, Wiredu, and Metz, since this fails to cohere with an integral picture of African relational metaphysics. See Molefe, "Rejection of Humanism," 59–60, 72–73.

cultural-contextual problems. However, I note that African medicine persons for instance, in their practice of medicine,[18] have closely guarded and passed-on resolute ethical codes as possible correlative moral activities known as "homeomorphic equivalents,"[19] which are observable in other traditions or close to principlism itself. However, principlism as it is applied by international biomedical ethicists in the African context seems not to adapt to realities that can be embraced as corresponding to local homeomorphic equivalents or, more specifically, what is genuinely impactful to African existential multifaceted problematic realities.[20] We possibly need to conceive these whole realities as likely homeomorphic equivalents resident in African socio-religious-cultural values, ideas, practices, and attitudes. These may include, for example, conceiving African approaches to "relational social autonomy" against merely individual autonomy as chiefly understood in principlism; justice understood as mutual aid/cost/assistance in the African sense against justice being simply taken as fairness in principlism; an African approach to reconciliation and healing understood in terms of some fraternal relational anamnetic solidarity—inclusive of the ancestors and the victimized of the past ("communal relational solidarity")—rather than only focusing on reconciliation with members of the living community common in principle-based approaches; being eco-friendly and eco-hospitable by extending respectful sympathy to all creation and its life forces rather than basically confining bioethics to interhuman relationships in clinics; as well as appreciating social informed consent than simply concentrating on personalistic consent. Atleast some of these African local approaches to bioethics would propose appurtenant answers, or at least some possible solutions and prescriptions to some global challenges in bioethics.

When some of the above local approaches are not appreciated as rich homeomorphic equivalents from global principlistic approaches which subscribe to the universalized common morality, a long-lasting enrichment and productive implementation becomes locally problematic. For a concrete illustration, this problematic is hinted at in different transnational groupings, like the international H3Africa (Human Heredity and Health in Africa) Ethics Consultation Meeting, which established

18. See Tangwa, *Contemporary Bioethics Problems*, 65.

19. The apparent question is: to what extent does the appreciation of common morality embrace homeomorphic equivalents? Panikkar, "Homeomorphic Equivalents," 21, 43, 47.

20. See Tangwa, "Leaders in Ethics Education," 103.

challenges with the implementation of African bioethical understanding of consent, community engagement, and the governance of sample and data access.[21] Such contemporary problematics refer us back to the previous paragraph that seems to open our eyes to unacknowledged realities of homeomorphic equivalents on local contexts. At this moment, I cannot help enlightening briefly on three of the mentioned approaches in the previous paragraph and how they can be well received in African contexts. *One,* an African equivalence of informed consent won't be evaluated on the basis of individual freedom and autonomy as understood in principle-based approaches, but rather on *communal relational solidarity* as social autonomy. *Two,* justice won't be understood as fairness in Sub-Saharan Africa (SSA), but more deeply as mutual aid (mutual cost or assistance) in solidarity with one's community or one's neighbor—as in cases of philanthropy and supererogation where there is doing more than what one's duty requires. Prescriptively, it is only by doing more than what one's moral duty requires that an African promotes community life, while if one abandons this pursuit, life will not be promoted. And so doing good, from a black African perspective, may not attract any extra benefits, but not doing it is considered not to promote life, and it so attracts sanctions.[22] Thusly, moral duty isn't considered any kind of extraordinary thing that promotes life in the African sense.

Third, an African would understand relational social autonomy as an antithesis to individual autonomy commonly appreciated in principlism, leading us to the *fourth critique* of principlism. In principlism, bioethicists over-prioritize personalistic autonomy, thereby overemphasizing the power of individual choices, but consequently underestimating, as critics sometimes aver, the role of the interconnectedness of the three-dimensional community as understood by Africans, and its respect for ecological life. African scholarships verily agree that the relational community is usually given precedence over the individual, but uttermost respect is accorded to individuality rather than individualism. It means Black Africans take superiority of the community common interests to override individualistic pursuits,[23] while simultaneously recognizing the

21. Tangwa et al., "Ethical issues in H3Africa," 3. See also H3Africa Consortium, "Research Capacity," 1346–48.

22. See Tangwa, *Contemporary Bioethics Problems,* 22–24.

23. If we pick out one artistic example, the Akan art motif depicts a crocodile with one stomach, but two heads locked up in a fight over food. This symbol captures both the problem of excessive fixation upon individual interests—symbolized by the two

uniqueness and creativity of the individuality of each human. As a result, many African scholars affirm that it is neither the autonomous individual (I) that achieves the uppermost good of the community nor is the individual held responsible for some wrong action; but rather, it is the collectivity of the communal members[24]—entailing the spiritual/religious interrelatedness of the living and nonliving—living out their communal relational solidarity (we), and who can unanimously increase or decrease the vitality of life inter-/intra-/trans-culturally.

In the African sense, atleast for Tangwa, autonomy-centered principlism—in spite of its positive values to modern medicine—exploits health as a marketable economic commodity.[25] It is one reason why economic determinism shouldn't rule bioethical pursuits "to such an extent that even ethical issues are couched and discussed in economic terms and language."[26] In addition to the aforementioned critiques, this is yet another bioethical inequality that leaves an African moral thought yearning for a bioethics that is *not commercialized*,[27] but is freely responsive to general as well as domestic relevant realities. It is why the dominant paradigms of bioethical principlism basically don't blend well with African local needs, values, specific conceptions, and religious/spiritual realities.

Precisely, the ethical inequalities lie in the fact that, on the one hand, bioethical principlism has not been shaped by a dynamic understanding of local cultural approaches to life, while on the other hand, principlism has not impacted the local socio-religious structural and institutional life as well as the concrete problems that challenge the local health realities. In this regard, just like Tangwa's studies show, we haven't made appropriate distinctions between the ethics of the developed world *in* the developing world. And although this is not a rejection of the general ethical

---

separate heads—and its solution when we share the same interest, namely, our common wellbeing—symbolized by the common stomach. See Wiredu, "Oral Philosophy of Personhood," 10.

24. Studies show how important it is for Black Africans to promote life fully in their daily lived life and throughout their expanded dynamic community life which cuts across all spheres of living and nonliving entities. See details in Bujo, *African Ethic*, 18–19, 60–61, 64, 66, 70–71, 86, 89; Bujo, "African Ethics," 430; Ilogu, *Christianity and Ibo Culture*, 24, 123–24, 128, 130.

25. Tangwa et al., "Global Health Inequalities," 242.

26. Tangwa, "African Perception of a Person," 43.

27. For the critique of commercializing bioethics, see Tangwa et al., "Ethics in Occupational Health," 9; Tangwa, "African Thought," 101–10; Andoh, "Bioethics and the Challenges," 67–75; Sebunnya, "Distinct African Bioethics," 22–31; Tangwa et al., "New Tuskegee," 15.

imperatives and obligations at all levels, different standards may and, in fact, cannot but be applied in different contexts at different levels in different ways.[28] This explains why some Africanists basically speak not of "African bioethics" but of "bioethics in Africa" since ethical guidelines for biomedical interventions remain foreign to an African context, and what is African is not merged with the universally proclaimed approaches to bioethics. And it is why, particularly for Tangwa, truly global bioethics seems almost impossible given that cultures aren't "even equally represented in contexts and fora where globalization decisions and projects are discussed"[29] by multinational Western agencies, yet we claim to be in a world that is branded as globalized.

We remain puzzled if we "could really honestly try to get into the spirit and swing of things"[30] of another culture. Some bioethicists like Tangwa are convinced that if African conventional "systems are looked upon, as they should, as viable alternatives rather than, as they usually are, as mere approximations to Western 'ideals', they stand to contribute great values to an emergent world culture."[31] And if Western culture is the "proprietor of modern bio and other technologies," it "should develop the habit of also standing up and moving around a bit, to view the dancing masquerade from different perspectives"[32] in order to be enriched from treasuries of knowledge from other cultures. It is one reason why Tangwa proposes "giving an increasing say to other communities, peoples and cultures and incorporating their views in international regulatory documents."[33]

In a nutshell, having initiated a critique to bioethical principlism in an African context, it has become clearer that both African moral thought and the universalized approach to bioethics need to enrich each

28. See Tangwa, "Ethics Committees," 156.

29. Tangwa, *African Bioethics in a Western Frame*, 43–44. I take cognizance of the different constitutive perspectives and the detailed discussions of globalization. These discussions have, in one way or the other, shaped the current global village and clash of civilizations. They may include: "Homogenisierung," "Heterogenisierung," "Glokalisierung," "Hybridisierung," "Kreolisierung," "Trans/Inter/Hyper/Multi/Pluri/kulturalität," as developed by Wolfgang Welsch (1944–), Johann Gottfried Herder (1744–1803), Samuel P. Huntington (1927–2008), Andreas Reckwitz (1970–), Naomi Robin Quinn (1939–2019) and others. Some scholars refer to "globalization" as a "cultural cloning." Bujo, "African Ethics," 436.

30. Tangwa, "Bioethics: An African Perspective," 198.

31. Tangwa, *African Bioethics in a Western Frame*, 143; see also 152.

32. Tangwa, "Morality and Culture," 20.

33. Tangwa, "Universalism and Relativism," 63–65.

other. And in an effort to close this gap and offer a contribution to the discourse of bioethics in this investigation, I am of the view that in our joint efforts to unearth a normative collaborative framework, we need to give room to the contribution of the local approaches too. For example, given that African moral thought appears hinged on an inclusive relational order, it would be unthinkable to determine any bioethical order (inclusive of binding normative codes) relevant to Black Bantu without appealing to their understanding of this relational social order. Though it may sound pre-logical, prescientific, or irrational to principlists, this investigation reveals how the Bantu approach to bioethics stems from their own conception of "the relational order of the three-dimensional community" along its wholesome active participation in promoting life. While taking explicitly such examples as from the Baganda, the Nso', Kom people, Swahili people, my contribution to bioethics remains focused at how we can enhance the "wholeness of life force" in a nonWestern principle-based rationality. More concretely, for Baganda, it is by necessarily and constantly increasing or decreasing a group's healthy "vitality and continuity," pursuing "solidary" participation in the total life, as well as respecting the "hierarchical" mode of existence of all beings.[34] Consequently, in this investigation, I ask and largely attempt to answer *how bioethical principlism can be complemented with the help of recipient local realities so we can lessen* local-global bioethical inequalities.

Critical Reactions to African Criticisms of Principlism and Initial Indications of a Solution

Despite the discussed criticisms of bioethical principlism in various approaches, there is doubt to what extent these Africanists fully interpret African moral thought and the extent to which we can either minimize local-global bioethical inequalities or complement principlism. Two issues come up in their overemphasis of the *Africanness* of bioethics. One relates to the discourse of *African bioethics*, questioning whether it flows from the relational ontological life of African communal moral thought, while the other relates to the *principlistic*[35] approach to *common morality*.

34. See Sempebwa, *Reality of a Bantu*, 126. "Ganda" people are the same as "Baganda."

35. Unless indicated otherwise, "principlists" refer to Tom L. Beauchamp and James F. Childress as the major proponents of principles of biomedical ethics together with adherents of their thought and Western principle-based theories to biomedical interventions.

If we pick out, for instance, Tangwa's bioethical approach to local-global bioethical challenges, it is eco-bio-communitarian, which approach *limitedly* includes shaping bioethics around human relationships (anthropocentrism) and interaction with fellow living beings (biocentrism). In the *first* (anthropocentric) instance, it is indisputable that African thought appreciates human relationships as central; consequently, advocating for bonding and caring for the well-being of the others and their worldviews.[36] Indeed, a functioning relationship with the *other*[37] remains significant for black Africans because humanity remains "a demand for a creative inter-subjective formation in which the 'other' becomes a mirror (but only a mirror) for my subjectivity. This idealism suggests that humanity is not embedded in my person solely as an individual; my humanity is co-substantively bestowed upon the other and me."[38] An additional emphatic African voice emanates from the African golden rule of "sympathetic impartiality,"[39] which focuses on promoting life for the welfare/well-being of each and all, though not for some creation types. The *second* (biocentric) instance, however, from the *other* and *me*, African relational ontology further suggests that bioethical life extends across the visible and invisible communities (*Other*), which can never only or limitedly mean responsibility for the "living"(other).[40] Worth to carefully note further, in this encounter of the *Other*[41] and *me*, Africans master the vital flux (energy) responsible for an all-inclusive harmony between the entire creation, or the tension opposing life in our normative life, for instance, fraternal solidarity. In wishing to secure life's victory over

36. See Wiredu, *Cultural Universals and Particulars*, 31, 182–85; Mkhize, "Ubuntu-Botho Approach to Ethics," 35–36, 39, 43; Wiredu, "Communalism and Communitarianism," 333; Ramose, *African Philosophy*; Molefe, "Partiality and Impartiality," 476; Jegede, "African Ethics," 247.

37. Though the "other" will later be defined, here it refers to living things, among them humans.

38. Eze, *Contemporary South Africa*, 190–91; also cited by Nicolaides, "African Ubuntu," 203–4; see also Beets, "Ubuntu," 68–83.

39. The *golden rule* of *sympathetic* impartiality according to Kwasi Wiredu promotes the equal well-being of all human beings. Africans believe that humans do not only exist as members of some cultural subgroups, but simultaneously as members of the whole global community. Wiredu, *Cultural Universals and Particulars*, 30–31; Jegede, "African Ethics," 246–48.

40. See Bujo, *African Ethic*, 1–2, 44–46.

41. Here, the "Other" refers to the living and the non-living, visible and invisible communities, inclusive of inanimate life and systems of collaboration.

death at all costs, Black Africans secure allies in the hierarchical world.[42] However, on reading charitably several scholarships on Africa south of the Sahara, we face challenges from African moral thought itself, in spite of the fact that these Africanists intend to protect African identity. This is because Black African relational social reality appreciates not only the living things, but the whole hierarchy of forces of creation—including the community of the future generation and the ancestors—as meaningful and purposeful in living out a good moral life.[43] This means it would be a moral cost if the vital energy within every creature isn't tapped as we work towards "promoting life," since for Black Africans nothing is known to be lifeless. And since all creation shares in the life principle—and on condition that this hierarchical participation of all phenomena aims at (conversion to) harmony—then we can say that life across all creation and systems of collaboration can be abundantly or fully promoted, first of all, by human beings. Thusly, the life principle as envisaged by African relational moral thought includes not only the living things, but rather everyone and everything wholly. So, for an African, we can reliably infer that human survival is implicated in the survival of all entities and functioning of all systems across the life principle. Also, this confirms the interdependence of all cosmic beings, besides human beings collaborating with all universal partners. It means we cannot do bioethics less of an "all-inclusive" tone that appreciates a harmonious relationship with the whole natural world as one functioning cooperation (unity).[44] Thus, the challenge to which Africanists are exposed remains—given the nature of the African attitude of the interconnectedness of life—an African concept of bioethics needs neither to be restricted to the local realities only (over-stressing the Africanness of bioethics), nor the community of the living only, but rather also embrace and share in the solidarity of the universal communal partners.

And whenever some scholarships stress the Africanness of bioethics (appealing to a cultural bioethics resident in each culture), they seem to deny common morality. Here common morality, according to the proponents of principlism, is the "set of universal norms shared by all persons

42. See Bujo, *African Ethic*, 2–3, 59, 64–66, 106. Ilogu, *Christianity and Ibo Culture*, 127–28; Bujo, "African Ethics," 425–26.

43. See Bujo, "African Ethics," 436.

44. Mkhize, "Ubuntu-Botho Approach to Ethics," 33–40; Sindima, *Ethics in Africa*, 169, 171; Bujo, *African Ethic*, 103, 167–74; Kelbessa, "Environmental Philosophy," 323; Behrens, *Environmental Ethics*; Ilogu, "Global Health Inequalities," 26, 128.

committed to morality," and "this morality is not merely a morality, in contrast to other moralities. It is applicable to all persons in all places, and we appropriately judge all human conduct by its standards."[45] Additionally, in the fourth and subsequent editions,[46] as well as in their recent article, Childress and Beauchamp emphatically argue that "principles can and should be regarded as deriving historically and philosophically from the common morality, that is, the *universal morality to which all morally committed persons subscribe* . . . Because our principles are universally applicable, we defend a global bioethics and not merely *customary, regional,* or *cultural rules.* Our principles correlate with basic human rights and establish what is *ethically acceptable for all societies.*"[47] To put it succinctly, they propose the principle of common morality while rejecting customary moralities and so insinuate that "acceptance of this principle could serve as an inclusion criterion, and nonacceptance as an exclusion criterion."[48] The onset of this bioethical discussion about the acceptance and criteriology of the two approaches, common and customary morality, hits right at the heart of the debate concerning cultural and universalized approaches to bioethics. However the question is, doesn't denying a universal or global bioethics translate into denying common morality as the ultimate source of moral norms and principles that are universally applicable and locally acceptable? And if the common morality is a set of norms that are shared by all morally committed persons, isn't the African approach short of a commitment to or excluded from morality?

Furthermore, Africanists' defense of cultural ethics exposes them further to a critique of whether cultural groups of people aim at different types of bioethics. Indeed some Africans question the extent to which one can establish the Africanness[49] or Asianness or Europeanness of bioethics and how different it may be to other bioethics or the universalized

45. *PBE*, 3.

46. To be more specific, the fifth ed. of Beauchamp and Childress's *PBE* embraces common morality as the ultimate source of moral norms. See DeGrazia, "Common Morality," 219–30, as well as the sixth ed. of *PBE*. See Herissone-Kelly, "Common Morality's Norms," 584–87.

47. Beauchamp and Childress, "PBE: Marking Its Fortieth Anniversary," 11. Emphasis added.

48. PBE, 451.While concluding this book (p. 451), they reiteratively define *common morality* in terms of "the set of norms shared by all persons committed to morality," and for them, "it is unimaginable that any morally committed person would reject this general principle."

49. See Fayemi, "African Bioethics," 98–106.

bioethics. It is the reason why some African scholars observe that "bio-ethics in Africa" could be new—since formal scholarly bioethics evolved recently in the twentieth century—but an emergence of a discussion of an "African bioethics" is slightly misleading because African bioethics has always been part of African culture—the peoples' genuine *way of life*.[50] Accordingly, we cannot dwell on the claim that the discipline of ethics in the practice of medicine is alien to Black Africans, just because they have had this informal practice as old as the human race. Coinage of the term "bioethics" is recent but its practice isn't recent.[51] Moreover, some Africans state that we shouldn't exceptionalize or rationalize an African, Asian, or European thought, since each cannot fully encompass every aspect of truth or exclusively offer the criterion for truth.[52] I am of the view that we ought not to emphasize the Africanness or Africanity of African bioethics (i.e., a *pro-tanto* bioethics), since it limits one culture from sharing in the global life of other cultures. But when we honestly open up to a genuine global discourse, there seems to be some room for the local approaches to market themselves together with their universalizable values such as healthy humanness, and so participate in a constructive global bioethical discourse, thereby allowing the African contribution to bioethics to translate into or permeate the universalized bioethical approach. Reciprocally, there are possibilities that the universalized approaches to bioethics will be acknowledged and embraced within the African cultural-contextual approach as they in return get eventually localized.

PROCEDURE OF THE INVESTIGATION AND METHOD

The investigation makes use of systematic and hermeneutical methods to reappraise scientific scholarships in order to bridge gaps within the critique of bioethical principlism and principlism itself. I draw my contribution from scientific scholarships as well as my own concrete observations

---

50. Culture is *the way of life* of a group of people, underpinned by adaption to a particular environment, a shared worldview, ideas, values, historical experiences, attitudes, expectations, practices, etc. Tangwa, *African Bioethics in a Western Frame*, 5; Tangwa, "African Thought," 104; Wasunna et al., "Bioethics in Sub-Saharan Africa," 525–30.

51. See Tangwa, *African Bioethics in a Western Frame*, 39.

52. See Bujo, *African Ethic*, 6–11; Larmore, *Morals of Modernity*, 60–62; Panikkar, "Homeomorphic Equivalents," 25, 59.

and experiences in the industrialized Global North and developing Global South.

In this first chapter, "In Search of a More Comprehensive and Inclusive Approach to Bioethics," I show the problem under investigation, the current discourse and state of research, and the suggested approach to a possible solution.

Further in the *first part*, the study investigates into the problem of bioethical principlism in an African context as perceived, with illustrations mainly based on the principles of autonomy and justice, in relation to how their principlistic interpretations don't sit well in diverse African contexts. By showing what misses for better functioning global bioethical relationships, this study remains familiar with the fact that although there is the universality of fundamental principles, relevance and applicability ought to be relevant and conform to concrete local realities since principles cannot be valid everywhere without a rich contextual concretization.

Taking various criticism of Africanists and African scholars seriously, in the *second part*, I investigate what according to African moral thought is essential for an appropriated interpretation of an African genuine approach to bioethics, and how bioethics can be unfolded so as to *bring together* rather than *take apart* the African approach to bioethics from a universally oriented principlism. After examining an African moral thought—for instance, in dialogue with works of Bujo, Mbiti, Magesa, Rwiza, Mulango, Ndombe, Sempebwa, Kasozi, Orobator Agbonkhianmeghe, Éla, Tempels, etc.—along their weaknesses and the weaknesses of African moral thought itself, this second part ultimately proposes a possible contribution and how we can unfold bioethics so that we can collectively enrich, add value, complement, unite, and harmonize an African bioethical approach to a universally oriented principlism. The last chapter advances findings and some recommendations ("Conclusion").

PART I

Critique of Bioethical Principlism in an African Context

ABSTRACT OF PART I

Part I reveals that principlism basically relies on the cultural approaches of the Global North as absolute while excluding rich approaches from the Global South. In their absolute nature, they seem not to fittingly sit in some African contexts. In contrast, if we are aiming at developing a suitable global language of bioethics befitting world citizens, principlism remains largely unsuccessful in some contexts south of the Sahara. This is partly because culturally tilted Global North principles seem to appear inadequate for addressing global concerns that are global in nature. This first part discloses how we need a new language of bioethics and a more comprehensive moral framework grounded in a foundation of universally shared outlooks which should appreciate pluralistic approaches to bioethics. Methodically, the chapter analyzes the North-South ethics dumpings, where corruption cuts across all global members. In this analysis, the risk of exploitation in bioethics is not limited to a single rubric (or to Africans and their immediate Western counterparts); it is not limited to an "inadequate informed consent process," "possessive individualism," "manipulation of non-human primates in research," or "exploitation through genetically modified foods" or "male circumcision trials or the candidate Ebola virus vaccine," but the lodged caveat cuts across biomedical exploitations as general weaknesses and inadequacies of an entire system, on the one hand. On the other hand, if the specific African socio-cultural reality is not accurately appropriated like in the exploitive case studies extensively discussed, then some will continue to basically speak not of "African bioethics" but of "bioethics in Africa," since ethical guidelines for health-related research are foreign to an African context, and what is African is being replaced with what is non-African in the guise of universalization or globalization of principles. So, the problem with bioethical principlism in an African context is not only an inadequacy of a well-intentioned collaborative structural-universal regulatory framework, a weak cosmic-ecological relationship, but also destabilized interpersonal-humanistic relationships across Global North-South cultures. If the entire system needs an entire conversion, does African moral thought supply some contributory approach to this total conversion? How does the African approach complement principlism with a more comprehensive approach to bioethics?

2

General Critique of Bioethical Principlism

This chapter assesses not only the general critique of bioethical principlism but also initiates the critique in the African context. Precisely, the chapter attempts to answer what the general problem with bioethical principlism is, in relation to an African context.

BIOETHICAL PRINCIPLISM AS COMMON MORALITY (UNIVERSALIZED PRINCIPLES)

An Approach to Bioethics in the Western and the African Sense

"Bioethics" is a term coined in the mid-1970s, synonymously referred to as "biomedical ethics."[1] On the one hand, the term "bioethics" was popularized by an American biochemist and cancer researcher Van Rensselaer Potter (1911–2001). Being uniquely concerned about the problem of the future of humankind and its survival, Potter conceived of "global bioethics" as a "bridge" between the present and the future, between science and values, nature and culture, and between humans and nature. On the other hand, numerous contemporary scholars associate the term "bioethics" to a Protestant German theologian, Fritz Jahr (1885–1953), who proposed a "bioethical imperative" that must *respect every living being as an end*

---

1. *PBE*, vii.

in itself and treat it, if possible, as such. Though the stress predominantly remained on the living beings, both Jahr and Potter *originally* held an integrative conception of bioethics, covering not only the human community, culture and health but all the physical and material aspects of our worldly life.[2] As of 1990s however, bioethics seems less of an integrative system, and more of a multidisciplinary mode of inquiry that is limited to *generic ethical questions.*[3]

Even in the Western principlistic sense, ethics can hardly be conceived outside generic terms to which also rules orient themselves in their concrete applications. Namely, before ethical theories identify and justify principles, rules, norms, rights, or virtues, the normative approaches to ethics demand guiding general moral norms. Even for practical or applied ethics' hunt for practical answers—for instance, in particular moral judgments, professional codes or practices, biomedical problems, and public policies—practical ethics appeals to moral concepts and general norms, precedents, or theories.[4] Nevertheless, in the mind of the proponents of principlism, both principles and rules are norms of obligation where moral doers draw only a loose distinction between them. Eventhough for justifying moral principles one requires a general comprehensive framework of norms, for rules one appeals to a more concrete and restricted framework of norms, without suggesting any distinctive dissimilarity between these norms of obligation. Yet, over and above that, not only are *prima facie* (whose infringement is contingently impermissible in contrast to *pro-tanto*) principles, rules, obligations, and rights *not absolute,* but also *all general moral norms can be justifiably overridden by other norms* with which they conflict,[5] at least in some contexts. Argued

2. See Tangwa, "Global Bioethics," 268.

3. See Harris, *Bioethics,* 4, 15; Gillon, "Bioethics, Overview," 4.

4. *PBE,* 1–2.

5. The infringement of rights may be contingently permissible (an example is a *pro-tanto* right or duty) or impermissible (an example is a *prima facie* right or duty). Beauchamp, "Rights Theory and Animal Rights," 220; see also *PBE,* 14–15. Examples of categories of rules: substantive rules—rules of truth telling, confidentiality, privacy, foregoing treatment, informed consent, and rationing health; authority rules—decisional authority, who may/should make authority; rules of surrogate authority—decision for incompetent persons; rules of professional authority—who should override patients' decision in professional ranks; rules of distributional authority—who should make decisions about allocating scarce resources; procedural rules—rules that establish procedures to be followed, e.g., for determining eligibility for organ transplantation and procedures for reporting grievances to higher authorities. Moreover, for principlists, "principles and rules of obligation have correlative rights and often corresponding virtues" (p. 14); A

differently, a right is *pro-tanto* if and only if there are exceptional contextual circumstances in which it is permissible to infringe it, even though it is impermissible to infringe it under normal circumstances.[6] This means that whenever we narrow down to specific contextual frameworks, we are undeniably faced with relentless corresponding questions regarding application of rules. For instance, although we generally appreciate human rights and liberty as inherent, in the recent COVID-19 pandemic, we have been witnessing multifarious restrictions of human rights and liberties through enforcement of quarantine (isolations or confinements) given the primacy of the principle of safety of life.

In the contemporary African ethical discussion, there is equally a reinvigorated *awareness*[7] of bioethical interventions to control moral challenges—related to recent life-threatening infectious diseases and waves of epidemics (some viral), such as cancer, measles, Ebola virus disease (EVD), malaria, cholera, smallpox, yellow fever, tuberculosis, poliomyelitis, HIV/AIDS and so on—which threaten each individual and the totality of communal life. From this awareness, Africanists like Tangwa conceive ethics as a systematic normative reflective science of morality applicable in concrete realities which can be morally evaluated as right or wrong within a given time, place, culture, context, and particular existential way of life. Here ethics presupposes that we already have a sense of morality, making ethics prescriptive rather than descriptive.[8] Such an approach appreciates contextual existential African realities than what applies in generic universal terms. Even if ethics were universal, it is not absolute since we have already admitted numerous defensible exceptions.

---

person who appears irrational or unreasonable to others might fail a psychiatric test, and as a result be declared incompetent (p. 113). An incompetent person is unable to formulate a preference; understand information and appreciate one's situation; reason through a consequential life decision (pp. 115–16).

6. Concerning the topics of moral rights and duties and moral permissibility and impermissibility, see Frederick, "Pro–Tanto versus Absolute Rights," 375–94; Beauchamp, "Rights Theory and Animal Rights," 220.

7. Narrowing down to Sub-Saharan Africa (Black Africa), bioethics awareness is consequent to the rise of the HIV/AIDS epidemic (mid 1980s), the Rio Earth Summit (1992), the Beijing Conference on Women (1995), Apartheid in South Africa following its collapse (1994), the realization that Africa was slowly and surely becoming the developed world's "septic tank" epitomized in the dumping of toxic waste in Côte d'Ivoire (2006), among others. Tangwa puts stress on biomedical research led by northern scientists, aimed at finding a cure or a vaccine against HIV/AIDS. "Leaders in Ethics Education," 100.

8. See Tangwa, *African Bioethics in a Western Frame*, 55; Tangwa, "Ethical Principles," S2–3.

Besides, making use of Tangwa's studies, he argues in defense of Kwasi Wiredu's context-oriented thesis of *sympathetic impartiality*, to which I will return in depth later. Suffice it here to say that Kwasi Wiredu argues exclusively for cultural universals, i.e., emphasizing universalism such as the biological unity of humans that is consistent everywhere. Wiredu argues: "Suppose there were no cultural universals, then intercultural communication would be impossible. But there is intercultural communication. Therefore, there are cultural universals."[9] It can be argued that sympathetic impartiality can be ascribed to everyone, proposing it further as an intercultural universal, making universalism to be upheld in this context. Tangwa argues more accurately "that all *ethical* norms or rules are cultural universals, because a rule or norm cannot properly be described as 'ethical' unless it is understood as having cross cultural validity, in the sense of being perceived as applying in all similar circumstances, irrespective of place and time."[10] However, this is not in any way affirming common morality overriding customary moralities. For Tangwa, even though these fundamental ethical principles as laid out in PBE may have cross-cultural validity, relevance, and applicability, in every real-life situation or concrete context remains unique and different from all others.[11] We can reliably deduce that the focus on contextual realities influences the framework of bioethics in Africa.

A context-oriented bioethics is further defended, for instance, in Tangwa's studies. He situates bioethics in the framework of "beliefs, attitudes, dispositions, outlooks and practices given which, any of the myriad specific ecological or bio-medical problems of the contemporary world might be very differently perceived, appreciated and evaluated."[12] For him, however, although the African social construction of knowledge is born from *lived life* within the existential cultural realities, the elements of African bioethics are to be found in its contextual cultural values, traditions, customs, and practices[13] that make up *what life is* for an African in some given village. It is from the lived life that we learn peacefully to "live and let live," a lifestyle and an attitude of coexistence that

9. Wiredu, *Cultural Universals and Particulars*, 21.

10. Tangwa, "Universalism and Relativism," 63–64. Emphasis original. See more in Wiredu, *Cultural Universals and Particulars*.

11. See Tangwa, "Ethical principles," S2.

12. Tangwa, *African Bioethics in a Western Frame*, 52.

13. See Tangwa, *African Bioethics in a Western Frame*; Tangwa, *Contemporary Bioethics Problems*, 16–17; Tangwa, "Leaders in Ethics Education," 91–105.

is eco-bio-communitarian. So what is central to an African approach to bioethics is the "peaceful coexistence, promoted by the idea of *live* and *let live*" that "is possible even where love in the active sense is absent, but it would not seem possible where *reverence* and *respect* are absent"; hence, Tangwa's "preference for defining *Bioethics as reverence and respect for life*."[14] What respect and life mean in African moral thought await us in this lengthy investigation. But the reader should note that even when bioethics is understood as respect or reverence for life relationally, suffice it now to say that in this outlook of a Black African bioethics is rather understood in integrative terms, which point to bioethics as part of an African daily totality of life or living, but not stressed in absolute generic terms—which can be justifiably overridden based on conflicting circumstantial principles, as it is in bioethical principlism.

The General Critique of Bioethical Principlism

In more than four decades, Beauchamp and Childress have strenuously defended their framework of bioethical principles against diverse criticisms till their latest edition of *PBE* in 2019, when they commemorated its fortieth anniversary. They claim that their cluster of a four-principle approach, like we already noted, is a universally appropriate framework of moral principles because it originates from the *common morality*. The common morality, a set of the universal norms about right and wrong human conduct, is "shared by all persons committed to morality . . . in all places."[15] Given that common morality is universally foundational;[16] by contrast, customary, regional, or cultural moralities only seek *specifications* from common morality but are not part of it.[17] The long-sustained challenge, however, has been to demonstrate how the four principles of common morality are foundational in their specifications and so globally applicable in this way.[18] When challenges arose to their unspecific

14. Tangwa, *African Bioethics in a Western Frame*, 186. Emphasis added.

15. *PBE*, 3; while concluding this book, they reiteratively define the *common morality* as "the set of norms shared by all persons committed to morality . . ." 451.

16. *PBE*, 5, 26 n. 5. The *morally committed persons* are sometimes referred to as *morally serious persons*. The novelty in the 5th ed. is to define the common morality as ". . . the set of norms that all morally serious persons share." *PBE*, 3.

17. Beauchamp and Childress, "PBE: Marking Its Fortieth Anniversary," 11; *PBE*, 445; Childress and Beauchamp, "Common Morality Principles," 164–76.

18. See Herissone-Kelly, "Common Morality's Norms," 584, 587.

framework of principles in the 1980s, critics such as Danner K. Clouser and Bernard Gert emerged as the most unsparing critics. They coined the term "principlism" to refer to the practice of using "principles" to supplant not only moral theory but also a plurality of potentially conflicting *prima facie* moral rules and ideals in handling biomedical specific problems.[19] The difficulty to date remained to demonstrate how principlism would conclusively handle specifications while bracketing out regional contextual moralities. Contrastingly, even though common morality is a shared universal product of human history and experience, its standards of truth cannot exist independently of a *combined history* of regional moralities.[20] We can all agree that since the Global South, just like the Global North, has a history rich in moral beliefs, it would be a moral inconsistency (according to the wisdom of *PBE*) if some "persons who accept a particular morality sometimes presume that they can use this morality to speak with an authoritative moral voice for all persons. They operate under the false belief that their particular convictions have the authority of the common morality. These persons may have morally acceptable and even praiseworthy beliefs, but their particular beliefs do not bind other persons or communities."[21] And if granted, then this principlistic emphasis of some beliefs, even when combined cultural histories make up the common morality, is inconsistent with their outright denial of regional or historical cultural moralities as part of the common morality. With this moral inconsistency, we already note how bioethical principlism may hardly respond to practical cultural contexts in the morally pluralistic world.

In some instances, common morality claims are parochially cultural imperialistic since they not only preclude moral diversity but also moral beliefs from other regions or cultures. It is yet another reason why for ethnographers and medical anthropologists principlism fails. It fails because our historical cultural social contexts deeply influence how we think about principles and the manner in which we respond or how we invent right and wrong in counteracting specific bioethical questions.[22] So to deny cultural moralities is to deny contextual existences in which

---

19. See Clouser and Gert, "Critique of Principlism," 219–36; *PBE*, 13, 428.

20. See Beauchamp and Childress, "PBE: Marking Its Fortieth Anniversary," 11; *PBE*, 4.

21. *PBE*, 6.

22. Biomedical ethics ought to be a situational creative process rather than being principlistic. Traphagan, *Critique of Principlism*, 145–45.

pluralistic moralities reign and in which moral doers repeatedly make their own decisions. To show further instances in which principlism remains abstract and doesn't respond to cultural practical contexts, similarly Clouser, Gert, and other critics have looked at "principles" as merely a collection of sometimes superficially abstract moral concepts, which are not only too general and vague to guide *moral action*—viz., when confronted with bioethical problems—but also too indeterminate to supply a more directive decision plan of moral action. Besides, if we are to rely on these principles, given that they are neither derived from nor give birth to one unified moral theory, the contemporary biomedical conflicts remain generally indeterminable. Beyond not determining conflicts, these principles not only often conflict with each other but also lack any systematic relationship to each other.[23] The same challenges of a non-unified determinate theory are for example consistent with even their earlier editions, whereby "an agent's adherence to non-maleficence alone would not be sufficient to establish that *she* is committed to morality" given that an "agent-fallibility means that only those who are committed to all the common morality's norms can qualify as morally committed."[24] This indeterminacy is for instance crystal clear in Beauchamp and Childress' own theory of justice presented in *PBE*. For instance, they categorically note that the construction of a unified theory of justice that captures their "diverse conceptions and principles of justice in biomedical ethics continues to be controversial and difficult to pin down."[25] And it is why, even when principlists themselves partially succeed in bringing coherence and comprehensiveness to their multilayered and sometimes fragmented conceptions of social justice, still different "commentators see these theories as having the weakness of Plato's ideal state in the Republic: They provide models but not truly practical instruments"[26] that can be

23. See Clouser and Gert, "Critique of Principlism," 219–36; Beauchamp and Childress, "PBE: Marking Its Fortieth Anniversary," 9–10; Principlism uses four freestanding principles not embedded in any system. Gert et al., "Principlism." Moral action can be categorized as: (1) actions that are right and obligatory (e.g., truth-telling); (2) actions that are wrong and prohibited (e.g., murder and rape); (3) actions that are optional and morally neutral, and so neither wrong nor obligatory (e.g., playing chess with a friend); and (4) actions that are optional but morally meritorious and praiseworthy (e.g., sending flowers to a hospitalized friend)." *PBE*, 45.

24. Herissone-Kelly, "Common Morality's Norms," 584, 587; Emphasis added. I use feminine pronouns to equally stand for other genders unless indicated otherwise.

25. *PBE*, 267.

26. *PBE*, 300.

relevantly impactful in other regional contexts. Given this impracticality coupled with such indeterminacy, scholarships establish that principlism qualifies neither as a global theory (where moral norms operate globally, not merely locally) nor as a statist theory (where normative requirements of the principle of justice operate locally, but not globally).[27] The claimed specification remains not specifying, and the promised balancing remains unbalancing even in local approaches.

However, to make the principles more determinate, in Beauchamp and Childress's latest article, they still stress specification or balancing or both, meaning we should not anticipate uniformity everywhere since the four principles are shared as inescapable starting points.[28] Nonetheless, principlism finds further difficulties in specification, for example, in settling two autonomous imperatives of first- and second-order desires.[29] With due competence, I might have a first-order desire to socialize with my gay closest friend in Uganda, and yet another second-order desire not to socialize with him fearing Uganda's new *anti-homosexuality legislation*,[30] or even a third-order desire to postpone our desired social time. The question that challenges specification here is whether the various practical decisions made can be proved as fulfilling the standards of autonomous actions, or I simply acted non-autonomously. Yet again, in further defense, if we are to succeed with specification in biomedical ethics, "we do not always appeal directly to moral principles or to derivative rules. We appeal to them primarily in deliberation and justification in novel situations (e.g., involving a new technology), in uncertain or ambiguous circumstances, and in outright moral conflicts . . . including

27. Outside the state, there is no justice; meaning, the absence of a global state, there can be no global justice. Nagel, "Global Justice," 113–47; Global justice seems to be all about "us" treating "them," especially "their" problem. Nili, "Global Justice," 629–53; Powers and Faden, *Social Justice*, especially chapters 1 and 4–7, and from p. 69 about inequalities from social institutions, and setting priorities in chapter 6.

28. Childress and Beauchamp, "Common Morality Principles," 164–76.

29. Analogously Tangwa talks more or less about moral challenges relating to victims of vulnerability (*dual loyalty*) in African occupational health practice. Dual Loyalty—the nature of vulnerability to workplace hazards is related to the inability of workers to avoid risk or to change situations where they (or their families) are faced with hazardous exposures and remain fearfully reluctant to act so as to avoid or limit exposures to victimization or job loss. Tangwa, "Ethics in Occupational Health," 4; "an exceptionally dedicated physician who has a first-order desire to work extraordinarily long hours in the hospital while also having a higher-order commitment to spend all her evening hours with her family." *PBE*, 100–102; see also 116–17.

30. Republic of Uganda, Anti-Homosexuality Act, 2023.

rules of veracity, confidentiality, privacy, and fidelity as an approach to professional ethics."[31] If this is granted, it means that specification and balancing work "down" by applying universalized normative principles in determining what is ethically acceptable in all particular and regional moralities. But this attempt still unfortunately fails. In *PBE*, this can be demonstrated in the conflict arising from the application of biomedical substantive rules of rights to privacy[32] and confidentiality. *PBE* distinguishes between the right to privacy and the right to confidentiality and demarcates the limits enshrined in both circumstances:

> The basic difference between the right to privacy and the right to confidentiality is that an infringement of a person's right to confidentiality occurs only if the person or institution to whom the information was disclosed in confidence fails to protect the information or deliberately discloses it to someone without first-party consent. By contrast, a person who, without authorization, obtains a hospital record or gains access to the computer database violates rights of privacy but does not violate rights of confidentiality. Only the person or institution that obtains information in a confidential relationship can be charged with violating rights of confidentiality.[33]

Put differently, principlists claim on the one hand that it would be an infringement (violation) of a person's right to confidentiality if a medical practitioner or a medical facility disclosed a patient's medical results without the patient's prior consent. On the other hand, "we might justifiably disclose confidential information about a person to protect the rights of another person."[34] Meaning, what confidentiality gives with one hand it regrettably takes it away with the other. Well, amidst this indeterminacy, specification fails once again on the European soil in Andreas Lubitz's Germanwings Airbus premeditated and suicidal ideated crash case, in the southern French Alps, murdering 144 passengers and five crew members on March 25, 2015. As a fact, on March 10, 2015, the 27-year-old copilot was admitted to a psychiatric department for a psychotic episode

31. Beauchamp and Childress, "PBE: Marking Its Fortieth Anniversary," 9–10; *PBE*, 362, 414–15.

32. Privacy is a foundational good. Popular privacy is the kind that people tend to want, believe they have a right to, and expect governments to secure it. They embrace privacy for home-life, telephone calls, e-mail, health records, and financial transactions. Allen, *Unpopular Privacy*.

33. *PBE*, 342–43; see also Powers and Faden, *Social Justice*, 80–82.

34. *PBE*, 15.

and hospitalized for two days. Medical leave was recommended, but this time the certificate was not sent by the pilot to Lufthansa's Aero-Medical Examiner, meaning that on the day of the crash he was on medical leave but the airline was not informed by him or his doctors.[35] Notably, Lubitz's flying was not brought to an end; just because, even the doctors were equally mandated to maintain article 9 (section 1) on confidentiality (§9 *Schweigepflicht, Abs 1*) of the professional code of the German Medical Association (GMA): "Physicians are obliged to maintain confidentiality regarding everything confided in them, or becoming known to them, in their capacity as a physician, including after the death of the patient. This also includes written communications from the patient, records concerning patients, X-ray images and other examination findings."[36] Such emphasis of individual autonomy enshrined in the principle of confidentiality in GMA is not much different to what is upheld in the Code of Medical Ethics of the American Medical Association: "Patients need to be able to trust that physicians will protect information shared in confidence. They should feel free to fully disclose sensitive personal information to enable their physician to most effectively provide needed services. Physicians in turn have an ethical obligation to preserve the confidentiality of information gathered in association with the care of the patient."[37] The challenge we are generally facing here and now with such principlistic approaches is the overemphasis of the mightiness of

35. In this murder-suicide technical report, on the same day, during the first flight from Dusseldorf to Barcelona, Lubitz had already tested how to block the autopilot system for descent after the chief pilot went to the toilet. In April 2008, he was given his first medical leave due to depression, while in August, his condition worsened with suicidal ideation, and he was admitted to hospital for a short time. He was permitted to fly again in July 2009 with the condition that if there were any signs of relapse, he would lose his pilot's license and insurance. Psychiatric consultations began in August 2009 and continued until he relapsed in December 2014. Hospitalization was suggested with the suspicion of psychotic symptoms. From November 2014, he consulted five different general practitioners, one psychiatrist, and one psychotherapist getting prescriptions of different mild psychotropic drugs. Soubrier, "Self-Crash Murder-Suicide," 399–401.

36. "Ärztinnen und Ärzte haben über das, was ihnen in ihrer Eigenschaft als Ärztin oder Arzt anvertraut oder bekannt geworden ist—auch über den Tod der Patientin oder des Patienten hinaus—zu schweigen. Dazu gehören auch schriftliche Mitteilungen der Patientin oder des Patienten, Aufzeichnungen über Patientinnen und Patienten, Röntgenaufnahmen und sonstige Untersuchungsbefunde." Even with the recent changes of this code in 2015, 2018, 2021, section 1 of this Article 9 was not affected. For instance, there were conclusive changes on the 14 December 2018 of section 3 and 4, but not the first section of Article 9. Article 9 on Confidentiality (§9 *Schweigepflicht*). The German Medical Association (Die Bundesärztekammer).

37. American Medical Association, "Confidentiality," 3.2.1.

individual autonomy, in which, for instance, both Lubitz and his doctors take refuge, but principlism still seeks further defense via specification. Even specification suffers in the exercise of autonomous individualism. What specification gives with the right hand it likewise cunningly takes it away with the left hand. The reader shouldn't forget that oftentimes when we criticize individualism, we implicitly incorporate the principle of respect for autonomy as the presiding moral principle, overriding each of the conflicting moral principles in context; i.e., we hierarchically prioritize the principle of respect for autonomy over other conflicting principles. In defense, however, the principlists stress "that it is indefensible to construe respect for autonomy as a principle with priority over all other moral principles."[38] That is, even though the principle of respect for autonomy is a relevant *prima facie* principle, along with other competing *prima facie* principles, it has no more and no less weight than the others and we should never use an *a priori* ranking of principles, virtues, or rights in biomedical ethics—as if morality is hierarchically structured. Accordingly, exercises of autonomy can justifiably be restricted or limited or overridden by various moral and legal weighty considerations, e.g., circumstances of emergency, incompetence, waiver, a physician having a therapeutic contract with a third party.[39] For example, there are cases when choosing not to choose is the only available choice a patient with decisional capacity is left with. Although it is the patient's right as an autonomous individual to choose her own goals autonomously, it is the physician's responsibility to inform the patient of the feasibility of the desired goal and the means to achieve it. In instances where the goal is not possible, with the headship of the physician, the patient (who cannot choose autonomously) and the family should jointly choose a feasible and cost-effective goal. At this moment, the patient and the family cannot choose for themselves alone, otherwise it would not only be abandoning the patient but it would be an insensitive form of violating the patient's autonomy.[40] This is how moral paternalism—"the intentional overriding of one person's preferences or actions by another person"[41]—controver-

38. *PBE*, 143.

39. See *PBE*, ix, 105, 135–36; Beauchamp and Childress, "PBE: Marking Its Fortieth Anniversary," 11.

40. See Loewy, "In Defense of Paternalism," 445–68.

41. *PBE*, 231–32; There is soft and hard paternalism. For soft paternalism, one prevents considerable nonvoluntary actions, like poorly informed consent or refusal, severe depression that precludes rational deliberation, and addiction that prevents free choice. Hard paternalism calls for interventions intended to prevent or alleviate

sially overrides the patients' rightful autonomy by appealing to the goal of mitigating suffering. Getting back to the Germanwings incident, no paternalism was talked about as a possible mitigation in line with the well-known health concerns of the copilot well before the plane crash. In this very scenario, even Lubitz's doctors, just like him, had to maintain confidentiality as autonomous entities when specification was needed most. Though studies prove how such use of disproportionate autonomy isn't historically new, we cannot disagree that it has been repeatedly confronted without suggesting an integral solution. Even when confidentiality, as one way to exercise autonomous choices, has not been an absolute moral imperative in biomedical ethics historically, in response to Lubitz's suicidal case, historical data on medical confidentiality shows that medical practices of secrecy have been regularly attacked. Part of the recommended solution to rigid autonomous rights of confidentiality and privacy have been infringements that remain a prerogative of a *competent community*, at least to protect morally *endangered at-risk third parties*, as well as fix social rules that will keep future lives from harm,[42] not simply the unguided individual autonomous choices based on principlism. Otherwise, we can still justifiably approximate that bioethical principlism isn't yet successful.

There are other ways to show how principlism is ethically indefensible. We note injuries suffered due to the exercise of autonomy in relation to therapeutic privileges. They are circumstances when medical practitioners may withhold pertinent medical information from patients without their knowledge or consent under the belief that disclosure is medically contraindicated (therapeutic privilege), which creates a conflict between the physician's obligations to promote patients' welfare and

---

harm to, or to benefit, a person, even though the person's risky choices and actions are informed, voluntary, and autonomous. It is not morally controversial if one restricts a hallucinogenic drug patient or a patient with a documented disorder from leaving a medical facility (hard paternalism) or protects a person from harm caused to her by factors beyond her control (soft paternalism). *PBE*, 233; Though antipaternalists, neopaternalists or libertarian paternalists still enforce freedom of choice. Sunstein and Thaler, "Libertarian Paternalism," 1159–202; *PBE*, 236.

42. Three moments of confidentiality in history: First, at the end of the sixteeth century, lay authorities put pressure on physicians to disclose the names of patients suffering from syphilis. Second, in the eighteenth century, physicians faced constant demands for information about patients' health from relatives and friends. Third, employers and insurance companies in the twentieth century requested medical data on sick employees. Phillip et al., "End of Medical Confidentiality?," 149–54.

respect for their autonomy.[43] In settling this charge, principlists argue that, in spite of the fact that patients do not have an absolute right to be told the truth, especially all at once, disclosure under some conditions can be delayed, though full disclosure must not be permanently delayed. So, it is ethically indefensible—as well as unacceptable—even if legally permissible, to invoke the therapeutic privilege merely on grounds that the disclosure of relevant information might lead a competent autonomous patient to refuse a proposed treatment. Otherwise, this would contradict the principle of autonomy. Besides, disclosure of relevant information is a necessary condition for obtaining informed consent, mitigating emergencies as well as reporting medical errors.[44] Put differently, even when the physician considers the patient's autonomous decision to be wrong, the principle of autonomy mandates the physician to value the patient's autonomy even if she rejects the offered treatment out of the physician's beneficence. But then, on the one side, intentional non-disclosure of information is occasionally supported since scientists—to establish the prevalence of a particular disease—barely conduct vital research in fields such as epidemiology if they always have to obtain consent from *human subjects* or *participants* for access to their medical records.[45] On the other side, limited disclosure of some central facts in oncological practices, or missing out one fact, may generally be inadequate for a valid consent or refusal.[46] Despite that, in an attempt *not* to override the general norm of respect for autonomy, controversial questions still arise as to whether all "medical errors"[47] should be disclosed or only harmful ones. How and how much of these errors, and when to disclose them? Should a physician benevolently deceive[48] given hindrances to prognostications (unforesee-

43. Bostick et al., "Therapeutic Privilege," 302–6.

44. Bostick et al., "Therapeutic Privilege," 302–6; *PBE*, 126–27, 330.

45. See *PBE*, 123–26, 129, 131; I take the term "human subjects" to synonymously mean those human beings studied or those enrolled in research as "participants." Although a research participant may highlight a voluntary cooperation, many are enrolled in research by others. In other studies, a subject suggests a being subjected to or under control of others. See *PBE*, 143 n. 1, and 370 n. 1.

46. Horikawa et al., "Changes in Disclosure," 37–42. On nondisclosure to selective disclosure, see Elwyn et al., "Cancer Disclosure in Japan," 1151–63. Fifty percent of cancer patients desired to know when the disease becomes terminal. Nwankwo et al., "Attitudes of Cancer Patients," 1829–33.

47. Lamb, "Hospital Disclosure," 73–83.

48. Deceptive reporting to insurers of procedure indications to obtain reimbursement for non-covered services creates ethical and legal problems for practitioners. The motive for deceptive reporting is rooted in the expectation that any medical

able recuperation), so as to fix monetary insurance reimbursements, so that they secure third-party payers' approval of medically indicated care? All these controversial moral questions in relation to autonomy show how "civil litigation has emerged over informed consent because of injuries, measured in terms of monetary damages, that physicians intentionally or negligently have caused by failures to disclose."[49] However the legal framework to fix such non-disclosure errors along autonomous informed consent is still barely inexistent in locally institutionalized codes in SSA, proving more how principle-based approaches remain indeterminate in specific local bioethical interventions.

Briefly, we switch from general critiques of bioethical principlism to more specific critiques of bioethical principlism in African contexts. But if we are to hint on the weakness of the legal framework, specifically for Black Africa, the principlistic ethical guidelines issued by international bodies—like from the World Health Organization (WHO), International Labor Organization (ILO), UNESCO's Universal Declaration on Bioethics and Human Rights (UDBHR), and other United Nations (UN) family organizations—frequently serve as the *only source* of principal ethical standards in the weak regulatory infrastructure of Global South countries.[50] Other reasons why only these declarations apply in other local contexts include limited institutionalization of ethical codes, fragile democratic and governance processes, a limited culture of accountability and respect for the rule of law. For example, the rationale for UDBHR, according to Henk ten Have, director of UNESCO's Division of Ethics of Science and Technology, was a clear lack of bioethical guidelines and national bioethics committees. He said: "Three-quarters of all our members don't have a national bioethics committee, so there's no body to advise the government what to do in this area."[51] Meaning the final text of the declaration must be negotiated to palatably fit all local contexts. This sounds like a far-reaching positive goal on the one hand, while on the other hand the final text of the declaration met global local resistance.

---

intervention recommended by the physician—even if of marginal benefit—should be covered. Cain, "Obstetrics and Gynecology," 475–78. Some physicians may resort to deception to secure third–party payer approval for patient procedures. Freeman et al., "Lying for Patients," 2263–70; Amid spiralling medical malpractice, insurance costs and large legal settlements, there is some recognition that open, honest, and timely disclosure may lower legal bills. Lamb, "Medical Error," 3–5.

49. *PBE*, 123.

50. Tangwa, "African Thought," 107; Wolinsky, "Bioethics for the World," 354–58.

51. Wolinsky, "Bioethics for the World," 355.

Even when African scholars prefer acting from moral conviction instead of legal constraints, the domestic legal situation of African countries remains complex, whereby some countries have for instance bi-jural while others tri-jural systems, depending on unfavorable French or English or some other colonial backgrounds. What is underlined is the fact that no matter how effective, a regulatory framework remains at best toothless outside of the cultural region of its application and jurisdiction. Therefore, even if ethical guidelines are universally applicable in all countries, they are subject only to home institutional review ethics committees and domestic contextual coloration.[52] By and by, not only do principlists "accept the criticism that *their* principle-based analysis fails to provide a general ethical theory,"[53] but also the bioethical principlistic framework, in its rejection of regional cultural moralities and simultaneously excluding them from the common morality, still remains generally abstract, too indistinct, giving birth to an unsystematic and non-unified guide to determining both global and regional conflicts in the world of biomedical ethics. Not only are critiques of bioethical principlism general as we have discussed above, but critiques are also pertinent to non-Western regional ways of life. The concrete open question remains how and why bioethical principlism doesn't sit well in non-Western contexts.

The General Critique of Bioethical Principlism: Influential Moral Theories

The earlier discourses seem to claim that bioethical principlism as it is applied in non-Western contexts seems to be predominantly based on Global North cultural approximations and theories. In retrospect, bioethical interventions appeal to the English physician and ethicist Thomas Percival's code—the early code of *Medical Ethics* in 1794 (expanded and published in 1803)—as the first comprehensive account of medical ethics. Childress and Beauchamp admit that Percival's book served as the spine of British medical ethics and indeed as the prototype for the American Medical Association's first code of ethics in 1847.[54] With an indubitably Global North–based background as the claimed true

52. See Tangwa, "Ethics in Occupational Health," 2; Tangwa, "Universalism and Relativism," 64; Tangwa and Nchangwi, "Bioethics in Cameroon," 356–66.
53. *PBE*, 13, 431. "Our" is in the original, but is here changed to "their."
54. See *PBE*, 13.

beginning of *PBE*,[55] the principlists explicate four influential theories: rights, virtue, Kantianism, and utilitarianism; whereby, for them, knowledge of these theories is indispensable for any reflective study in biomedical ethics and the consequential development of *PBE*.[56] One challenge open to the proponents of principlism is that they rely basically (consciously or unconsciously) on the *cultural approaches and theories of the Global North as absolute*, excluding rich approaches, the would-be alternatives, from the Global South, to generally sort global biomedical problems or rather propose common morality for other cultural contexts. In this section, I discuss the theories briefly while maintaining greater emphasis to consequentialism.

A Critique of Virtue and Rights Theories

The first of the four theories is the theory of virtues. Although some scholarship claims that principlism discounts the virtues, at least we can strongly argue below that it appreciates differently how virtues and rights are pronounced in practical moral life. In the first place, even though a number of critics maintain that "principlism overlooks or even discounts the virtues,"[57] this misunderstanding stands not only in contrast to the principlistic understanding of a virtue, at least in Aristotelian terms, as a deeply entrenched commended dispositional ("ἐπαινοῦμεν δὲ καὶ τὸν σοφὸν κατὰ τὴν ἕξιν") trait of character[58] that is socially valuable and

55. They so clearly report how their proposed global approach is purely Global North based: "deontology and consequentialism were irreconcilably opposed theories between which one had to choose. Tom said . . . if pushed to make a choice . . . he favored a consequentialist over a Kantian or deontological approach . . . Jim favored rule deontology. We quickly realized that our different approaches could generate and sustain a common set of ethical principles for bioethical discourse and practice. This insight is probably the true beginning of Principles of Biomedical Ethics." Beauchamp and Childress, "PBE: Marking Its Fortieth Anniversary," 9. Here "Tom" is Tom Beauchamp while "Jim" is James Childress.

56. *PBE*, 31–32, 385.

57. *PBE*, ix.

58. We can reliably say that the principlistic virtue ethics relies on Aristotelean virtue ethics. "Some forms of virtue are called intellectual virtues, others moral virtues: Wisdom or intelligence and Prudence are intellectual, Liberality and Temperance are moral virtues. When describing a man's moral character we do not say that he is wise or intelligent, but gentle or temperate; but a wise man also is praised for his disposition, and praiseworthy dispositions we term virtues." (λέγομεν γὰρ αὐτῶν τὰς μὲν διανοητικὰς τὰς δὲ ἠθικάς, σοφίαν μὲν καὶ σύνεσιν καὶ φρόνησιν διανοητικάς, ἐλευθεριότητα δὲ καὶ σωφροσύνην ἠθικάς. λέγοντες γὰρ περὶ τοῦ ἤθους οὐ λέγομεν ὅτι σοφὸς ἢ συνετὸς ἀλλ᾽ ὅτι

morally reliably present in a person,[59] but also contradicts the broadly expanded part of the principlistic theory of "moral character," part of accounts of "moral norms," "moral status," "moral virtues," "moral theories," "moral ideals," and "moral excellence." These all reflect a considerable appreciation of the classic virtue accounts by Aristotle and David Hume.[60] In contrast to other studies, indeed Aristotle's rich contribution toward the ethical marriage between one's "external performance" and "internal state" as used in principlism is also still immaculately relevant today. Acts done are in conformity with the virtues *only if the agent also is in a certain state of mind* (ἐὰν ὁ πράττων πῶς ἔχων πράττῃ).[61] For instance, a kindhearted biomedical researcher down in an African village will appreciate an honest dialogue as she secures informed consent, a scenario that pronounces an interrelationship between right action (external performance) and proper motive (internal state). Indeed Aristotle is not suggesting that it is so worthily ethical for her to act kindly and loyally by

---

πρᾶος ἢ σώφρων· ἐπαινοῦμεν δὲ καὶ τὸν σοφὸν κατὰ τὴν ἕξιν· τῶν ἕξεων δὲ τὰς ἐπαινετὰς ἀρετὰς λέγομεν). Aristotle, *Eudemian Ethics*; Aristotle, *Ethica Nicomachea/Nicomachean Ethics*, 1103a32–1103b.

59. *PBE*, 411–12.

60. Beauchamp and Childress, "PBE: Marking Its Fortieth Anniversary," 9–12. In Aristotelian terms, virtues are moral excellences that lead to finest actions: "Since we hold that the truly good and wise man will bear all kinds of fortune in a seemly way, and will always act in the noblest manner that the circumstances allow; even as a good general makes the most effective use of the forces at his disposal, and a good shoemaker makes the finest shoe possible out of the leather supplied him, and so on with all the other crafts and professions." (τὸν γὰρ ὡς ἀληθῶς ἀγαθὸν καὶ ἔμφρονα πάσας οἰόμεθα τὰς τύχας εὐσχημόνως φέρειν καὶ ἐκ τῶν ὑπαρχόντων ἀεὶ τὰ κάλλιστα πράττειν, καθάπερ καὶ στρατηγὸν ἀγαθὸν τῷ παρόντι στρατοπέδῳ χρῆσθαι πολεμικώτατα καὶ σκυτοτόμον ἐκ τῶν δοθέντων σκυτῶν κάλλιστον ὑπόδημα ποιεῖ· τὸν αὐτὸν δὲ τρόπον καὶ τοὺς ἄλλους τεχνίτας ἅπαντας), Aristotle, *Ethica Nicomachea/Nicomachean Ethics*, 1101a1–5; For moral character and achievement, just like in principlism, in Aristotelian virtue ethics, self-cultivation and aspiration are necessary. *PBE*, 51.

61. ". . . but acts done in conformity with the virtues are not done justly or temperately if they themselves are of a certain sort, *but only if the agent also is in a certain state of mind* when he does them: first he must act with knowledge; secondly he must deliberately choose the act, and choose it for its own sake; and thirdly the act must spring from a fixed and permanent disposition of character." For a fuller Greek version: ". . . τὰ δὲ κατὰ τὰς ἀρετὰς γινόμενα οὐκ ἐὰν αὐτά πως ἔχῃ, δικαίως ἢ σωφρόνως πράττεται, ἀλλὰ καὶ ἐὰν ὁ πράττων πῶς ἔχων πράττῃ, πρῶτον μὲν ἐὰν εἰδώς, ἔπειτ' ἐὰν προαιρούμενος, καὶ προαιρούμενος δι' αὐτά, τὸ δὲ τρίτον ἐὰν καὶ βεβαίως καὶ ἀμετακινήτως ἔχων πράττῃ." *Nicomachean Ethics*, 1105a28–33; ". . . there is a state of mind in which a man may do these various acts with the result that he really is a good man . . ." (οὕτως, ὡς ἔοικεν, ἔστι τὸ πῶς ἔχοντα πράττειν ἕκαστα ὥστ' εἶναι ἀγαθόν, λέγω δ' οἷον διὰ προαίρεσιν καὶ αὐτῶν ἕνεκα τῶν πραττομένων), Aristotle, *Ethica Nicomachea/Nicomachean Ethics*, 1144a18–19. In the above citations, and in both language versions, emphasis is added.

not reporting the ineptitude of a fellow colleague who manipulates consents, nor is he suggesting that such a failure to report unethical research elements propose loyalty and kindness as not virtues! Aristotle's point is that a researcher's action in biomedical research among these Africans in some village can be virtuous only if performed in the right state of mind since both the right action and the right motive are present in a truly virtuous action.[62] Just like Aristotelian virtue ethics, principlism richly has it that "virtues may guide professional conduct better than obligations or rights."[63] Virtues and principles jointly work in applied practices of medicine. We can further think of some careful disclosure of HIV or cancer positive test results to some client. The disclosure necessarily demands exceptional care whereby one may even delay this disclosure (therapeutic privilege) so as to carefully prepare the client or exercise utmost respect of the client's autonomy. Notwithstanding the fact that both the right action and right motive are present in a truly virtuous action, absoluteness of virtues isn't implied. It is one reason why for Aristotle a virtuous person hits a middle point in actions (appropriate feelings), qualifying a *moral virtue as a mean* (ἀρετὴ ἡ ἠθικὴ μεσότης) between two vices (one of excess and the other of defect). This is what he calls elsewhere the *"mean amount,"* and *"the best amount is of course the mark of virtue"* (καὶ ὡς δεῖ, μέσον τε καὶ ἄριστον, ὅπερ ἐστὶ τῆς ἀρετῆς).[64] So we can scarcely discriminate the right action from the right motive as existent in a truly mean amount, i.e., the best amount or the mark of virtue. Besides, African

---

62. *PBE*, 35, 410–11; The five focal virtues—compassion, discernment (Aristotelian practical wisdom or *phronesis*), trustworthiness, integrity, and conscientiousness—are important for the virtue of caring (pp. 38–41).

63. *PBE*, 330. "A passionate, or a compassionate, engagement with others can blind reason and prevent impartial reflection" (p. 39).

64. ". . . to feel these feelings at the right time, on the right occasion, towards the right people, for the right purpose and in the right manner, is to feel the best amount of them, which is the *mean amount*—and the best amount is of course the *mark of virtue*." (οἷον καὶ φοβηθῆναι καὶ θαρρῆσαι καὶ ἐπιθυμῆσαι καὶ ὀργισθῆναι καὶ ἐλεῆσαι καὶ ὅλως ἡσθῆναι καὶ λυπηθῆναι ἔστι καὶ μᾶλλον καὶ ἧττον, καὶ ἀμφότερα οὐκ εὖ· τὸ δ' ὅτε δεῖ καὶ ἐφ' οἷς καὶ πρὸς οὓς καὶ οὗ ἕνεκα καὶ ὡς δεῖ, μέσον τε καὶ ἄριστον, ὅπερ ἐστὶ τῆς ἀρετῆς.). Aristotle, *Ethica Nicomachea/Nicomachean Ethics*, 1106b21–23. "Enough has now been said to show that *moral virtue is a mean*, and in what sense this is so, namely that it is a mean between two vices, one of excess and the other of defect; and that it is such a mean because it aims at hitting the middle point in feelings and in actions." (ὅτι μὲν οὖν ἐστὶν ἡ ἀρετὴ ἡ ἠθικὴ μεσότης, καὶ πῶς, καὶ ὅτι μεσότης δύο κακιῶν, τῆς μὲν καθ' ὑπερβολὴν τῆς δὲ κατ' ἔλλειψιν, καὶ ὅτι τοιαύτη ἐστὶ διὰ τὸ στοχαστικὴ τοῦ μέσου εἶναι τοῦ ἐν τοῖς πάθεσι καὶ ἐν ταῖς πράξεσιν, ἱκανῶς εἴρηται.) Aristotle, *Ethica Nicomachea/Nicomachean Ethics*, 1109a17–20. Emphasis added.

moral thought doesn't discount an ethic of virtues, and we likewise note numerous Aristotelian perspectives widely applied in *PBE*, suggesting it would be uncharitable for scholarships to continuously claim that principlism discounts the virtues. Indeed, in several biomedical research interventions in Africa, if a proportionate use of virtues transpired, professional conduct would have been richly exhibited.

The second of the four theories is rights theory, with the claim of absoluteness. For principlism, a right gives its holder a justified legitimate entitlement to determine by one's choices what others morally must or must not do.[65] One gets an impression that these entitlements are absolute and cannot be justifiably or permissibly overridden. But this claim is unsuccessful in various legitimate circumstances. For example, we recently had justified Ebola Virus Disease (EVD) infringements of movement rights in Sub-Saharan Africa. Given these legitimate infringements, the question is whether we can morally state that rights aren't absolute and so can be morally vetoed. Let us say the infringements of these rights may be contingently impermissible (an example is a *prima facie* right) or permissible (an example is a *pro-tanto* right).[66] Danny Frederick is convinced that if and only if it is impermissible to infringe a right in any possible circumstances, then the right is absolute, meaning an absolute right always overrides any competing moral consideration. Though we may argue that all rights are constantly absolute, it is necessarily permissible to override[67] or justifiably infringe a right due to the EVD. Such an infringement shows that right entitlements aren't in some circumstances absolutely held and can be justifiably or permissibly overruled or disregarded.

Contemporary local approaches to rights mostly refer to globalized Global North declarations as their sources. But since the understanding of globalized rights barely casts homeomorphic equivalents in the Global South, the Global North rights language remains unpersuasive and not suitable to manage local biomedical conflicts. These principlistic declarations seem to build their framework upon a Westernized rights theory approach.[68] On examining, for instance, both historical manifestations of the declaration of human rights of the French National Assembly of

65. *PBE*, 401.

66. See Frederick, "Pro-Tanto versus Absolute Rights," 375–94; Beauchamp, "Rights Theory and Animal Rights," 220.

67. See Frederick, "Pro-Tanto versus Absolute Rights," 375–94.

68. See *PBE*, 402.

1789 and the one approved by the General Assembly of the United Nations of 1948, it is the unconvincing Global North rights language at the apex of cultivating global principles, as Bénézet Bujo writes: "Through their consent, the member states of UNO, which signed the 1948 declaration, wanted to make it known that the human rights, which were agreed upon with the consent of all participants, had a binding character for themselves. At the same time this binding character was extended to all cultures although the point of departure of the discussion and the agreement was purely western and reflected Western culture. Against the reproach of a certain ethnocentrism, this approach doesn't fit non-Western cultures."[69]

The context in non-Western cultures sometimes presents dissimilar meaning to that of Western cultures in which the universalized rights language is wrought. The approach of human rights doesn't fit non-Western cultures, for instance, along its association with property. Article 17, 1–2 of the Universal Declaration of Human Rights of 1948 states: "Everyone has the right to own property alone as well as in association with others. No one shall be arbitrarily deprived of his property."[70] The approach to property—in which the drafters distinguished three types of property: personal property, real property (land), and profit-making enterprises—already suggests a Global North outlook. Besides, since a right gives its *holder* a justified legitimate entitlement to determine by *her choice* what others morally must or must not do, property here suggests individualism and privacy. In the African genuine worldview, property is not typically private since an African *lived life* cannot be privatized for oneself and cannot be accomplished without consulting with the communal life force—the extended clan family relatives without whose collective approval (lineage relationality) it is not permissible to sell or rent. Ignoring a people's valuable culture, therefore, as the new form of property, is the first violation of human rights.[71] We essentially need a new *lingua franca* to suit a global discourse in bioethics.[72] And if we are trying to develop a

69. Bujo, *Ethical Dimension of Community*, 146–47.

70. Peterson, "Universal Declaration of Human Rights," 43; Gudmundur and Eide, *Universal Declaration of Human Rights*.

71. See Bujo, *Ethical Dimension of Community*, 142, 149–50.

72. See Fenton and Arras, "Bioethics and Human Rights," 127–33; a merger between bioethics and human rights. Arras and Fenton, "Bioethics Human Rights," 27–38. The Universal Declaration on Bioethics and Human Rights, adopted by UNESCO on 19 October 2005 (hereafter UNESCO Declaration on Bioethics 2005), is an important step in the search for global minimum standards in biomedical research and clinical

global language, we already notice that using the culturally slanted rights theories in bioethics is still unsuccessful in developing a global strategy befitting a world community. Culturally tilted Global North principles are inadequate on their own in addressing global bioethical concerns without involving other cultural voices. We need a language and a moral framework grounded in a foundation of universally shared, transcultural perspectives that will also genuinely recognize bioethical moral pluralism (epistemological pluralism).

A Critique of Kantianism: A Nonconsequentialist/Deontological Theory

The third theory of the four is Kantianism. Principlism is grounded in autonomous use of reason associated with Kantian theory—the theory of duty, holding that some features of actions other than or in addition to consequences (as I will put it later in *consequentialism*) make actions right or wrong. In interpreting the German philosopher Immanuel Kant (1724–1804), for instance, principlists take the donation of the kidney to have moral worth that should be based on a generalizable/universalizable obligation done from duty of beneficence, without lying or coercing one's will, and not out of passion, self-desire, ambition, desire, compassion, or affection. If we consider a donation from a child, who can barely consent individually, Kantians would understand this as using someone as a means. Such an action will not only be against one's "autonomy of the will" but also against the universally valid moral principles. In this sense, it is only the rational will acting morally that chooses autonomously. It means Kantianism is grounded not in heteronomous reason (like I will argue later for Black Africans) but rather in autonomous reason.[73] Accordingly, in Kantian ethics, the principle of autonomy is such an exclusive principle of morals on which, for instance, human value, respect, and consent is grounded.

Before highlighting on consent, its foundational principle of respect of autonomy has two ethically weighty words: "respect" and "autonomy."

---

practice. Andorno, "Global Bioethics," 150–4. Human rights cannot constitute the whole of global health justice. Tasioulas et al., "Human Rights," 365–82. Human rights as an older notion of natural rights. Pagden, "Human Rights," 171–99. On the relationship between human right theories, theories of justice and bioethics, see Powers and Faden, *Social Justice*.

73. See *PBE*, 395–97.

"Respect involves acknowledging the value and decision-making rights of autonomous persons and enabling them to act autonomously."[74] The word "autonomy" here derives from ancient Greek roots: αὐτονομία, which was anglicized as "autonomy." From αὐτόνομος, *autonomos*, we get αὐτο–*auto*–"self" and νόμος–*nomos*–"law," to signify "one who gives oneself one's own law," or rather "having its own laws." Originally, *autos* ("self") and *nomos* ("rule," "governance," or "law") meant the self-rule or self-governance of independent city states. Humanizing this etymological sense, the autonomous individual acts independently as a master of herself, obliging others to respect her self-determination, analogous to the way an autonomous government independently manages its ministries. To put it more clearly, an autonomous competent person is expected to perform *a range of tasks independently*, meaning mental skills (competence) are closely related to cognitive skills (autonomy). In contrast, however, a person of *diminished autonomy* can considerably be controlled. In these cases, it is not only the person's autonomy controlled, but also the person lacks the competence that would give a valid consent or refusal, since she is not adequately capable of legally, psychologically, and cognitively taking part in decision-making.[75] So to give a valid

---

74. *PBE*, 104.

75. Cases of *diminished autonomy* include persons who are cognitively impaired, mentally incapacitated, incarcerated in jail, epileptic, having pancreatic cancer or amyotrophic lateral sclerosis (ALS), Alzheimer's disease, children born with hypoplastic left-heart syndrome, psychosis, dementia, transient ischemic attack (TIA), transient global amnesia (TGA), cases of immaturity, infants, alcoholics, those suffering from stroke, Parkinson's disease, chronic depression affecting cognitive function, senility, etc. Moreover, the principlists prefer a three-condition theory of autonomy: 1) Internationality: Intentional actions require plans in the form of representations of series of events proposed for the execution of an action, corresponding to the actor's conception of the act in question even when the planned outcome may not be materialized, given conflicting wants and desires. 2) Understanding: An action is not autonomous if the actor does not adequately understand it. Conditions that limit understanding include illness, irrationality, and immaturity. Deficiencies in a communication process also can hamper understanding. 3) Noncontrol: "free of controls exerted either by external sources or by internal states that rob the person of self-directedness. Not all influences exerted on another person are controlling. See *PBE*, 99–100, 102, 112–15, 136. Others mention three types of autonomy: autonomy of thought (thinking for oneself); autonomy of will, or the capacity to deliberate; and autonomy of action (examples include those paralyzed by curariform drugs). Autonomy is viewed as a prerequisite for all the virtues. Gillon, "Autonomy," 1806–8. Other principlists concentrate on autonomous choice rather than general capacities for self–governance and self-management. Gaylin, "Competence of Children," 33–38; Miller et al., "Children's Competence," 255–95. A child's decision-making competence—a strong legislative presumption that persons older than a certain age are competent, whereas younger persons are not. Hein et al., "Children's

consent or refusal (like in instances of informed consent), discretionary and volitional acts that are free from considerable controlling conditions need to be in place as a requisite condition not to preclude autonomous choices, since the principle of autonomy remains the exclusive principle of morals in the Kantian sense, which presupposes the moral doer's ability to act according to the moral law.

Instances in biomedical ethics where we vividly identify preclusion of autonomous choices were exposed in the Nazi's horrific medical experiments. Not until in 1970s, decades after the Nuremberg Trials held in the late 1940s, did ethics in medicine significantly make use of the term "informed consent" ("explicit consent'); besides, it protected the quality of a subject's understanding, autonomous choices, and any morally presumed perilous consent. Moreover, there have been circumstances when consent has been inferable from actions as implicit (or implied) or even tacit consent (occurring silently or passively through omissions).[76] Such an inferable consent sights uncoordinated applications of principles. For example, think of a patient who is believed to have donated her organs before her death. When an objection lacks, this lack represents consent in some cultures, at least an implicit form, since it is validly presumed based on lack of objections. Could we then go ahead to save a life that needs a transplant by relying on the deceased's intimate communal relationships? Or rather emphasize active principlism (autonomous choices) that won't increase organ donation rates from the dead since some cultural moralities don't entertain presumed consent which may rely on intimate communal relationships? It is one reason among others why founding bioethics on Kantian ethics, which emphasizes autonomous reason than heteronomous reason, is challenged as abstract within African scholarships that defend rather an *inclusively practical* collective approach rested on the competence of an extended clan lineage fellowship as we shall still emphasize later.

The debate between abstract and practical as well as the Kantian principlistic approach and African thought relates well with African scholarship. The reader should note that Africanists like Tangwa defend an African approach they claim not to be theoretical but rather practical. For Tangwa, it is why "in the domain of morality correct practice without

---

Decision–making Competence," 1; see also Hein, "Children's Competence to Consent," 74–74; Tangwa, "Ethical Principles," s5.

76. See *PBE*, 108, 118.

theory is preferable to correct theory without practice."[77] Here practice isn't tantamount to principlism, at least as far as implicit consent is concerned. For instance, as Tangwa specifically notes, "within Nso' culture, in the face of a question or suggestion, silence does not signify consent as it might within other cultures, but rather refusal or disapproval."[78] Tangwa finds difficulties in grounding ethics on moral principles *a priori*—morality as founded on the universality of autonomous reason—without taking individual contextual experiences into account. Likewise, other scholars like Bujo, in his masterpiece *Foundations of an African Ethic: Beyond the Universal Claims of Western Morality*, find this principlistic foundation abstract and noncommittal to practical contexts. Similarly, other scholars (like Bujo and Tangwa) think, as did the British philosopher Ludwig Wittgenstein (1889–1951)—in his scrutiny of language and its connection to the world in *Tractatus Logico Philosophicus* of 1921 and *Philosophical Investigations* of 1953—that the discourse of morality should not be founded on principles *a priori*, because all discourse on morality is subject to the language games in context and life forms of individual peoples. So in positing and justifying norms, whereas principlism builds and proceeds from a Kantian abstract rational manner, argumentations that seek an absolute "ultimate justification," the African approach embraces rather also a communitarian dimension.[79] Regardless, Kantian ethics contributes not only to the principlistic understanding of informed consent, the obligation to respect the reasoned choices of others and to the principle of respect of autonomy, but also to the quality of valued relationships we owe to others.[80] However, building bioethics not on a collectively inclusive practical dimension but rather on Kantian duty-based ethics leads us to tease out more challenges. Primarily, since for Kant moral rules are categorical, we are faced with unforeseen conflicts of two bioethical imperatives or laws. As an example, on the one hand medical practitioners have (often) "due care" obligations or "obligatory therapeutic contracts"

77. Tangwa, "Biomedical and Environmental Ethics," 387.

78. Tangwa, "Bioethics: An African Perspective," 196.

79. See Bujo, *African Ethic*, 22.

80. Therefore, it would be a mistake, as Christine M. Korsgaard also argues, to present Kantian theory as defending deontological constraints on the goal of promoting the good. In her interpretation, Kant does not claim that there is a general duty to promote the good that then must be constrained. But Korsgaard argues rather that "Kantians see the subject matter as the quality of relationships, what we owe to others . . . the norm that one should produce good outcomes derives from norms of proper relationships." *PBE*, 398.

to their patients under a third-party arrangement,[81] *inter alia*, police entrants, parents caring for an ill child, military applicants, candidates to religious institutions, athletes joining sport clubs, who send examinees to physicians; while on the other hand, care of the autonomous interests of the examinees conflicts with institutional codes to which medical practitioners are likewise committed (since they have "due care" obligations or "obligatory therapeutic contracts"). The question remains: whose needs should take moral precedence in the medical practitioner's eyes of fidelity since there is an unanticipated conflict between the medical practitioner's commitment to respect the patient's autonomy and the medical practitioner's obligation to the third party (ill child, athlete)? This question nevertheless fails in the practical application of principles. In our previous case of Andreas Lubitz's Germanwings Airbus crash case, conflict arose since doctors were unwaveringly accountable to two categorical codes. Would they inform Lufthansa's Aero-Medical Examiner (airline) about the medical leave in advance, or strictly adhere to the confidentiality (*Schweigepflicht*) professional code of the German Medical Association? Moreover, Kant discourages moral actions not executed from the duty of obeying or respecting the moral law, but from sympathy or emotions, since they have no moral worth. Again, his emphasis on obligations from moral imperatives (law) is barely ever practical amongst intimate relations or moral friendships or in the communitarian life of Black Africans, where sympathetic impartiality extends even further than the human communities to include other creatures to whom humans ought to extend respectful sympathy.

A Critique of Consequentialism

Here is the fourth and last theory. I turn to consequentialist accounts which majorly lay a ground in subsequently explaining numerous cases of Global North–Global South exploitative collaborations in biomedical interventions as evidenced in numerous scholarships. Some of these consequentialistic theories include utilitarian and justice accounts.

---

81. See *PBE*, 354–55.

Utilitarianism

Boethical principlism relies on theories that appreciate how actions are right or wrong in relation to their balance of good and bad consequences. The renowned consequentialists (welfarists) are utilitarians—with their basic principle of beneficence, best known as the principle of utility, the principle that balances positive value over disvalue. Among proponents are the hedonistic utilitarians Jeremy Bentham (1784–1832) and John Stuart Mill (1806–1873),[82] who perceive utility depending on whether moral actions consequently promote happiness (or pleasure). Contrarily, for principlists, principles of beneficence (utility) are neither sufficiently foundational nor absolute to ground all other moral principles and rules in the way many utilitarians have maintained.[83] Like we saw in the case of other principles before, principlists treat utility as only one among other equally central *prima facie* principles that can legitimately or justifiably be overridden in contextualistic circumstances, or even override other weighty *prima facie* moral principles.

Justice and Social-Economic Prioritizations

Another central consequentialist account is that of justice. For principlists, the term "fairness," what is deserved (entitlement), explicates the term "justice" in light of what is due or owed to affected individuals and groups. Likewise, the term "distributive justice," the general abstract material principle of justice, alludes to fair, equitable, and appropriate distribution of benefits and burdens, meaning fairness is when every person is treated according to rules and actions that maximize not simply the human welfare but more inclusively the public (social) utility. Moreover, the essential components of the theory of justice (material principles) identify further morally fair properties, for instance, the principle of need, that persons must possess to qualify for particular distributions. Though the discourse of justice focuses on questions of fairness, it is not only fairness that has been considered by other scholars, for instance the libertarians, as the only determining factor for just health policies.[84] Numerous discrepancies in concretization of fairness arise when the moral

---

82. Mill, *Utilitarianism*. For a deeper discussion of utilitarianism, see Frey, "Utilitarianism and Animals," 172–97.

83. See *PBE*, 218, 388–90.

84. Powers and Faden, *Social Justice*; *PBE*, 267–73.

focus shifts from maximizing the public (social) utility or welfare to articulating principles of *individual liberty rights* and participation in fair free-market exchanges. Among libertarians are John Locke and Robert Nozick. In Nozick's theory, justice consists in the operation of just procedures, *but not* in the production of just outcomes, such as an equal distribution of health resources.[85] Articulating a libertarian fair free-market explains why principlists are concerned that individual "expenditures for treatment, rather than prevention, are far higher in the current health care systems of most industrialized nations."[86] On top of that, amidst the limited resources, prioritization as a means to fairness remains dictated by "rationing,"[87] "responsibilisation,"[88] "*triage*,"[89] which are all majorly determined not only by fair socioeconomic factors,[90] but they largely favor the individual flourishing.

If we specifically think of the *triage* model of rationing, it has basically appealed to medical utility during the COVID-19 ICU (intensive care unit) choices of admitting (or denying access) and discharging patients amidst the available scarce resources, to an extent of withdrawing ECMO

85. *PBE*, 270–71, 274.

86. *PBE*, 303.

87. As of August 5, 2005, there were 96,189 patients awaiting organ transplantation. During 2004 only 27,036 transplants were performed, highlighting a growing disparity between patients listed and organs available. Smedira, "Allocating Hearts," 775–76. Need for organs and need in distribution decisions. Kamm, *Intricate Rights*, 207–20. A major problem is the growing gap between the demand for and supply of transplantable human organs. Dhooper, "Organ Transplantation," 322–27.

88. On encouraging "responsibilisation" in healthcare, see Brown, "Moral Responsibility," 695–98. Patients with alcoholic cirrhosis should not receive equal priority for scarce transplantable organs. Ubel et al., "Transplantable Organs," 600–7; Scarce resources necessarily force uncomfortable selection. Thornton, "Liver Transplant," 739–42.

89. A French term meaning "sorting," "picking," or "choosing." The term "triage" has been applied to sorting items such as wool and coffee beans depending on their quality. In biomedical ethics, triage is a process of prioritization. It has been used in war, in community disasters, and in emergency rooms where injured persons have been sorted for medical treatment. Triage decisions usually appeal to medical utility rather than social utility. For example, disaster victims are generally sorted or ranked according to medical need. White et al., "Life Support," 132–38; Rasita et al., "Ethics of ICU Triage," 5–15. Limited resources necessitate rationing and triage (prioritization) decisions in South Africa. Joynt, "ICU Triage and Rationing," 613–29; *PBE*, 312.

90. Few developing countries can afford dialysis programs and those that do, ration this scarce resource. In South Africa, rationing has been practiced, and almost 60 percent of patients were denied renal replacement treatment because of social factors related to poverty. Socioeconomic factors influence decision to accept patients more profoundly than medical ones. Moosa and Kidd, "Rationing Dialysis Treatment," 1107–14.

(extracorporeal membrane oxygenation, a technology for treatment of respiratory failure).[91] The moral question of criteria—whether we had to adhere to medical prognosis, age, life expectancy, quality of life, *skin color/ race*, or the *fair innings argument*—remained ethically controversial in recent COVID-19 cases of bioethical rationing. Considering the last two points, *first*, regarding race and color, there have been and are still several disparities between "black-white patients"[92] as far as rationing of scarce resources is concerned. Some studies "demonstrate that beliefs about biological differences between blacks and whites—beliefs dating back to slavery—are associated with the perception that black people feel less pain than do white people."[93] Consequently, some rationing disparities in health care arise from such conscious (explicit) or unconscious (implicit) racial biases among physicians, which contribute to color/racial/ ethnic disparities in the use of medical procedures.[94] Basically, relying

91. Some patients in ICUs are too sick to derive much benefit from them, while others are too well to require the technology and skills offered. Truog, "Triage in the ICU," 13–17. Since children are unable to consent, the provision of standard therapy may be legally and ethically obligatory. Frader and Lantos, "ECMO and the Ethics," 409–13. ECMO is a technology for the treatment of respiratory failure in newborns. Mike et al., "ECMO," 212–18. ECMO has been used to support cardiorespiratory function during pediatric cardiopulmonary resuscitation (CPR). Thiagarajan, et al., "Resuscitation in Infants," 1693–700. Life support systems in acute fatal illness has problems of logistics, ethics, and consent. Bartlett, "Acute Fatal Illness," 456–65.

92. Disparities between white patients and Black patients have not substantially improved during the past decade or so. Lurie, "Health Disparities," 727–29. African American patients had significant access problems in obtaining coronary artery bypass graft (CABG) surgery. Hannan et al., "Coronary Artery Bypass Surgery," 68–77. For the decade of the 1990s, efforts to eliminate racial disparities in the use of high-cost surgical procedures were unsuccessful. Jha et al., "Racial trends," 683–91. Disparities in the management of coronary artery disease differ for Hispanic and black populations. Freund et al., "Acute Coronary Syndromes," 126–32. African-Americans continue to have the worst heart disease survival rates of all racial groups. Capers and Sharalaya, "Cardiovascular Care," 171–80, Women and black patients eligible for a primary prevention implantable defibrillator device (ICD) are less likely than men or white patients to receive one. Hess, "Cardioverter-Defibrillator Counseling," 517–26, Disparities in outcomes for African American kidney transplant recipients have persisted for forty years after analyzing 202,085 transplantations. Taber, "Kidney Transplant," 878–87.

93. Hoffman, "Racial Bias," 4296–301, especially 4300. The study claims that Black Americans are systematically undertreated for pain relative to white Americans. It examines whether this racial bias is related to false beliefs about biological differences between Blacks and whites (e.g., that Black people's skin is thicker than white people's skin).

94. Green et al., "Implicit Bias," 1231–38. Moral on disparities, see Blair et al., "Unconscious (Implicit)," 71–78. Although many physicians demonstrate an implicit (unconscious) preference for white people, this bias does not appear to impact their

on principlism may neither promote justice nor morally ameliorate such cultural and social beliefs. *Second*, in rationing of scarce resources, the fair-innings argument (FIA) is frequently put forward as a justification for denying elderly patients treatment when they are in competition with younger patients, whereby some find it tragic if a young person dies but only unfortunate if an elderly person does.[95] In the FIA—a fair period or amount of time in one's whole life-time experience—each has an equal chance to get fair innings. Alan Williams, the proponent, argues that everyone is entitled to some "normal" span of health in one's entire lifetime, and anyone failing to achieve this equality in health care has been cheated, whilst anyone getting more than this is "living on borrowed time"—that is above the threshold, say of seventy years of age, and no longer entitled to receive social health care support.[96] In other words, the FIA converges around priority for younger patients given that medical utility favors them for more quality-adjusted life-years ahead. This is one reason why egalitarian justice also prioritizes young people since they have not yet had their fair innings (they have neither maximized their appropriate chances nor moved through stages in the life cycles).[97] Interpreting the FIA argument, that greater longevity will impose an unfair burden on younger generations, is to continuously attack an anti-ageist stance. If we consider the current increase in life expectancy and rapid increases in the older-age population, which is generally considered a positive development, it already raises consequential health care burden concerns given the United Nations' projected growth in the older-age population of threefold by 2050.[98] The "burden concerns" reiterate a language of individualistic socioeconomic terms that basically dictate libertarian prioritization in bioethics and enforces intergenerational bias against the elderly, thereby not enforcing an integral social utility, which must not only cater for the present but the past as well as the future generation without discrimination.

The dictates of individualistic socioeconomic factors at the heart of bioethical principlism can equally be explained, for instance, amongst

---

clinical decision making. Dehon et al., "Physician Implicit Racial Bias," 895–904.

95. Rivlin, "Fair Innings Argument."

96. Intergenerational equity requires greater discrimination against the elderly. Williams, "Intergenerational Equity," 117–32.

97. *PBE*, 306, 310. Though Beauchamp and Childress' priority, an allocation leading to specification emphasizes not only an egalitarian strategy of equal worth of persons and fair opportunity, but also emphasizes a utilitarian strategy of overall maximum benefit for both patients (especially the young) and society. *PBE*, 308.

98. Hazra et al., "Fair Innings," 209–17.

uninsured persons who donate organs, but rarely receive them.[99] Studies note, however, that in allocating such scarce organs or even vaccines, no single principle (be it an economically oriented principle) has sufficiently and inclusively incorporated all morally relevant considerations, and therefore bioethical principlism must be combined into multi-principle allocation systems.[100] It becomes clear that in rationing organs, principles that place economic terms at the forefront for freely donated organs are ethically ineffective and incur a moral cost. The argument of equitable access finds it morally inconsistent to ration freely donated organs according to a patient's economic ability to pay. Yet if life was to be rescued, it shouldn't be a marketable piece of merchandise like commercializing donated heart transplants. Accordingly, the principlistic arguments of prioritization through rationing, the triage model, FIA, free-market economies, and commercialization of bioethical interventions, rightly promise fairness in allocation of scarce health care resources, but unfortunately fail to practically deliver it (in consequence). And if we presently speak with a language of free-market economies and commercialization of bioethical interventions, what becomes apparent are increased socioeconomic exploitative consequentialist collaborations, which have steadily escalated in the Global South.

In this chapter, basically principlism has relied on the cultural approaches of the Global North to dictate a common morality suitable for the Global South. In such an absolute nature, bioethical principlism seems to encounter more challenges than it can solve in numerous African contexts. Such culturally tilted Global North principles seem to appear insufficient for addressing concrete Sub-Saharan bioethical concerns that bear a global mark. Otherwise the risks of exploitation in bioethics seem to cut across as general weaknesses and inadequacies of an entire bioethical system. What new comprehensive language of bioethics do we need to address the current trend of an exploitative moral framework? Won't it be constructive to respond to specific non-Western sociocultural realities so that we can accurately speak of an appropriated approach to bioethics within an African context without being complacent with the current wave of universalization or globalization of principles?

99. Herring et al., "Organ Donors," 641–52.
100. Persad et al., "Scarce Medical Interventions," 423–31.

3

Critique of Bioethical Principlism in an African Context

African cultures, world-views, philosophies, customs and practices which, traditionally, were, in the general sense, evidently bioethically very sensitive, have not yet awoken to the new bioethical challenges for fairly obvious reasons, the more palpable of which are that, since the phenomenon of western colonization, Africa has remained a mere consumer of western ideas and products and that the developments which make bioethics and biolaw contemporary imperatives have not yet been truly felt in Africa. And yet, African traditional systems might indicate or suggest better or more viable ways of tackling bioethical challenges in a world whose cultural frontiers are rapidly shrinking. If African traditional systems are looked upon, as they should, as viable *alternatives* rather than, as they usually are, as mere *approximations* to western "ideals," they stand to contribute great values to an emergent world culture made possible and inescapable by western advances in technology and communication.[1]

In spite of the fact that African scholarship acknowledges the universalized fundamental biomedical principles, their relevance and applicability seem not to conform to concrete local realities. Besides, studies disclose

1. Tangwa, *African Bioethics in a Western Frame*, 143. Emphasis in original.

that the universalized principles cannot be absolutely valid everywhere in their current form in all non-Western localities. To show how principlism doesn't sit well in various African contexts, this chapter presents manifold illustrations of specific case studies of exploitative North-South consequentialistic collaborations, which are more pertinent to the principles of autonomy and justice as variedly understood in the African outlook, in bioethical principlism and in numerous principlistic Western studies. And since these exploitations are largely controlled in Western countries, the parallel continual increase in cases in the developing countries presents them, not as viable *alternatives* but rather as mere *approximations*. I make use of distinctive cases to show how bioethical principlism does not concretely sit well in contextual African realities. In this section, having briefly discussed theories in the previous section, I now maintain greater emphasis to consequentialism and its case-based challenges in African settings.

AN AFRICAN CRITIQUE OF BIOETHICAL PRINCIPLISM: SPECIFIC CASE STUDIES OF EXPLOITATIVE NORTH-SOUTH CONSEQUENTIALISTIC COLLABORATIONS

Developing world populations, particularly those of Sub-Saharan Africa, are highly vulnerable in medical research which has witnessed an exponential increase in recent years for various reasons. While much of this research is easily justifiable as aiming to serve the urgent health needs of the populations subjected to it, care must be taken to avoid harming, exploiting or otherwise treating research subjects in ethically unacceptable ways. The research scandals that have been witnessed in recent times in countries such as Nigeria, Kenya, Tanzania, Cameroon, etc. show that the protection of research subjects, particularly in Sub-Saharan Africa, is an important ethical imperative whose observance cannot be taken for granted.[2]

The underlying ethical denominator here are vulnerabilities exploited in the bioethical interventions as reported from various scholarships. In his

2. Tangwa, "Vulnerable Human Beings," S19. Pages indicated thus.

study *Research with Vulnerable Human Beings*, Tangwa strongly indicates how world populations necessarily require protection from the possible outrageous abuses of research that have repetitively transpired in recent world history, stemming more specifically from vulnerabilities inclined towards the helpless needy-weak, the poor-sick, and those economically disadvantaged. Bioethical interventions in SSA, for instance, reveal how several putative ethical imperatives in biomedical research were not fulfilled, even when these Global North–funded studies, though abusively disreputable, progressed through and through as I will give evidence in the following case studies.

Tangwa notes that "the exploitation of human vulnerabilities has been abundantly manifested in coercive research, deceptive research and inducive research."[3] Coercive research achieves its scandalous objectives unethically through unashamed harm towards innocent unknowledgeable human subjects, while deceptive research misleads these participating subjects along multifarious circumstances of inadequate or less-informed consents. Induced research elicits the poverty-stricken or ignorant subjects to do what they would not have done or rather act against their would-be judgements. In point of fact, configuring biomedical ethics in economic terms doesn't sit well in African vulnerable settings as Tangwa claims. His thought contextualizes a moral doer, the human subject or participant, in the known distinctive characteristics of her world. For him, "since moral obligations are obligations only from the point of view of a particular *moral agent*, we need to make appropriate distinctions between the ethics of developed world research in the developing world, collaborative or cooperative research between the developed and developing worlds, developed world research in the developed world and developing world research in the developing world."[4] Making this distinction confirms the consequences of vulnerability, whereby in many Western driven biomedical research projects in developing Africa there are controversial moral problems associated with enrolling the economically underprivileged, those in constraining or coercive situations (like acute illnesses without medication or food), or those who are homeless. Such participants seem vulnerable to exploitive

3. Tangwa, "Vulnerable Human Beings," S18.

4. Tangwa, "Ethics Committees," 156. Emphasis added. *A moral agent* can be morally evaluated, praised, or blamed for its motive and actions. See DeGrazia, *Taking Animals Seriously*, 203.

abuse.[5] These morally constraining or abusive situations are oftentimes shaped in monetary offers, which consequently generate moral problems related to injustices of exploiting vulnerable situations, which are also rightly captured in *PBE*. They occur "from a distributive injustice of too small a payment to subjects by contrast to an irresistibly attractive, large payment. In the undue-profit situation, the subjects in research receive an unfairly low payment, while the sponsor of research garners more than is justified . . . two moral problems of exploitation—undue inducement (unduly large and irresistible payments) and undue profit (unduly small and unfair payments)."[6] These undue inducements make use of all accessible vulnerable situations.

To explicate these undue inducements, Tangwa notes within the Global South volunteers and desperately poor study participants who are offered irresistible gifts or amounts of cash as compensation.[7] The corruptive and profit-oriented commerce conquers sacred vulnerable human relationships through inducements. This unholy relationship of inducements turns out to be manipulatively destructive rather than constructive to the only God-given life that all human beings have equally. Here the bait used by the Global North to effect these inducements toward the Global South is still in economic terms. The ethical challenge of undue inducements arising from commercialized (market-oriented and profit-driven) bioethical interventions is linked to less respect for others, their culture, and their whole vulnerable way of life as Tangwa notes:

> Industrialized world medical research is not only based on physicalism and materialism, it is also generally market-oriented and profit-driven, and tends to be susceptible to morally blind economic forces. As a consequence, it engages to a high degree in ad hoc rationalizations and justifications. For these reasons, the ethical challenges of industrialized world research, especially in non-Western contexts, are many and varied. Is it possible to combine commercial motives with philanthropic altruism or ethical imperatives in general? Can high-tech medical research avoid harming the vulnerable or exploiting the desperately weak, poor, and ill? Can it avoid undue-inducement . . . ?[8]

5. Hawkins and Ezekiel, *Exploitation and Developing Countries*.

6. *PBE*, 289.

7. Tangwa et al., "Ebola Vaccine Trials," 49–60; Tangwa and Nchangwi, "COVID–19," 7.

8. Tangwa, "African Thought," 105; see also Tangwa, "Bioethics, Biotechnology and Culture," 136–38.

Tangwa further summarizes these economically driven exploitative inducements with a broadened distinction between "due" and "undue" inducement.[9] For him, reimbursements to research participants for both monetary and non-monetary expenses incurred may be so attractive to cause them to join the research study against their would-be sober moral judgments as I noted above. But the legitimate moral question remains to establish when an inducement is rightly due to a study subject and how distinctive it would be from any compensatory fees. The moral paradox presented by bioethical principlism remains if a *small inducement* may be permissible while only a *big inducement* is impermissible, and more so, to which extent would both inducements be objectionable or beneficial to the study subject or participant! Although some studies defend undue inducement on grounds of mutual benefit,[10] there is something morally repugnant about offering monetary payments to relatively impecunious people for the purposes of taking part in research. In actual fact, the monetary inducement does not only add to their difficulty—as vulnerable targets of bioethical interventions—in adequately estimating the risks of participating, but also the acceptance of inducement is likely to increase the inequity of research for the benefit of the "well-off" or "funder."[11] African moral thought already notices the difficulties with principlistic dualistic tendencies that dichotomize between *due* and *undue moral inducements*. How can such dichotomized inducements be *due* to study subjects since inducements will forever remain inducements!

To be more specific, African scholarship refers to these twenty-first-century exploitive unethical principlistic practices from high-income to lower-income vulnerable settings as "ethics dumping,"[12] the term introduced by the European Commission. Under intensified "globalization"[13] of research activities, Doris Schroeder et al. note that *"research with sensitive ethical issues is conducted by European organizations outside the EU in a way that would not be accepted in Europe from an ethical point of view.* This exportation of these non-compliant research practices is called

---

9. Tangwa, "Vulnerable Human Beings," S19; Tangwa, "Universalism and Relativism," 64; Tangwa, *African Bioethics in a Western Frame*, 163.

10. See Wilkinson and Moore, "Inducement in Research," 373–89.

11. See Russell et al., "Paying Research Subjects," 126–30.

12. Tangwa et al., "Ebola Vaccine Trials," 49–60; Tangwa and Nchangwi, "COVID-19," 7. Other studies understand *ethics dumping* as *off-shoring*. Marouf and Esplin, "Care in Clinical Trials."

13. Sørensen, "Globalization," 143–48.

ethics dumping."[14] This interest in ethics dumping targeted at vulnerable African countries has been more than often linked to biomedical research ethics and practices that are in many cases incompatible with African cultural and social diversity as UNAIDS notes.[15] Precisely, ethics dumping occurs mostly in two vulnerable areas. "First, when research participants and/or resources in low- and middle-income countries (LMICs) are exploited intentionally, for instance because research can be undertaken in an LMIC that would be prohibited in a high-income country. Second, exploitation can occur due to insufficient ethics awareness on the part of the researcher, or low research governance capacity in the host nation."[16] Here exploitation manipulates the whole vulnerable system, which awakens significant moral glitches associated with exploitive consequentialist ethical dumping in LMICs[17] disguised lately as "modernization" or "globalization." For instance, as a way to show environmental injustices in the seemingly globalized bioethics, Richard N. Rwiza notes how it is common to globally trade in toxic substances by exporting hazardous wastes from developed to developing countries.[18] And if principles of globalization mean the intensification of economic, political, social, biotechnological, and cultural connections across borders,[19] but not the kind of globalization that "began with the slave trade, since slaves were persons without culture, who could be bought and sold,"[20] Tangwa finds nothing wrong with it, in and of itself. For Tangwa, however, globalization which "has turned the world into a veritable global village, is increasingly bringing both the benefits and hazards of Western technology and, especially, biotechnology to all parts of the globe. Some of the most urgent hazards involve human health and the global physical environment."[21] Moreover, the dichotomized "globalization creates insiders and outsiders, integration and fragmentation, and winners and losers. Those who are able to take advantage of globalization, usually those already with power and

14. Schroeder et al., "Ethics Dumping," 2. Emphasis added.

15. UNAIDS, "Good Participatory Practices"

16. Schroeder et al., "Ethics Dumping," 2. There are numerous studies on research misconduct in Africa proving how underdeveloped policies in Africa are misused. Nortjé et al., *Research in Africa*, 117–21.

17. See Tangwa, *Contemporary Bioethics Problems*, 47–48.

18. Rwiza, *Environmental Ethics*, 150; see also 137–60.

19. See Sørensen, "Globalization," 143–48.

20. Bujo, *African Ethic*, xii.

21. Tangwa, "Biomedical and Environmental Ethics," 394.

resources, gain at the expense of those unable to modernize or use opportunities, usually those who are already marginalized, who become even more powerless."[22] This seems generally consistent with the principlistic ethics dumping of the twenty-first epoch, where some people or cultures develop an attitude of being worthier of healthy surroundings in the global village than others. So, we end up having a global village that is dichotomized into the have and have nots, who have to appeal to the Global North based international organizations for fair legislative declarations. For example, amid the present-day globalized declarations—via the United Nations, the World Trade Organization, pharmaceutical pricing and marketing boards, and the World Health Organization—"some groups have called for a world ethos that would work against the negative effects of globalization."[23] Within the cries of these groups, what needs to be known, we should not simply brush off the negative effects of a global monoculture within pluralistic ethical insights, which global efforts have destroyed the world by levelling down unique historical cultures.[24] And if we consider a worthwhile life as necessarily requiring certain minimum living conditions for all cultures as never optional, they are living realities that should never be neglected between the Global North and Global South bioethical approaches and collaborations. Given this evident ethics dumping, I have suggested three areas of morally challenged realities to which I am to subsequently supply evidence in the next sections: 1) exploited interpersonal and humanistic collaborations (male circumcision trials and Ebola virus vaccine trial); 2) exploited structural-universal collaborations (lack of cooperative values and corruptive collaborations); and 3) exploited cosmic-ecological relationships (genetic manipulation in non-human primates and genetically modified human food trials).

Exploited Humanistic and Interpersonal Relationships

In this foremost category are exploited humanistic and interpersonal relationships. First is an example of exploitive rampant circumcision trials. Though circumcision rites have been generally practiced in Africa prior to the arrival of colonial-era missionaries, the same missionaries described these practices as barbaric, backward, and incompatible with

22. Tangwa, "Ethics in Occupational Health," 3.
23. Bujo, *African Ethic*, xii.
24. See Bujo, *African Ethic*, xii, 74.

Christianity, consequently spearheading their abandonment.[25] It is only later on that circumcision received a new impetus basing on some studies that appeared to support its effectiveness as a prevention strategy in the fight against HIV infections[26]—paradoxically, in a twenty-first century where girls and women circumcision is a highly contested human rights' issue world over. In contrast to this global campaign, Tangwa's studies reveal that in 2007 the WHO and the Joint United Nations Programme on HIV and AIDS (UNAIDS) approved a campaign to circumcise millions of African boys and men.[27] To this Tangwa notes, "although the cohorts were not deliberately singled out by race or class, the 'subjects' across the four circumcision trials were virtually all poor African men and women, raising concerns about vulnerabilities that are consistent with a history of medical exploitation."[28] It is in these historical and geopolitical African contexts where biomedical shortcomings like exploitations and disparities have remained open to unending ethical questions till today. Any researcher would wonder, just like Tangwa does, why these affected communities of color in the Global South cannot assume a more central role in medical decision-making, especially about their own life or bodies: "The inclusion of African men and women in decision making regarding their own bodies, cultures, and health was—and remains—overshadowed by a Western hegemonic model promising HIV reduction

25. As a fact, in every country in Africa, there is an ethnic group that practices circumcision for cultural or religious reasons. See Gwandure, "Male Circumcision," 89. Conversely, male circumcision has been later imposed by powerful groups onto the marginalized. In Kenya, a tradition of forcible circumcisions of non-circumcising Luo tribal minorities is longstanding. Tangwa et al., "A new Tuskegee," 9; Muraya, "Female Circumcision," 26–45. Anthropological engagements with questions of illness and healing in Africa at the dawn of colonialists. Upton, "Illness and Healing," 97–117; Camilla and Yonatan, "Female Genital Mutilation," 356–82; Glass, "Forced Circumcision of Men," 567–71; Lamont, "Forced Male Circumcision," 293–311. Circumcision dates immemorial in Africa. Colonial explorers like John Hanning Speke were circumcised on the field of battle during the search for the source of the Nile by local Somali men: "One of them shrieked, charging at Speke–Speke parried a sharp blow that snapped off the blade of his sword. Speke was in a daze when one of them, pressing a long knife to his throat. Lashed out 'Circumcision or death, you Christian dog.' He pulled Speke's foreskin and stretched it tight, then sliced it off with his razor edge blade." Edwardes, *Death Rides a Camel*, cited from Aggleton, "History of Male Circumcision," 17.

26. Fox and Thomson, "HIV/AIDS and Male Circumcision," 107. Male circumcision reduces HIV incidence in African men; see Gray et al., "Circumcision for HIV Prevention," 657–66; and Bailey et al., "Circumcision for HIV Prevention," 643–56.

27. Tangwa, et al., "New Tuskegee," 1; WHO, "Traditional Male Circumcision."

28. Tangwa, et al., "New Tuskegee," 4.

to African men through circumcision."[29] I take interest in the use of the word "hegemonic."

Here the adjective "hegemonic" (from the Greek word *hēgemonikos*) originally indicates the capability to command as a leader (*hēgemōn*). Yet for an African this capability legally binds only if it involves the leader (chief) sharing and owning the well-intentioned message of circumcision (from any authoritative official) with the local people; otherwise it would be a mysterious decision that simply comes down to them as a monologue, i.e., through some hegemonic superficiality. Opposed to hegemony, in response to Safe Male Circumcision (SMC) funded through the US Centers for Disease Control and Prevention, in collaboration with Africa Comprehensive HIV/AIDS Partnership (funded by the Gates Foundation) while teaming up with Botswana's Ministry of Health, one elder is reported to say: "We have expressed that we thought Ministry of Health was going to help. Now they send us . . . young people to insult us. They should respect our culture and not confuse it. Go and tell them to come and talk to our chief."[30] Since the chief is their voice and custodian of their ancestral moral tradition, the seemingly superficial government programs cannot be uplifted at the expense of promoting life through cherished local traditional values, but simply remain a hegemonic model that is characteristically a *monologue* with less room for a genuine *dialogue*.

Promoting life across human communities demands we proceed from cultural monologue to bicultural, inter-, intra-, and transcultural forms of dialogue.[31] If we are to talk about transcultural forms of dialogue in bioethics, at least from an African approach to ethics, Onwubiko recommends that we have to necessarily presuppose "cultural parity," i.e., the "culture at the export level" should remember that the vehicles of its exportation are only elements of a culture, its products, and not the

29. Tangwa, et al., "New Tuskegee," 9.

30. Mageso and Marguerite, "Safe Male Circumcision," 748.

31. In the analysis of Achebe's famous *Things Fall Apart* of 1958, while arguing against hostilities directed towards the African way of life (intracultural manipulations) as burdened with racism and *hegemonic Eurocentrism* (in its three forms: colonialism, civilization, and Christianity), Onwubiko suggests a dialogical transcendence that implies a form of biculturalism (an interplay of "acculturation" in a two-directional nature), leading on to transculturalism. Onwubiko, "Re-Encountering African Culture," 91, 95, 100; see also Van Binsbergen, "Hermeneutics of Race," 8. Tangwa also stresses "Eurocentric hegemony" as "derived from colonialism and colonial indoctrination cum proselytization." Tangwa, "African Thought," 101.

culture itself. "Dialogue as charity"—dialogue understood as *bicultural-ity* (interlocution)—must go beyond today's mediated "monologue" in which no genuine communication is actually happening.[32] For genuine communication to happen, the African approach to ethics cannot simply be subsumed under other ethical systems like bioethical principlism, but demands to be taken genuinely as a *true dialogue partner*. In accepting the confrontation with the universalized principlism, it hopes for a recip-rocal give and take that can enrich both dialogue partners.[33] Moreover, to enrich the participating partners in the discourse of bioethics, the dialogue "between the West and Africa, as well as between industrialized countries and the Third World, has to be based on a dialogue of partners with equal rights."[34] The question is, if it is not a hegemonic model of operation, where is the genuine dialogue in our example of circumcision?

Like Tangwa, some African medical practitioners simply perceive this as a hegemonic model used in circumcision trials to imply "colonial paternalism."[35] One reason is that it is not a genuine relationship of life between the two dialogue partners, the international generous funders and the affected local subjects. Tangwa takes it to be cultural imperialism, an instance where a "benevolent dictator" makes decisions in the "best interests" of the locals, without their relevant extensive input on their own terms (their collective consent). The locals seem to have insufficient regard for their own version and meaning of life (limited awareness), as well as their own experience of health priorities. If not extremely guarded against, reoccurrence of the same mistakes may indeed evoke the errors and harms of the colonial past, and reawaken traumatic memories of co-lonial discriminatory brutality, marginalization, experimentation,[36] and

32. Onwubiko, "Re-Encountering African Culture," 96–97.

33. Bujo, *African Ethic*, xiv.

34. Bujo, *Ethical Dimension of Community*, 179.

35. Mallinson and Sibandze, "HIV and Male Circumcision," 133.

36. See Tangwa et al., "New Tuskegee," 12–15. "The brutal and uncontrollable pas-sion of the Negro . . . ungoverned sexual passion . . . the legal enforcement of cir-cumcision among the negro race would effectually remedy the predisposition to raping inherent in this race." "Circumcision for the Correction," 346, quoted from Tangwa et al., "New Tuskegee," 3. Discriminatory statements about Black people repeatedly appear in George W. F. Hegel's work *The Philosophy of History*, 91–99: Africa is not only closed upon itself but a gold-land confused within itself and enveloped in the dark mantle of night (p. 91). In Negro life, consciousness has not yet attained to the realization of any substantial objective existence—man is in a wild and untamed state. No reverence or morality—his character has no harmony with humanity (p. 93). Africa is "no historical part of the world . . . the unhistorical, undeveloped Spirit" (p. 99).

viewing Negros as sexual predators that need circumcision as a medical correction[37] to their supposed excessive and primitive drive against the females of the white race.[38] It is one reason Tangwa looks at *circumcision as a new Tuskegee.* It is therefore not simply an act of health care that increases humanness, but it's "enacted upon others by those with power, who fund programs to circumcise those with less power."[39]

And indeed, if Black Negros are still taken as sexual predators and so require circumcision as a medical rectification, it rightly resonates with funded studies from LMICs that appear (at the beginning of this section) to support the prophylactic effect of male circumcision as a health response to sexually transmitted diseases/infections (STDs/STIs) including HIV. Contrastingly, however, evidence from Global North before the WHO approval of large-scale circumcision in African (2007), as well as evidence from Global South after, finds the prophylactic effectiveness mild and weak. For instance, a UK scientific study conclusively reported: "We did not find any significant differences in the proportion of circumcised and uncircumcised British men reporting ever being diagnosed with any STI . . . We also found no significant associations between circumcision and being diagnosed with any one of the seven specific STIs."[40] In similar fashion, the findings from an analysis of data from the National Health and Social Life Survey of United States informs the existing debates on the utility of circumcision thus: "With respect to STDs, we found no evidence of a prophylactic role for circumcision and a slight tendency in the opposite direction . . . These results suggest a reexamination of the prevailing wisdom regarding the prophylactic effect of circumcision."[41] Comparable Global North reports are evident in the Global South. Recent studies from the Global South decisively detect "a weak association between male

37. "The enforced mass circumcision of the colored race was to correct of sexual crimes among their race, like the Negro rape problem, an issue that was presented at the annual convention of the Colored Physicians Association in 1889. Such an approach (Negros as sexual predators) seemed like a medical solution to ignorance, bad hygiene, and low morals in the blacks towards White southern womanhood and White male sexual hegemony." "Circumcision for the Correction," 345–46, quoted from Tangwa et al., "New Tuskegee," 4.

38. Since the Black male sexuality is excessive and primitive, the legitimate scientific way to manage this drive against the females of the white race is circumcision. Fox and Thomson, "HIV/AIDS and Male Circumcision," 103.

39. Tangwa, et al., "New Tuskegee," 15.

40. Dave et al., "Male Circumcision in Britain," 499–500.

41. Laumann et al., "Circumcision in the United States," 1052–57.

circumcision and HIV infection in Malawi. Twelve percent of circumcised men were HIV positive compared to almost 10% of uncircumcised men."[42] In agreement, President Yoweri Museveni, one of the foremost champions of the fight against HIV/AIDS in the world, equally castigates circumcision as a way to prevent HIV/AIDS: "I have always heard people and partners saying that when you are circumcised you don't contract the virus . . . before we started this campaign the HIV prevalence was at 18% then it dropped to 6% but when they started this talk of circumcision it confused the masses then it went up to now 7.3% . . ."[43] But this is not to generally discard circumcision, and more so, traditional African means of circumcision as locally effective. For example, Malawi has majorly had a successful traditional circumcision as in Lesotho, where traditional male circumcision is undertaken by more than 90 percent of circumcised men.[44] So coupled with the discussed necessity to involve locals and their chiefs, it is a clear indication that principlistic biomedical interventions from above may fail in advancing humanity and inter-personality if there is substantial sociocultural resistance. Of course, suggesting inculturation or enculturation of universalized principles within African cultural heterogeneity and nuance[45] isn't anywhere close to turning down the prophylactic effectiveness of safe male circumcision.

The second example, where humanness has been put to the test in Africa, is disguised in vaccine trial on human subjects or participants. The Ebola epidemic (2013–2015) that broke out in West Africa raised serious moral controversies.

> On September 29–30, 2014, the WHO organized a "consultation to assess the state of the art work to test and eventually license candidate Ebola vaccines," in which more than 70 experts, including many from affected countries in West Africa, are supposed to have taken part. This panel considered and apparently approved urgently carrying out clinical tests on humans of some candidate Ebola vaccines in the development pipeline, including Chimpanzee adenovirus serotype 3 (Chad 3) by Glaxo Smith Kline and others and Recombinant vesicular stomatitis

---

42. Namuunda et al., "Circumcision and HIV Infection."

43. Bob, "Museveni Castigates Circumcision."

44. Maffioli, "Male Circumcision Effective."

45. Tangwa et al., "New Tuskegee," 10.

virus (rVSV) vaccine by a consortium involving Canadian Public Health plus others.[46]

Controversies leaked out at phase I/II of the clinical trial (testing for safety and immunogenicity) of this candidate Ebola virus vaccine in 2015. This happened in a Sub-Saharan African country that had not registered any cases of the Ebola virus disease. The trial was sponsored and funded by one of the biggest Global North multinational pharmaceutical companies. The protocol received ethics clearance from the relevant national ethics committee. The study trial aimed at recruiting a total of three thousand research participants (adults and children) across four or five Sub-Saharan African countries. Participants received a monetary payment of approximately US$20. But before the study commenced at the second site, some members of (the public) raised an alarm that the government was carelessly risking the health, safety, and lives of its citizens in the cause of an unproven vaccine. Consequently, the study was immediately suspended while the trial involving children was withdrawn. The protesting general public (the research participants, their families, and communities) knew little about the study.[47] Before turning to the biomedical challenges related to this vaccine trial, let us comparably observe what the leading international regulatory declaration on research involving human beings—the World Medical Association (WMA) of the Declaration of Helsinki (DoH)—pronounces:

> The research protocol must be submitted for consideration, comment, guidance and approval to the concerned research ethics committee before the study begins. This *committee must be transparent in its functioning*, must be independent of the researcher, the sponsor and any other undue influence and must be duly qualified. It *must take into consideration the laws and regulations of the country or countries* in which the research is to be performed as well as *applicable international norms and standards* but these must not be allowed to reduce or eliminate any of the protections for research subjects set forth in this Declaration.[48]

Let us summarize atleast three significant minimum requirements that fell short of the above WMA guidelines in this vaccine trial. *First*, a

---

46. Tangwa, "African Thought," 106.

47. Tangwa et al., "Ebola Vaccine Trials," 49–59, especially 49–51.

48. WMA, 64th General Assembly. Emphasis added.

functioning transparent "ethics committee" lacked. In Tangwa's analysis of WMA principles for medical research involving human subjects, although there was a highly impressive committee for science review, it was not necessarily for ethics review, since there was no research ethics training that had been provided. And even if such training had been provided, well knowing that every research ethics committee needs an ethics expert as a member, there was none in the protocol. Over and above that, suspending the study before the recruitment of children not only ethically questions the inclusion of vulnerable children in a clinical study—like numerous other twenty-first-century abusive study malpractices on vulnerable children in Africa—designed for testing safety and immunogenicity, but it also solemnly questions the competence of the ethics committee (its expertise, independence, possible pressure from sponsors, etc.). Undoubtedly the whole study involved feigned structures and procedures that, on the surface, appeared to conform to ethics demands but gravely violated them,[49] just like several horrifying abuses and malpractices of unethical experiments on Africans[50] that are not even co-authored in native languages[51] suitable to homegrown ways of life (ways

49. See Tangwa et al., "Ebola Vaccine Trials," 52.

50. Exploitative *human research abuses and malpractices* prove unethical experimentation and unethical medical/clinical trials such as on vulnerable children, should remind us several other examples in the developing world, including, a clinical trial designed to test an antibiotic drug called trovafloxacine (Trovan) in 2001, which was carried out on children in northern Nigeria (Kano) in 1996 by a Pfizer team of doctors from the USA during a meningitis epidemic, and resulted in eleven deaths, while a further 200 of the children became blind, deaf, or lame, while some survivors suffered permanent brain damage and paralysis. Thirty families sued the Pfizer pharmaceutical company over these trials. Though the trial compared Trovan with the recommended drug Ceftriaxone, unfortunately children in the control arm allegedly got Ceftriaxone at an inadequate dose. The investigations revealed that the clinical trial had not been approved by a local research ethics committee, and that the families concerned were not adequately informed that their children were research participants. The families sued Pfizer through the US courts, resulting in an out-of-court settlement, with Pfizer having to pay a substantial compensation. Another example can be traced in Zimbabwe, during the early 1990s, where Dr. Richard Gladwell McGown, a British anaesthetist was working. He was arrested and charged on allegations of having carried out dangerous medical experiments on five hundred patients. Courts found him to have conducted interventional studies using new drugs and anaesthetics, without the approval of the National Drugs Authority, and knowledge of his patients, of whom six died. A Harare court found him guilty of professional negligence. See Ndebele et al., "Research Ethics in Africa," 4; Tangwa, "Universalism and Relativism," 64 and *African Bioethics in a Western Frame*, 155–56.

51. Other controversies surrounding the tenofovir trials are among the liveliest in recent research ethics in SSA, at least in Cameroon. They were carried out on

that are less of academic intellectual bioethics). In all these unfortunate trials on vulnerable Africans, exploitive violations have abused both international and local procedures, in which lawful and collaborative efforts must be renewed. In one response, the H3Africa policy documents and guidelines on informed consent and confidentiality basically recommend a harmonious regulatory procedural framework of RECs oriented from bottom to broader guidelines, for instance, in genomic research and biobanking,[52] where the local intimate relationships (bottom) are more informed of the genetic disorders of some of its community members than the broader universalized system. These are ethical distinctions that must be grasped by well-trained and updated ethics committees, which oftentimes is not the true contextual reality. In some empirical studies, to which Tangwa was part, ethical review committee processes that are poorly recognized in Africa were majorly challenged with inadequacy of resources, inadequate expertise or training of members, dependency on paper-based data management systems (lacked electronic or online systems), and lacked or had outdated SOPs.[53] Moreover, updating SOPs continuously needs a parallel capacity building in health research ethics (HRE), whereby even if the protocols legitimately and validly receive ethics clearance from the relevant national ethics committee, responsible implementation of approved research protocol is bound to encounter several practical challenges if not closely regulated and monitored by well-trained and equipped ethical review committees (ERCs) to fit local[54] contexts as stressed from thirty-one African ERCs. "Thus, ethical approval alone does not necessarily ensure protection of the safety and welfare of research participants throughout the research; hence the need for approved research to be monitored by ERCs."[55] Even if in some later

---

commercial sex workers and were prematurely suspended in February 2005 with several reasons, one being, according to the principal investigator, the trial documents pertaining to research participants such as informed consent forms were only in English whereas the city of Douala where the trial was taking place is predominantly French speaking. Tangwa, "Cameroon," 954.

52. Tindana et al., "Research Ethics Committees"; Staunton and Keymanthri, "Biobank Governance." For emphasis of collaborative research, see Hyder et al., "Ethical Review," 68–72; Mendy et al., "Biobanking," 252–60.

53. See Tangwa et al., "Ethics Review Committees," 149–56, especially 149 and 156.

54. See Tangwa et al., "Research Ethics Committee," 26–31; Tangwa et al., "Research Ethics in Central Africa," 4–11.

55. Tangwa et al., "Independence of Ethics Review Committees," 189. The thirty-one ERCs from eighteen African countries include: Burkina Faso (2), Cameroon (2), Ethiopia (2), Gabon (1), Gambia (1), Ghana (5), Kenya (1), Malawi (1), Mali (1),

studies Tangwa proposes monitoring by ERCs to be more effective along simplified succinct SOPs suitable to local contexts,[56] we necessarily need not only to renew local competences of ERCs in Africa, but also work towards a broader and deeper ethical conversion that will sustain functioning transparent ethics committees if bioethical principlism is to sit well in local settings. Here we see a necessity of mutual enrichment from both the global and the local competencies in their approach to ethics.

Second, "the laws and regulations of the country" were not respected, viz., adhering to the local legislation and regulatory infrastructure. In interpretation of World Medical Association guidelines, it should have included a genuinely participatory non-authoritarian and legal framework that is overseen by a well-constituted, qualified, and genuinely independent ERC. The absence of such infrastructure or doubt about its genuineness or impartiality, despite appearances, delimits a no-go area for biomedical interventions. It is because the verifiable existence of such a transparent infrastructure should be a precondition for any human subject research, especially in resource-destitute settings like Black Africa,[57] at least if we are to refer to the guidelines of the WMA of the Declaration of Helsinki as stipulated above. Over and above that, Tangwa's analysis of the informed consent documentation—just like we saw above where research is not co-authored in the familiar local languages—reveals a notable ignorance of local legal requirements (e.g., researchers seeming unaware of the local age of consent in this country of vaccine trial). The lawful age of consent is twenty-one, not eighteen, as it was noted. It was not explained that those below twenty-one required (additionally) the proxy consent of their parents or legal guardians.[58] Worse still, to show how such an ethical flaw scarcely fulfilled the local requirements of informed consent, the study physicians (or their agents) simply gave out information sheets to human research subjects to take home, asked them back the next day only to sign the informed consent form and then the procedural instructions, before a monetary payment of US$20. Moreover, there was no process of community engagement beyond the media

---

Mozambique (1), Nigeria (4), Rwanda (1), Senegal (1), Sudan (1), Tanzania (4), Uganda (1), Zambia (1) and Zimbabwe (1).

56. See Tangwa et al., "Small Is Beautiful."

57. See Tangwa et al., "Ebola Vaccine Trials," 49–59, specifically 57–59. There is "lack of adequate legislation governing research and bioethics." Tangwa, "Cameroon," 950, with "no national framework for the operation of RECs" (p. 953).

58. See Tangwa et al., "Ebola Vaccine Trials," 59; Schoeman, "Research Ethics Governance," 1–15, especially 13.

announcement by the minister of public health emphasizing that the study had been approved by the WHO and was simultaneously taking place in many countries.[59] The whole background shows not only injustices of non-disclosure of information (since disclosure is a component of informed consent) but also several non-compliances to a would-be acceptable universal framework in the process of making a valid informed consent within a given local context. We note carefully that one of the non-complying elements in pursuing an informed consent in the discussed African sense is a misguided reliance on the individual autonomy of the research subjects outside their *broader family relationships*.

Although individual autonomy is a component of informed consent in biomedical ethics,[60] need demands us to get beyond this individual autonomy concealed in bioethical interventions if we are to address global bioethical problems such as those in vaccine trials in Africa. Borrowing the wording of Tangwa—in Henk ten Have's book *Global Bioethics: An Introduction*, a renowned Western bioethicist—he ascertains that the emergence of "global problems" ought to shake

> Western bioethics from its static position and dogmatic slumber, anchored on the ideology of possessive individualism and an exaggerated emphasis on the moral value and principle of individual autonomy, on which it had developed for over half a century, to the realization that a new and broader perspective is required in addressing these problems. He therefore affirms the necessity of *a "global bioethics" as distinct from "Western bioethics"* which recognizes differences and cultural diversities and moves towards a convergence of shared values and a common global perspective as the only way of adequately addressing global problems.[61]

Affirming the necessity of "global bioethics" as distinct from "Western bioethics" demands our renewed approach to bioethics, for instance,

59. See Tangwa et al., "Ebola Vaccine Trials," 53–57. "Each potential subject must be adequately informed of the aims, methods, sources of funding, any possible conflicts of interest, institutional affiliations of the researcher, the anticipated benefits and potential risks of the study and the discomfort it may entail, post-study provisions and any other relevant aspects of the study. The potential subject must be informed of the right to refuse to participate in the study or to withdraw consent to participate at any time without reprisal." WMA, 64th General Assembly.

60. O'Neill, "Some Limits of Informed Consent," 4–7; Dickert et al., "Reframing Consent for Clinical Research," 3–11.

61. Tangwa, "Introduction by Henk," 268; see also Tangwa, *African Bioethics in a Western Frame*, 42, 59.

in our broader understanding of an inclusive process of consent as understood in an African context. We need to restructure the priority of individual autonomy, which overemphasizes the power of individual choice, by appreciating the role of the family and community relationships.[62] Basically, we are far from these restructurings. Beyond Kant's deontological emphasis of individual autonomy as discussed in principlistic theories, there are other obvious Western accounts in biomedical ethics emphasizing the same. They include the Belmont Report of 1979, which concerns ethics and health care research based on the principles of beneficence, justice, and respect for persons. The report takes the ethical function of informed consent as protecting individual autonomy, treating individual subjects and participants in clinical trials and research as autonomous agents.[63] Another account is that of globalization, which we have already hinted on. Firstly, founding bioethics on the principles of globalization remains problematic since it currently emphasizes individual pursuits of neoliberal market ideologies that seemingly complicate more, quite fast if not so fast, than simplifying more global challenges like environmental degradation, pandemics, and organ trafficking. In all ways, empowering individuals is no longer sufficient if we inevitably must implement a unifying universal framework of bioethics.[64] Secondly, globalization as espoused in an agency of the United Nations, WHO, fails—since it had nothing to say or even said nothing, for instance, in guiding the global pharmaceutical production of Prof. Victor Anomah Ngu's HIV/AIDS virus vaccine, VANHIVAX as I will soon discuss his case—to fulfill its oversight role for global health since it directly engages "in commercial drug discovery, development, and clinical tests, while at the same time issuing guidelines and directives for all"[65] global partners. In dialogue with Tangwa, we are compelled to deduce that the WHO's simultaneous issuance of universal guidelines coupled with its general oversight role may be perceived as paradoxically unethical and unresponsive to local bioethical challenges. Given that whenever it is less emphatic with its supervisory role, it may implicitly grant a fertile ground for the dominant bioethical principlism to exploit local moral thoughts as it wills. In short,

62. Tangwa et al., "Global Health Inequalities," 242.

63. The standard response has been that obtaining informed consent is a way of treating individuals as autonomous agents: Kantian autonomy and not individual autonomy. Kristinsson, "Autonomy and Informed Consent," 253–64.

64. See Henk, "Global Bioethics," 600.

65. Tangwa, "African Thought," 106.

overemphasis of individual autonomies (autonomous perspectives) and neoliberal market ideologies reminiscent of principle-based globalization and other similar bioethical interventions seem to be the largely dominant threatening tools employed by the industrialized Global North countries along international organizations at the expense of the developing Global South countries.

It is one reason an African ethic appreciates each one's creative individuality (innovativeness, inventiveness, resourcefulness, ingenuity) but only within a sociality of community members. This communal moral hope, thinking in terms of an unending future, is enshrined at least in a valid consent or refusal that is *relationally determined*. As far the relationship between the individual and social community, the individual decision is more concretely embedded in a social context. It is different from the principle-based autonomy that does not envisage the plurality of existing understandings of autonomy. This is because an informed consent is not only an individual's autonomous authorization of a medical intervention or of participation in research through a consent agreement, but also a conformity to the social rules of consent, very much so, as *PBE* would rightly state.[66] So forgetting the public's sociality of life is to fail in implementing any biomedical ethics in an African context. In this Ebola trial vaccine, political officials, or sometimes prominent people, independently (with autonomous interests) authorized research, giving their individual consent, without the general public's ultimate authorization of the same project. Also, it is ethically manipulative and deceptive because high-profile people or politicians are neither necessarily medically qualified to speak about the efficacy of a medical procedure[67] nor hold the community approval (social competence). Certainly the public wouldn't have protested if the project competently held together with the necessary social rules, as the ultimate form of social/communal/collective informed consent, which is the antithesis of individual consent. It is an ultimate form of consent because it relationally incorporates the plurality of the existing understandings and the extended community life. Therefore, even when an individual informed consent addresses the *micro*-level issues, there are broader moral issues at the *meso* and *macro* levels—what some African scholarships have referred to as "tiered-informed-consent" processes that comprehensively encompass community consultations—relating to

---

66. *PBE*, 120, 122.

67. Gwandure, "Male Circumcision," 93. See details in Bridges et al., "Male Circumcision Services," 60–76.

families, communities, religious groups, ancestral groups, legal sages and the broader citizenry—making all to be collectively affected by genomic data from individuals. This is because the ancestry-based perceptions have previously fueled life-threatening discord, so valid consent processes must by necessity be fixed in this broader setting.[68] This is in line with the pertinent issues that were revealed in the report of the H3Africa Ethics Consultation Meeting. Although we already indicated challenges from that report in our introduction relating to consent, community engagement and governance framework, most clearly, the report points to the centrality of trust that can be warranted within a robust community engagement.[69] Though in recent studies concerning public health emergencies (PHEs), reciprocity, transparency, and accountability between communal stakeholders or collaborators[70] are more emphasized, trust remains re-echoed as the stronghold of communal life. It is within this broader community perspective that genomic knowledge can easily identify with the ethics of hereditary chronic diseases like sickle cell disease (SCD) and individual autonomy turned into communal autonomy, namely, informed consent translating into social/collective autonomy.

Consequently, in analyzing the Ebola trial vaccine, a collective consent lacked because some facts that the community needed in deciding whether to refuse or consent to the lodged biomedical investigation remained undisclosed. It is because the adequacy of the information to be disclosed goes beyond the individual person. A family historical health complexity (hereditary diseases) and unconventional beliefs are familiar with the extended family community not the individual. The community determines the adequacy of any disclosure and establishes the body of information to be disclosed in reasonably simple non-technical language, not a foreign language. It is why the reader should carefully note that biomedical research in African settings requires a broader community engagement that assesses not only the aim and the investigator's personal interests, but also demarcates the communal right to withdraw without penalty; the applicable methods; the anticipated benefits to the past,

---

68. Tangwa et al., "Tiered Informed Consent"; Treadwell et al., "Genomics Applications for Sickle Cell," 323–32.

69. Tangwa et al., "Ethical Issues in H3Africa." The report engages members from research ethics committees and national ethics councils from eighteen countries: Benin, Botswana, Burkina Faso, Cameroon, Cote d'Ivoire, Ethiopia, Ghana, Guinea, Kenya, Malawi, Mali, Mozambique, Nigeria, South Africa, Sudan, Tanzania, Uganda, Zambia. See also H3Africa Consortium, "Research Capacity," 1346–48.

70. Tangwa et al., "Health Emergencies in Sub-Saharan Africa."

current, and future community, not simply some individuals; the risks to the past, present, and future community; and more so the limits of authorizing any consent as fixed by the previous community of ancestors, etc. Otherwise breaching this requirement of acquiring an informed social consent will, for example, evoke the wrath of the ancestors and so it will not sit well in an African setting.

Third, even when universalized principles (international norms and standards) were to be strictly applied in this Sub-Saharan African country, there are several ethical flaws to prove that the Declaration of Helsinki (inclusive of WMA) was neither fulfilled, satisfied, nor respected. Retrospectively, we know that research ethics governance in biomedical ethics began as a reaction to the abuse of human research participants in the Western world. Examples that shaped ethical governance include the Tuskegee syphilis study (1932–1972) and Nazi experiments that resulted in the Nuremberg Trials. In consequence, the Nuremberg Code was established in 1949 in the aftermath of the trials against twenty-three German physicians who conducted unethical medical experiments during the Second World War. The Nuremberg Code, the first ever international document on medical experimentation on human subjects, advocated for the principles of voluntary participation and informed consent in research. It became ethically clearer to exercise freely the power of informed choice without constraint or coercion.[71] However, in analyzing these vaccine candidate trials, the Ebola vaccine trials, manipulative control is generically reported, though neither persuasive nor coercive, but rather in varied forms of withholding information, lying, exaggeration, and framing information positively or negatively about the efficacy of the Ebola vaccine, which are all morally incompatible to a successful process of achieving a valid informed consent or refusal. Actually, in the analysis

71. The Nuremberg Code included the ethical principles of "do no harm" and the right to withdraw from a study. Moreover, researchers should be qualified for a study. Principles have to guide RECs (research ethics committees) to review the ethicality of projects. The Nuremberg Trials resulted into not only the promulgation of the Declaration of Helsinki (1964) but also the development of the International Ethics Guidelines for Biomedical Research Involving Human Subjects (CIOMS) in 1982. The regulatory framework of the Nuremberg Code determined that a comprehensive research protocol should be developed for each study and approved by an independent REC—assessing compliance with the principles, funding, sponsors, institutional affiliation, incentives for participants, any potential conflicts of interest. RECs have the right to monitor studies. See Schoeman, "Research Ethics Governance," 3–4, WMA, "Declaration of Helsinki," 373–74; *Trials of War Criminals before the Nuremberg Military Tribunals*; Tangwa and Nchangwi, "Sprinting Research," 356–66.

of the information sheets, Tangwa et al. show inappropriate or mislead-ing language. For instance, what was promised was "to test a vaccine against Ebola," whereas "there are no vaccines or treatment against the Ebola virus"[72] by this time, but rather a candidate vaccine. The failure to explain in simpler ordinary language technical medical terms ("screening visit," "blood factors," "extra study procedures," "muscle of your upper arm," "serious medical conditions") or jargon ("dummy vaccine," "ac-tive components" and "placebo") remains insignificant to a prospective subject and ethically problematic especially in LMIC setting.[73] This un-ethical way of biomedical interventions, interventions that override valid communal ways of obtaining informed consent, contribute to intolerable distrust—there by undermining trust in scientific research that would be an investment tool used to build up African local capacity[74]—between communal human relationships, given that it incapacitates the poor for the sake of forcing consent from them, making it an *exploitative consent*[75] of participants since the poor human subjects are unduly enticed, a ma-nipulative exploitation that contradicts the very essence of a would-be global bioethics. In agreement with *PBE*, indeed when biomedical rela-tionships and trust are reduced, decision-making as well as the process of a valid informed consent get tainted. Inevitably, here arises "problems of nonacceptance and false belief, a single false belief or a breakdown in a person's ability to accept information as true or untainted, even if he or she adequately comprehends the information."[76] Breakdown in the ability to accept information due to illegal flows leads to marginalization, depri-vation, stigmatization, and discrimination in LMIC settings. Even when

72. Tangwa et al., "Ebola Vaccine Trials," 53.

73. See Tangwa et al., "Ebola Vaccine Trials," especially 54–57; If we point out a placebo in principlism, it is a substance or intervention that the clinician believes to be pharmacologically or biomedically inert or inactive for the state of health being handled. While "pure" placebos, such as a sugar pill, are pharmacologically inactive, active medications are sometimes used as "impure" placebos. The therapeutic use of placebos (for example, the prescription of an antibiotic for a common cold) has injured some clinical relationships and hugely reduced trust. The use of placebos typically, but not always or necessarily, involves limited transparency, incomplete disclosure, or even intentional deception. *PBE*, 127.

74. See Emerson et al., "Use of Human Tissues."

75. We can think of an exploitive consent in terms of a coercive consent from barely unfree victims of transcontinental organ trafficking, transborder prostitution and the rampant dehumanizing transcontinental exportation of human eggs and sperms from vulnerable populations without their would-be-sober will.

76. *PBE*, 134.

we all agree that there is nothing justifying discrimination of the economically disadvantaged persons or groups from participation in legally approved research, there are increased risks of exploitation experienced by the economically distressed, indicating an unjust and paternalistic form of discrimination that serves to further marginalize, deprive, or stigmatize the less advantaged.[77] By way of narrowing down, I note that one of the two moral problems of exploitation we mentioned—undue profit (unduly small and unfair payments)—suffered significant injury in the whole course of an unjustified offer (US$20) during these candidate Ebola virus vaccine trials, an attractive offer *akin* to stimulating of the subjects' interests during the Tuskegee syphilis study, where participants received free burial assistance and insurance, free transportation, free medicines, free hot meals, free stops in town on the return trips, and so on. I am of the view that biomedical interventions between the Global North and Global South must pursue a genuine social autonomy. For in both cases of circumcision and vaccine trials, they would be more globally fruitful if locally inculturated and culturally friendly than being economically couched from funders and researchers, individualistic, less conforming to local broad-based requirements of informed consent and so exploitive of African vulnerabilities. Embracing a true social autonomy necessarily implicates responsibility of both the genuine efforts of the funding Global North as well as the genuine efforts of the funded Global South.

Exploited Structural-Universal Collaborations

The second instance is of exploited structural-universal relationships between the Global South and the Global North. These are suggestive of not only instances of *weak cooperative values* between the Global South and Global North but also shows how *economic international collaborations can be exploitive.* The relationship of these exploitive collaborations to principlism lies in the fact that the internationally proclaimed principle-based approaches or guidelines seem not to concretely impact the local contexts but rather serve Western interests in the internationally decreed guidelines. These guidelines are further largely overridden in local contexts by corruptive collaborations, sometimes agents from local

77. See *PBE*, 287. Africa's vulnerabilities expose research participants to possible exploitation by researchers and sponsors, thus calling for improvement in regulatory frameworks. Kilama, "From Research to Control," S91–101. Pages indicated thus.

administrations in conjunction with foreign agents. Let me mention a couple of examples.

First, during the World Congress of Bioethics (2000) in the UK, Tangwa jointly presented with Cameroon's late foremost medical scientist and oncologist, Victor Anomah Ngu, on "the Effective Vaccine against and Immunotherapy of the HIV: Scientific Report and Ethical Considerations from Cameroon." The problematic that sparked off Ngu's investigation was to answer why for long HIV provoked immune responses that failed to exterminate the HIV virus from the human body. It is from here that the immune system could be subsequently induced so as to yield immune responses that eradicate the virus from the body. It is again from here that their presentation drew attention to a candidate vaccine against the HIV/AIDS virus, VANHIVAX (VANVAHIV), discovered by Ngu—who was no neophyte in this field since he had won the Lasker Prize for Cancer Research in 1972—and so far tested on some desperate HIV-positive patients with remarkably promising results, in terms of reduction in viral load, increase in CD4 count, gain in weight, and even sero-conversion from positive to negative.[78] *At that time,* Ngu reasoned that HIV being an "enveloped" virus,[79] which "is 'perceived' by the host immune system as 'partly-self' because of the presence of host cell wall membrane on the viral envelope and that this situation leads to an ineffective response by the body's immune system to the virus," and determined that the "viral core antigens without the envelope would be 'perceived' as 'non-self' by the host immune system, they would elicit an appropriate and effective immune response," whereby Ngu asserted that "in a normal uninfected person, therefore, such 'unenveloped' core

---

78. Victor Anomah Ngu was an ecologist and professor of surgery at the University of Ibadan (1965–1971) and Université de Yaoundé (1971–1974); vice-chancellor of the Université de Yaoundé (1974–1982); president of the Association of African Universities (1981–1982); Minister of Public Health for Cameroon (1984–1988); director of the Cancer Research Laboratory at the Université de Yaoundé (1984). Tangwa, "Leaders in Ethics Education," 96–98; Tangwa, "African Thought," 106–7; Ngu and Tangwa, "Immunotherapy of the HIV," 77–84, especially 80.

79. The envelope of HIV is derived from the host CD4 cell wall, and killing the virus with its envelope will also kill host CD4 cells, leading to an autoimmune disease. The virus uses this threat to blackmail and to block, so to speak, effective immune responses that alone can kill the virus. Ngu and Tangwa, "Immunotherapy of the HIV," 80; Tangwa, "HIV/AIDS Pandemic," 223; Sriwilaijaroen and Suzuki, "Sialoglycovirology of Lectins," 483–45.

antigens . . . would serve as an effective vaccine."[80] In his own voice, Ngu puts across his research results succinctly:

> The HIV is perceived by the host immune system as *partly-self* because its envelope is derived from CD4 cells. All things being equal, immune responses to a *non-self* virus will be considerably stronger and more effective in killing such a completely alien virus than that to a virus, like the HIV, which is perceived as somehow kindred or partly-self.
>
> The ideal solution for the body is one where the immune responses kill only the virus without any damage to its envelope and therefore the CD4 cells. To provoke such ideal immune responses in an uninfected person, i.e., by vaccination, one must start with HIV antigens from which the viral envelope has been destroyed beforehand, transforming it into *non-self* antigens. Such immune responses will kill an *invading virus*, sparing its envelope and CD4 because the envelope was absent from the viral antigens used to vaccinate the uninfected person. The procedure is effective because, from the start, it is directed against the virus only.
>
> *All HIV antigens, minus its envelope, would thus be an effective vaccine for the HIV!*[81]

Returning to the World Congress of Bioethics (2000) in the UK, the presenters, Ngu and Tangwa, substantiated the above convictions. To put it differently in Tangwa's words, in order to provoke a new immune response to HIV that is without an envelope and will effectively kill only the virus, sparing the viral envelope and CD4, one started with HIV antigens, tested on patients' own auto-vaccine since testing would not happen on normal healthy people, with only the envelope destroyed to constitute an effective HIV vaccine. In the testing, the auto-vaccine provoked immune responses that killed the virus, confirming its effectiveness as a vaccine. This procedure of using vaccines to induce the killing of the virus in patients constituted a form of immunotherapy for HIV. The presenters were convinced that the vaccine was completely effective and an affordable (US$ 0.10 per shot if produced on an industrial scale) way to produce an HIV vaccine for the vulnerable using a local, simple approach as opposed to complex modern principlistic methods. Retrospectively,

---

80. Tangwa, "HIV/AIDS Pandemic," 222. Tangwa relies on Professor Victor Anomah Ngu's seminal hypothesis published in *Medical Hypotheses* as "The Viral Envelope in the Evolution of HIV."

81. Ngu and Tangwa, "Immunotherapy of the HIV," 80–81. Emphasis original.

the reader ought not forget that after the presentation—which drew a huge audience and during which Ngu handled questions and comments related to the science and Tangwa those related to the ethics, governance, and overarching issues—the BBC science journalist invited both for an interview, which lasted about an hour. Responding to several questions, Tangwa and Ngu felt the case for VANHIVAX/VANVAHIV had been even better expressed than during the presentation. Surprisingly, when they were back to Cameroon after the Congress, on emailing the BBC journalist requesting a transcript of what might have been broadcast, he responded that he was very sorry the interview had not registered on the recording machine and could not therefore be broadcast! As if that was not bad enough, several research teams from some prestigious institutions from the USA and Europe came in pretense to Yaoundé, proposing to Ngu collaborations in his vaccine project; but when they got the scientific details of his protocol to go and study closely, Ngu never heard from them again.[82] Consequently, Tangwa concludes their World Congress of Bioethics presentation on the HIV candidate vaccine, to which many remained surprised till today: "First of all, no one seemed to have entertained the possibility of a 'homegrown' candidate vaccine proposal coming from Africa. What seemed to have been expected were proposals of how Western researchers could carry out ethical and acceptable HIV/AIDS vaccine research in Africa, with or without the collaboration of local scientists. Secondly, surprise was expressed as to why, with the interesting clinical results presented, no Western researchers or funding agencies had shown interest in VANVAHIV."[83]

Like such ethical bewilderments, more unraveled and some never ended. Astonishingly, close to two decades later, Ngu's "idea of an immunotherapeutic vaccine, which seems to have gained scientific currency today, was dismissed by some as an impossible and contradictory concept."[84] But even though his scientific basis was disputed by some international elements as unscientific, Ngu, through his own meager financial resources, and the basic science of that time, confirmed his hypothesis by pre-testing the vaccine as an *autologous vaccine* for each sero-positive patient in 1999, and more so produced convincing results

---

82. Tangwa, "Leaders in Ethics Education," 96–97; Ngu and Ambe, "Immunotherapy of the HIV," 2–8; Tangwa, "African Thought," 106–7; Tangwa, "HIV/AIDS Pandemic," 224.

83. Tangwa, "HIV/AIDS Pandemic," 224.

84. Tangwa, "African Thought," 107.

of measuring the viral load from 2001.[85] Without forgetting the exploitative attitudes from the journalist, the failed international collaborations with world scientists who disappeared with Ngu's protocol and never appeared again, which in my reading represent principlistic attitudes, need demands cultivating a better harmonious way to approach global bioethics. To challenge this exploitative attitude, in the 2002 World Congress of Bioethics in Brasilia, Brazil, under the general theme "Bioethics, Power and Injustice," Tangwa's presentation—titled "The HIV/AIDS Pandemic and the Ethics and Politics of Vaccine Research in Africa"—"drew attention to the pathetic situation of HIV/AIDS infections and deaths and the need for a collaboration between the North and South, devoid of a predatory and exploitative attitude, transparently well-intentioned and unquestionably ethical."[86] Tangwa reiterates not only the limitations of the bioethical principlism to biomedical solutions within an African context, where he cautions us to be very careful not to confuse global commerce or economic interests with the ethics of healthcare, but he also journeys with us in reappraising African communitarian values, knowledge systems, and practices.[87] In my reading of these African studies, though these practices still remain threatened in preference to principlistic practices, which practices are jumbled with economic and commerce ill interests, African approaches could aid the search for not only a possible global solution to biomedical exploitive interventions at large, but also solutions that can ably ameliorate a Black African moral thought in particular, towards its holy ethical marriage with principlism. An ethical marriage is significant because it brings together Western thought, understood as principle-based approaches, and African thought in an ethical mutual friendship that is not only more globally appealing in achieving mutual needs but also locally attractive. Most importantly, the need to safeguard against exploitive collaborations as well as cultivating cooperative values through well-intentioned structural-universal collaborations in the practice of biomedical ethics remains necessarily mutually wanting for both the Global South and Global North.

Nonetheless, I am not ruling out some functioning structural-universal economic collaborations, though buried in corruption. Here corruption, like I mentioned above, disrupts a would-be functioning

---

85. Ngu, "Vaccines for the HIV"; Ngu and Ambe, "Immunotherapy of the HIV," 2–8; Ngu and Tangwa, "Immunotherapy of the HIV," 81.

86. Tangwa, "Leaders in Ethics Education," 97.

87. Tangwa, "HIV/AIDS Pandemic," 217–18, 229.

implementation of ethical guidelines in local and global circumstances. It means that if we have to emphasize a conversion in bioethics, the Global South is as equally ethically genuine as the Global North. Since corruption presupposes two conniving partners at least, Global South countries, especially dictatorial governments and corruptive local chiefs, for example, by the same token have other patterns in crime in the Global North—viz., kleptocracy, whereby the elite of African countries collaborate with foreign ideologies (dating back from slavery) in order to recklessly exploit the rich resources of their own people.[88] Several moral problems are not only of connivance, but also depict an immorally structured world economic order between the Global South and the Global North as Bujo questions. From an economic and moral perspective, we can for example ordinarily ask whether the supply of weaponry to countries sitting in debt does not advance exploitation between local and global relationships. Firstly, powerful weaponry ceaselessly sold to poor countries implies not only that the taxes meant for the country's population needs are poorly used for dictators to shield their long hold on power, but also the importation of weapons leads to an increase in indebtedness of these poor countries. Secondly, in the final analysis, the great beneficiaries of weaponry sales are the Global North industrialized countries, which for profit-oriented economic reasons are not willing to stop the murderous export of military arms, even when dictators are rapidly killing their own citizens.[89] The presupposition of two conniving corrupt partners in a bioethical crime is very clear. The Global North and Global South are in an unholy ethical alliance! But we indispensably need

88. Almost everywhere on the Black continent, there is some sort of kleptocracy; i.e., the upper class (e.g., government officials of the ruling party) becomes wealthy at the expense of stealing the people, or rather through the taxpayer's money. The loot is transferred to foreign countries. See Bujo, *Ethical Dimension of Community*, 168. Some villages or clans or ethnic groups raided others in search of slaves at the demand of foreign elements. Here arises questions of ethnic superiority and the unwillingness to confer equality on some people who are not of one's own clan family. See Odozor, "African Moral Theology," 594–95.

89. Moreso, Western criticism of African dictators lacks credibility, if one recalls that many politicians on the continent are supported by Western states and industries to carry out their inhuman enterprises. Examples include Jean Bedel Bokassa, the former president of the Central African Republic, incredibly supported by the French government; and Yamoussoukro in Ivory Coast with the construction of the often-criticized giant "Dome of St. Peter," whose building material was partly ordered from Europe. See Bujo, *Ethical Dimension of Community*, 175, 191–92, on Yoweri Kaguta Museveni of Uganda with his bloodthirsty presidential campaigns and almost four decades of cruel dictatorship.

both for a global as well as a local approach to bioethics to constructively materialize.

Second, the unethical manipulative connivance between the Global North and the Global South can be further demonstrated using the outcomes of the well-known concluded Truth and Reconciliation Commission (TRC) under the chairmanship of Archbishop Desmond Tutu, in which unethical dealings with the chemical and biological weapons projects of the Apartheid regime's army is explicit.[90] It is here that Tangwa wonders at the silence of the International Association of Bioethics amid weakened structural-universal cooperative values. He points out some findings as reported by Wendy Orr, a militant medical doctor from South Africa who was a member of TRC. Wendy talked precisely about South Africa's Chemical and Biological Warfare (CBW) programme.

> According to her, the work of the scientists involved in the CBW programme was covert and secretive in every imaginable way. It was conducted under the guise of private 'front companies' supposedly conducting commercial research and development work; it was funded from undeclared, unaudited secret government coffers; research and experimentation were conducted on a strict 'need to know' basis with little or no communication or interaction between scientists working on different projects or between scientists inside and outside the programme. The aim of the programme was to develop agents harmful to individuals, groups and communities. The chemists, engineers, physicists and veterinarians, didn't see themselves as bound by any particular ethical codes of conduct, making the present conventional approaches to Bioethics/ Medical Ethics exclude ethical obligations. Some secretive and non-interactive scientific research projects include:
> - The production of thousands of kilograms of street drugs (like Ecstasy and Mandrax), supposedly for use as crowd control agents.
> - Research into the use of various carrying agents for organophosphate poisoning, for example, beer, whiskey, chocolate, shampoo.
> - The development of "applicators" which were, in effect, murder weapons, for example, screwdrivers which could inject poison into a chosen victim and leave an almost undetectable external puncture wound.

90. Tangwa and Nchangwi, "COVID-19," 5.

- Research into toxic agents which are easily administered, le-
  thal, tasteless, odourless and undetectable in the body.
- Research into and the stockpiling of millions of drug resistant
  cholera, anthrax, plague and botulinum organisms.
- In spite of the fact that South Africa was supposedly subject to
  international sanctions, the South African Defense Forces received support for this programme from a number of foreign
  countries, including the USA, the UK, France, Israel, China
  and Germany.[91]

Here the camouflaged approaches were covert, secretive, and commercialized without attached moral obligations. The conniving participating partners seemed not bound by any particular ethical codes of
conduct, either universal or local. Harmful codes were set on economic
terms to injure both the individual, institutions, groups, and the globe,
exhibiting the highest forms of interpersonal-structural-universal ethical
injustices cutting equally across the Global North as well as the Global
South. Correspondingly, other bioethicists reaffirm imbalanced inter-
continental trade and unfair commercialization of drugs that has led not
only to huge profit maximizations and disreputable corruption of the
vulnerable but also continually systematized dependance of the vulnerable populations on the industrialized West. Even when these vulnerable
populations supply low-priced raw materials of their herbal medicine
to Western pharmaceuticals, unethically, the vulnerable people end up
purchasing the controlled finished drugs so expensively.[92] To some, these
moral exploitations seem endemic and dictated by Western economic
terms (economic determinism), just like Tangwa and Schroeder note:

> Western medicine and medical technologies, like Western cul-
> ture and technology in general, have the fatal weakness of being
> driven and controlled by apparently morally blind economic
> forces and interests. Economic determinism rules the Western
> world and culture to such an extent that even ethical issues are
> couched and discussed in economic terms and language. The
> most imperative of moral categorical imperatives seems to
> leave the Western world unimpressed and unruffled unless it is
> somehow stated in terms of or connected with economic and

91. Tangwa, "Leaders in Ethics Education," 95–96; Tangwa, *African Bioethics in a Western Frame*, 172–73.

92. The study argues in the global context of indigenous peoples' rights, consent and benefit-sharing. Wynberg, Schroeder, and Chennells, *Indigenous Peoples, Consent Consent and Benefit Sharing*.

marketing considerations and cost-benefit analyses. Even the moral philosopher is constantly prompted in the Western world to demonstrate a position's "cash value."[93]

That bioethical interventions are disguised in an economic language which generally exploits vulnerabilities inclusive of old colonial systems in other cultural contexts, which weakness cannot be stressed enough. But even if these "exploitative North-South research collaborations often follow patterns established in colonial times,"[94] like Schroeder believes, corruptive ventures involve atleast two moral criminals. In addition to the studies discussed, we substantiate this position with a fresh case of an intensified post-COVID-19 organ transplant tourism on the African continent, which reveal these two or more conniving moral criminals. Organ tourism, according to an Interpol report, can only be done in the framework of complex networks, due to the required skills (medical specialists, surgeons, nurses), logistics (compatibility of the patient and donor), and well-equipped healthcare facilities (analytical laboratories, clinics, operating rooms). Most often, victim-donors receive a smaller payment than had been agreed with the recruiter (or broker), while in some cases they may never get the promised payment.[95] The common denominator in all these exploitations as we have largely stressed is economic determinism. It is the reason Tangwa suggests that we need not couch bioethics as a marketable commodity, which makes it fail. An African *authentic* account doesn't consider health resources as marketable commodities controlled by the forces of demand and supply, even if it were securing transplant organs. It doesn't favor only the wealthy, who have the financial resources to choose between multifarious medical options.[96] And so the escalating organ tourism south of the Sahara portrays, like we initially stated, *a distributive injustice of too small a payment to subjects (victim-donors) in contrast to an irresistibly attractive large payment paid by the donor.* Moreover, I argue, in such an undue-profit exploitation, the victim receives an unfairly low payment, while the broker/recruiter garners more than is justified. Like so, the two major moral problems of exploitation

93. Tangwa, "African Perception of a Person," 43.

94. Schroeder et al., "Ethics Dumping," 1–2.

95. Recently, Interpol assessed the problem of trafficking in human beings for organ removal, especially from impoverished desperate communities of displaced populations, the unemployed, migrants, asylum seekers, refugees sexual laborers, etc., who are at greater risk of exploitation. ENACT, "Trafficking of Human Beings."

96. Tangwa et al., "Global Health Inequalities," 242.

among two or three moral criminals, *undue inducement* (unduly large and irresistible payments to the broker) and *undue profit* (unduly small and unfair payment or no payment to the victim-donor), still demand us a well-intentioned normative codification in our collaboration, at least in biomedical interventions. Neutralizing the existing unholy collaborations in biomedical ethics between the Global North and Global South requires a genuine consensus in strict regulations that apply equally to all, with the same intensity to all structural-universal dimensions of life.

Exploited Cosmic-Ecological Relationships

> Due to their genetic proximity to human beings and to their highly developed social skills, the use of non-human primates in scientific procedures raises specific ethical and practical problems in terms of meeting their behavioral, environmental and social needs in a laboratory environment. Furthermore, the use of non-human primates is of the greatest concern to the public.[97]

In the third instance are exploited cosmic-ecological relationships, relationships that extend to nature. In some bioethical debates, we note significant ethical discrepancies in cultivating a functioning consensus (whether local or global) within various attempted collaborations relating to the scientific use of non-human primates. In this first example, the beneficial impact of the above directive (2010/63/EU) has regulated and led to decline in the exploitive manipulation of non-human primates in research within the EU, on the one hand. On the other hand, biomedical procedures with non-human primates are increasingly difficult in the EU, making European scientists to seek collaborations outside of Europe. Because of this, the decline in EU has been paralleled by an increase in numbers used elsewhere. Researchers from high-income countries (HICs), where directives on the use of non-human primates are strict, are conducting similar experiments in countries of collaboration where regulations are less strict.[98] To be specific, Kate Chatfield and David Morton report an incident in 2013 where an academic from

97. European Union, "Directive 2010/63/EU."
98. Chatfield and Morton, "Non-Human Primates in Research," 81–89, especially 81–84.

a UK university bypassed British law in his research with wild-caught baboons in Nairobi. A professor of movement neuroscience, part of a team investigating methods to treat conditions affecting the brain such as stroke, spinal cord injury, and motor neuron disease, was accused of exploiting a cheap and plentiful source of animals in the Sub-Saharan country of Kenya. The accusation followed an undercover investigation by the British Union for the Abolition of Vivisection (BUAV), which had covertly obtained photos and video footage of the baboons at the relevant institute in Nairobi. The UK professor was quoted as saying that animal welfare standards were not as high in Nairobi as in the UK. Moreover, he accepted that the experiments would not be permitted in the UK but argued that it was better to capture wild baboons, who had lived for four or five years in the wild, rather than breed them in captivity. On asking the institute in Nairobi, it claimed that the studies were aimed at advancing medical research for the benefit of Kenya and the world. And far from being endangered, baboons were considered pests in the wild.[99] It is a general perception from this report that the conceived standards of animal welfare in Kenya (Africa) are lower than those legally required in UK (EU). On top of that, Africans treat baboons as pests, while owing to principlism they are treated essentially as utilitarian means to the ends of harmful experimentations or human science. Such an exploitative relationship raises pertinent ethical questions regarding the moral status of non-human animals. This already proves not only a weakened bioethical relationship between humans and *"cosmic-ecological life,"*[100] but also ethically violates the *EU directive* (2010/63/EU) from which the researcher originated from, as well as confirms the absence of a strong legal system of enforcement and institutionalization of ethical codes that is evident in Black Africa, as Tangwa equally illustrates:

99. Chatfield and Morton, "Non-Human Primates in Research," 85.

100. In this investigation, "cosmos" and "ecology" are used as synonyms. I use "ecology" as a noun—a thing created. I find creation summarized in ecology. The Greek sense of the word οἶκος, *oikos*: house, home, a place to live—depicts similarly what creation is. The German zoologist Ernst Haeckel applied the term *ökologie* to the "relation of the animal both to its organic as well as its inorganic environment." Thus, ecology deals with the organism and its physical surrounding environment. It involves relationships between individuals within a population and between individuals of different populations. These interactions between individuals, between populations, and between organisms and their environment form ecological systems, or ecosystems. Ecology has been defined variously as "the study of the interrelationships of organisms with their environment and each other," as "the economy of nature," and as "the biology of ecosystems." See Pimm and Smith, "Ecology."

This is one reason that many researchers from the advanced countries strictly respect and comply with ethics regulations at home but violate them easily abroad. The research scandals that have recently been recorded in the developing world, particularly in trials in sub-Saharan Africa, involving Trovan and Tenofovir, are sufficiently illustrative of this fact. Such scandals today hardly occur in the developed world, not because the temptations to do unethical research are absent, but most likely because of fear of legal consequences. In any case, it is a fact that the majority of African countries do not have legally binding and enforceable frame-works for medical practice or research on humans and some of those which do exist are weak or lack enforcement capacity.[101]

Similar ethical violations can be traced in the second example of human food trials. In this example, a poor bioethical relationship between humans and the cosmic-ecological life is again noticeable. There is exploitation through philanthropic organizations as they determine research priorities without necessarily involving the affected LMICs. In relation to the "Human Food Trial of a Transgenic Fruit" research, Jaci van Niekerk and Rachel Wynberg present concerns about developing a genetically modified "vitamin-enriched" banana for cultivation in Uganda through a proposed trial with a North American university.[102] Since East Africa is the secondary center of banana diversity (after India), with Uganda being the largest producer and consumer in this region, already there are ethical risks of undermining natural local food systems and erroneously reducing banana agro-biodiversity. Over and above, Uganda is home to banana varieties that are already higher in beta-carotene than the proposed GM banana. In addition to the country being situated in a fertile rift valley and tropical zone, it cultivates foods naturally rich in beta-carotene, such as sweet potatoes, leafy vegetables, and various types of fruits, which are affordable, healthy, locally acceptable, and highly nutritious. Such research in LMICs driven by philanthropic donors from HICs should be sensitive to local peoples' self-determination.[103] That being the

101. Tangwa, "Ethics in occupational health," 2.

102. Schroeder et al., "Ethics Dumping," 6.

103. In the words of a concerned Ugandan, "Just because the GM banana has been developed in Australia and is being tested in the US does not make it super! Ugandans know what is super because we have been eating home-grown GM-free bananas for centuries. This GM banana is an insult to our food, to our culture, to us a nation, and we strongly condemn it." See Niekerk and Wynberg, "Transgenic Fruit," 91–98, especially 95–96; Rwiza, *Environmental Ethics*, 181–82.

case, in cultivation of such an ethical sensitivity, we anticipate founding a functioning friendship not only between *global partners with African local natural ecological environment*, but also *locals and their cosmic-ecological environment*. Analogously, it is more effective to relationally convince ourselves that turning plastic garbage into a biodegradable product—or let us specifically think of eco-friendly agricultural mechanization in the development of genetically modified bananas—is more environmentally friendly not only to the benefit of global and local partners alike, but equally also to the cosmic-ecological environment that sustains the global community of both the local people and the foreign researchers.

RÉSUMÉ OF PART I: CRITIQUE OF BIOETHICAL PRINCIPLISM IN THE AFRICAN CONTEXT

Precisely, Part I has established the problem with bioethical principlism in an African context. In chapter 2, we have noted how the bioethical principlistic framework remains generally abstract, too general, and indistinct, giving birth to an unsystematic and non-unified guide to resolving regional and global conflicts in biomedical ethics. In chapter 3, we have noted that one of the difficulties remains: the language of principlism rests on socioeconomic terms that basically dictate libertarian prioritization (rationing) without enforcing an integral social utility. Yet if life was to be rescued, it shouldn't be a marketable piece of merchandise.

Our discussion in chapter 3 has shown how principlism basically relies on the cultural approaches of the Global North as absolute (excluding rich approaches from the Global South), which approaches do not fittingly sit in some African contexts. But if we are aiming at developing *a global language of bioethics befitting world citizens, principlism remains largely unsuccessful*. Culturally tilted Global North principles are inadequate for addressing global concerns that are global in nature. We need a new language of bioethics and a comprehensive moral framework grounded in a foundation of universally shared outlooks that appreciate bioethical moral pluralism.

In analyzing the North-South ethics dumpings in chapter 3, where corruption cuts accross all global members, the risk of exploitation is not limited to a single rubric (or to Africans and their immediate Western counterparts); it is not limited to an "inadequate informed consent process," "possessive individualism," "manipulation of non-human primates in research," or "exploitation through genetically modified foods" or "male circumcision trials or the candidate Ebola virus vaccine," but the lodged caveat cuts across biomedical exploitations as *general weaknesses and inadequacies of an entire system*, on the one hand. On the other hand, if the specific African sociocultural reality is not accurately *appropriated* like in the exploitive case studies discussed, then some will continue to basically speak not of "African bioethics," but of "bioethics in Africa" since ethical guidelines for health-related research are foreign to an African context, and what is African is being replaced with what is non-African in the guise of universalization of principles or globalization.[104] So, in answering the major posed question of Part I, the problem with bioethical principlism in an African context is not only an inadequacy of a well-intentioned collaborative structural-universal regulatory framework,[105] or a weak cosmic-ecological relationship, but also destabilized interpersonal-humanistic relationships across Global North/South cultures. If the entire system needs conversion as African moral thought seems to suggest, does African moral thought supply some contributory approach to this conversion or rather complement principlism? What is this African moral thought? In Part II, without bracketing off what some critics aver, I present and assess African moral thought in our pursuit of a more comprehensive approach to bioethics, as well as offer various significant interpretations in relation to bioethics.

104. See Andoh, "Bioethics and the Challenges," 67–75; Compare with Metz, "African and Western Moral Theories," 49–58.

105. Tangwa et al., "Ebola Vaccine Trials," 51.

PART II

African Moral Thought: An African Interpretation of Bioethics

ABSTRACT OF PART II

In Part II, we attempt to reconstruct an African interpretation of bioethics as already presented by numerous Africanists in Part I. One prominent perspective is Tangwa's *eco-bio-communitarian* approach. For Tangwa, as we saw in chapters 1 and 2, central to bioethics is the "peaceful coexistence, promoted by the idea of *live* and *let live*," hence preferring to define *bioethics as reverence and respect for life*.

Further on, taking the criticism of other various Africanists and African scholars seriously, Part II attempts to answer how bioethics can be unfolded so as to *bring together* rather than *take apart* African approaches to bioethics and a universally oriented principlism. We attempt to establish what could aid us in enriching and complementing a universally oriented bioethical principlism from an African outlook so that we can collectively unite both approaches along their own weaknesses.

Besides, Part II disassociates itself from the dangers of preoccupying ourselves with which moral thought should be prioritized in biomedical ethics, since African and Western thought each have weaknesses and are at least perfectible. So I find no necessity of stressing an Africanness of bioethics and then separating it from the general understanding of bioethics. Though Tangwa's *eco-bio-communitarian* approach is plausible, I find it over-stressing the Africanness of bioethics. Though the chapter picks something from Tangwa's approach, it offers a systematic approach beyond Tangwa's *eco-bio-communitarian* approach in the light of African moral thought.

4

An African Interpretation of Bioethics

CRITIQUING EMPHASIS OF AFRICANNESS: DECOLONIZING ETHICS

As my point of departure, I begin with this critique in order to journey well with the reader since emphasizing the Africanness of bioethics does not broaden the background of this chapter and doesn't unveil my rhythm of thought. I initially wonder if we can speak of an African approach to bioethics (an African moral thought) without wholly welcoming an authentic African reality, that is, a truly *appropriated* African way of life. In the reading of Tangwa, the traditional understanding of African reality has no difference with the traditional four-principle approach (bioethical principlism) of Childress and Beauchamp. This traditional perspective implies neither a lower status at all or a static tradition (culture) or a life that is not currently defensible, but rather an *appropriated* way of living out life. In the appropriated essentials of Africans, cultures, metaphysics, attitudes, customs, and the totality of life—reality is at least very similar in many ways. Although this may quite be compatible with some observable nuances and dissimilarities, my focus on Black Bantu of Sub-Saharan Africa proves commonality in *cultural realities* and *mores*[1] like I strongly elaborated in chapter 1. Nonetheless, I do not completely rule out the cultural differences among African realities, and neither should

1. See Tangwa, "African Perception of a Person," 41; Tangwa et al., "Global Health Inequalities," 245.

87

my generalizations in this investigation need to be flawlessly accurate to be truly satisfying to anyone who has previously associated herself with some African social reality. And if that is granted, then African social reality is not a mere myth,[2] but rather a specific conception of reality, in which *lived reality* may not only aid us in proposing solutions and prescriptions to local problems in bioethics,[3] but also support our universal pursuit of science and technology based on, for example, the African eco-bio-centric attitude of "live and let live" as Tangwa suggests. But if the specific African sociocultural reality is not accurately *appropriated* like we have discussed in Part I, then implementation of globalized bioethical princiflism will remain foreign to an African context. But here again my emphasis is also neither the "Africanness of bioethics" nor "African bioethics," but rather the lack of a genuine African approach to bioethics in preference to principle-based foreign values. It is here that the global community hasn't given ample room for an African contribution to fully dialogue with the universalized discourse in bioethics.

In emphasizing the Africanness of African bioethics, Tangwa takes "African bioethics in the same sense in which we talk of American bioethics,"[4] which is contrary to my appreciation of an African approach to bioethics. Tangwa's emphasis of Africanness or Americanness of bioethics may not entertain a productive bioethical dialogue. The question is: what is exactly the Africanness or Africanity of African bioethics? My concern is that even when we have different methods or approaches to arrive at the *goal of bioethics* as *life*, I barely see any would-be-difference in these said bioethics, because I think of *only one bioethics*, which we all pursue, bioethics that is life oriented, whether in Africa or in non-African contexts, in Western or non-Western contexts. Thusly, I see only differences in the approaches to *bioethics* or *life* but not a difference in bioethics itself. And so, it would be more constructive to emphasize genuine dialogue along these different approaches than emphasize the mightiness of an African bioethics (Africanness) or Western bioethics (Westernness).

Besides, in interpreting Tangwa's writings in the most charitable way, I come across an inadequate epistemological clarification between the approach (method), the sources and elements of African bioethics

2. See Tangwa, *Contemporary Bioethics Problems*, 16.

3. See Tangwa, "Leaders in Ethics Education," 103; Tangwa, "African Thought," 101–10; Andoh, "Bioethics and the Challenges," 67–75.

4. Tangwa, "Leaders in Ethics Education," 102–3.

in his book *Elements of African Bioethics in a Western Frame*, which I partly treated in Part I. Before defining bioethics as reverence and respect for life, he presented bioethics as the framework of beliefs, attitudes, and dispositions of African cultures, understanding them to be the constituent elements of African bioethics, rather than what we would mostly understand as the raw materials or sources for reflecting on an African approach to bioethics. It can be rightly objected that I have interpreted Tangwa's perspective of African moral thought unfavorably, but I have found no other alternative interpretation on reading him. This is why I am not exploring the Africanness or Africanity of bioethics, but rather an African approach to bioethics as I conceive of it in light of African moral thought, in line with Tangwa's studies plus numerous interpretations of African scholarships. I am of the view that we should *relax* (ease) the overemphasis of Africanness of bioethics in broader scholarship by realizing a sufficient mental decolonization—by exorcising our minds of colonial conceptions of thought—and political will on the part of both the Global North and Global South, colonized and colonizing peoples, the universalized and localized senses of bioethics. Although many colonized peoples have been able to outgrow some negative effects and productively integrated themselves toward positive transformative localization of foreign values, for integrative bioethics that I argue for in this investigation to triumph, we must overthrow the mentality of "my country first" or "Africanness" or "Americanness" or "Europeanness."[5] This is in no way to suggest that we ought not domesticate what is foreign. Already Tangwa believes we can purposefully domesticate colonial legacies as vehicles of unity and integrative interaction. For instance, it would be undesirable for Africans to continue bemoaning the imposed European languages on the African continent as well as colonial systems of education in foreign languages, where the advantages of modern writing at that time were obvious over the orality of African tradition. Contemporarily, however, modern writing and education belong to no culture any more than to the single human culture since human cultures, as Tangwa reiterates, form intersecting and overlapping concentric circles within a circle—think of tinted spectacles or lenses through which we culturally view reality differently (cultural tint). But such a cultural tint shouldn't blur us from recognizing that "objective reality" is variedly multicolored.[6]

5. See Tangwa, "Introduction by Henk," 269; Onwubiko, "Re-Encountering African Culture," 93–94.

6. See Tangwa, "Colonialism and Linguistic Dilemmas," 4 and 9–10; Tangwa,

Multicolored morality here means "no human community is perfect, and hence, no system of morality is perfect."[7] This implies not only a possibility of *gradual perfection* since no rationality, tradition, or morality is fixed or static, but rather also a chance *to enrich* and *be enriched* from other moral traditions. In one way, it is not only to mean that human cultures are both fallible and perfectible, but also, what is implicitly suggested in these cultural lenses is to outgrow negative attitudes towards colonialism and move positively towards a productive, transformative indigenization. Indigenizing principlism, according to Puleng Segalo and Lien Molobela, is not to emphasize the Africanness of bioethics, but rather to socially cultivate responsible *decolonial ethics*[8] that speak to the value of acknowledging the multiple spaces of reality people occupy and the treasuries of knowledge they employ in running their daily lives.

One way to decolonize ethics is to give a voice to an African thought in medical research ethics as Tangwa puts it.[9] Moreover, conceptual decolonization compels not only Africans but equally also non-Africans in equal measure. Indeed, to decolonize ethics is to embrace these rich multiplicities of worldviews springing from concretely lived social realities of how each culture interacts with its environment. It is from this perspective that each must nurse doubts and cultivate the humility to withdraw from or deny any superimposition of infallible ethical principles that don't relevantly tally with local social realities, or which alienate people from their rich and peaceful histories of life. It is why my perspective of African moral thought goes further than these limitations to embrace the whole, i.e., enriching others and having the humility to be enriched by them without imposing my Africanness on them. Let me lay down some preliminary elements of media of African moral thought. It is from here that I systematically appropriate the epistemological characteristics that systematically shape not the Africanness of African bioethics that some Africanists underscore, but rather an African approach to bioethics—that even goes further on than Tangwa's limited eco-bioethics. In the following section, it seemed practical to approximate some elements (because space cannot allow me realizing a comprehensive exhaustive list here) of media of African moral thought in order that I can structure the characteristics of their moral thought.

African Bioethics in a Western Frame, 96, 108–9.

7. See Tangwa, *African Bioethics in a Western Frame*, 142.

8. See Segalo and Molobela, "Africanist Research Ethics," 45.

9. Tangwa, "African Thought," 103.

SOME PRELIMINARY ELEMENTS OF THE MEDIA (CHANNELS) OF BANTU MORAL THOUGHT

At some level, I do journey with African studies that establish an African approach to bioethics in relation to African cultural values, traditions, customs, and practices that are often uninterrupted,[10] while at another level I consider various dissimilar scholarships on African moral thought to achieve the aim of this section. To a slight extent, let me mention a few preliminary elements of the media of Bantu moral thought that will guide us in structuring the characteristics of African moral thought.

Bantu Languages and Culture

Scholarship generally affirms the unifying and linking role of languages not only to the people but also to the existent treasures of moral wisdom embedded in peoples' way of life. For Martin Heidegger, "language already hides in itself a developed way of conceiving,"[11] while for Hannah Arendt, "language is the medium of thinking."[12] Language as the medium of thinking and conceiving of people realizes its unifying role through its relationship to culture.[13] This coupling relationship is captured by Hubertus C. G. Merkies in the presentation of Giovanni Battista Vico (1668–1744), the founder of ethnopsychology (the psychology of races and peoples): "Understanding culture is like understanding language, because human ways of life are embedded in human way of speaking."[14] Culture as a particular peoples' way of life is grasped by a particular language. In a similar arithmetical sense, "language can be compared to the visible one-tenth of an iceberg, the other nine-tenths of which are submerged under water. The nine-tenths of the iceberg could be said to be the culture of the people who use the language of the people."[15] This points to a biologically deeply rooted need—so to say a cultural universal—of the individual to have optimal contact with her cultural surroundings,

10. Tangwa, *African Bioethics in a Western Frame*, 58.

11. Heidegger, *Being and Time*, 199.

12. Arendt, *Life of the Mind*, 174.

13. See Levi-Strauss, *Linguistique et en Anthropologie*. "Langue et culture sont deux modalités parallèles d'une activité plus fondamentale . . . l'esprit humain." Cited from Merkies, *Ganda Classification*, 9.

14. Merkies, *Ganda Classification*, 273 n. 1.

15. Yotsukura, "Ethnolinguistic," 269.

just like plants optimally need contact with their milieu in relation to light (before they wilt),[16] through the language. It is no different to "Bantu people."[17] Their languages unify people with their cultural surroundings. For instance, the Baganda confirm the uniqueness of each group: *Buli ggwanga ne byalyo*, meaning, every tribe (surrounding) has its own rules of conduct[18] shaped by its language. The human surrounding as "the 'real world' is to a large extent unconsciously built on the language habits of the group."[19] Even though language remains a cultural universal in all human beings, its particular use in every cultural surrounding remains unique. So, we can reliably deduce that there can never be any pluralistic languages (two or three) that are indeed adequately alike and so denoting an identical social reality.

Over and above that, we are led to construe that though we may experience syntactical and grammatical differences, there is no sharp difference between other languages and "Bantu languages"[20] in their

16. Merkies, *Ganda Classification*, 43 n. 18.

17. *Bantu* or *Abantu* (the plural form of *muntu*) means people. The Bantu groups include the Eastern Bantu who live in Kenya, Tanzania, Rwanda, Burundi, Democratic Republic of Congo, Somalia, Comoros, and Uganda (in addition to several examples given from other Bantu people, *this study familiarizes itself more with practical examples of Bantu people of the Baganda (Ganda) ethnic group* in the southern part of Uganda around the northern shore of Lake Victoria). The Southwestern Bantu are found in Zambia, Malawi, Zimbabwe, Mozambique, Swaziland, Namibia, South Africa, Botswana, Angola, and Lesotho. The Western Bantu are found in Equatorial Guinea, Gabon, and Cameroon. See Kasozi, *Ntu'ology of the Baganda*, 19–20; see also footnotes 2–4, 6. For more on Bantu peoples, see Sempebwa, *Reality of a Bantu*, 1–5.

18. See Nsimbi, *Olulimi Oluganda*, 31.

19. Merkies, *Ganda Classification*, n. 17.

20. "Es ist wie alle Bantu-Sprachen, eine Klassensprache, das heißt, die Hauptwörter werden nicht wie etwa in Deutschen durch ein grammatikalisches Geschlecht in männliche, weibliche und sächliche Hauptwörter eingeteilt, sondern sie werden in Artformen gruppiert, in Klassen." Jahn, *Umrisse der neoafrikanischen Kultur*, 104, "'Bantu' bezeichnet zunächst einen Sammelbegriff für ein sprachliches Gebiet in Schwarzafrika, in dem viele so genannte 'Bantu Sprachen' linguistische Gemeinsamkeiten und auffallende Verwandtschaftsverhältnisse (genealogische) aufweisen, die durch das Auftreten eines regelmäßigen *nominalen Klassensystems* (=grammatikalische Kategorien wie die lateinischen Deklinationen) als Grundlage gekennzeichnet sind. Die Bezeichnung 'Classes nominales' besagt, dass die Übereinstimmung durch das Substantiv (Nomen), das Subjekt der Aussage erfolgt (z.B. in Kiswahili: *Kintu kikubwa kile kimeaguka* = Dieses große Ding ist gefallen). Da Präfix *ki* wird beim Adjektiv und beim Verb wiederholt. Im Allgemeinen dienen die Präfixe zur Unterscheidung der Kategorien (Menschen-Lebewesen végétaux), etc.)." *Encyclopedia Universalis*, 278. Quoted from Ndombe, *Verständnis der negro-afrikanischen (Bantu-) Weltanschauung*, 137, see footnote 412; "Die Gruppe der Bantusprachen in Afrika teilt ihre Substantiva in zahlreiche Klassen ein und kennzeichnet sie durch Präfixe." Dempwold, "Sprachforschung und Mission,"

significant characteristics of identifying, linking, and unifying peoples. Some Luganda (the language of Baganda) proverbs indicate this role: "Omuganda engeriye ogitegeerera ku lulumi lwe," i.e., the characteristics of a member of the Baganda ethnic group are betrayed by the language;[21] "Ozaayanga omubiri n'otazaaya lulimi," one may expatriate the body, but not the language. Literally, language is crucial for personal and cultural identity.[22] We may further construe that the role of identifying, linking, and unifying of a language can be better understood in the triadic inter-relationship of one's culture as one's particular way of life and as grasped by one's particular language: the individual in a social network, the culture and the linguistic interrelatedness, which all remain in one unified accord.

Bantu Contextual Symbols

The totality of African moral life can be clarified through the pre-dominance of an infinite variety of symbols, as Jean-Marc Éla states : "L'homme africain s'exprime à travers une infinie variété de symboles, puisés dans l'univers concret de son expérience, là où le langage trouve ses signifiants."[23] Such a boundless diversity of symbols is collateral to their symbolic thinking.[24] And if one is to understand Bantu moral life, then

---

152; Other languages like Luganda, have verifiable inflexions with a tendency towards agglutination, with static roots, which gain meaning, value or status only by use of initial vowels, prefixes, infixes, suffixes, etc. An example is the static root "sajja": with the prefix: "mu" *musajja* = man; add to that word the vowel "o" *omusajja* = the or a man (though the added initial vowel "o" doesn't change the meaning of the word); with an infix: *kusajjakula* = to become a man (by age); with a suffix: *musajjaggere* = a genuine man. Take the "historical" overtones of the suffix "wawu," in the word "Muganda-wawu" to mean a genuine "muganda," a genuine or pure member of this folk, whose lineage can be traced back to the ancestors. Nonetheless, nouns aren't classified in accordance with a masculine, feminine or neutral grammatical gender, rather, nouns are distinc-tively classified by prefixes along corresponding pronouns. See Kasozi, *Ntu'ology of the Baganda*, 21–22; see also footnotes 8–13, 14, 17. They are also small classes like Li-, MA class to express a collective idea—things found in quantities, or in sets or things which do not exist singularly but in pairs or quantities, e.g., *Malagala* (leaves), *Mayinja* (stones), *Liiso* (eye), *Maaso* (eyes). Kasozi, *Ntu'ology of the Baganda*, 29 nn. 11–12. See also Sempebwa, *Reality of a Bantu*, 48–51.

21. Kasozi, *Ntu'ology of the Baganda*, 22 n. 15.

22. Kasozi, *Ntu'ology of the Baganda*, 21–22; see also footnotes 8–13, 14, 17.

23. Éla, "Symbolique africaine," 93–97. About Black African symbolism and sym-bolism in initiation rites, see also Éla, *My Faith as an African*, 34–41.

24. According to Ndombe, symbolic thought is key for understanding the African

the *muntu*'s (individual human being) *participation in the life principle* has to be necessarily understood in relation to the function of the symbol at the center of her life. The widely known Congolese thinker Abbé Vincent Mulago's theory of *"lebensnotwendige Partizipation"*[25] indeed gives credence to African symbols.[26] In Mulago's interpretation of African thought, the invisible world manifests itself outwardly in the *muntu*'s participation in the life principle centrally through symbols. But if the civilization of bioethical principlism denies itself from the world of symbols, then an African ethic that relates naturally with the visible as well as with the invisible realities would have no place, as the Cameroonian thinker Éla argues. For him the immensity of symbols amongst Bantu interrelates between the invisible community and visible community of *muntu* endlessly and transcendently.[27] Transcendence of African ethic embraces the symbolic media with an extension to the created totality of nature. For Black Africans, "all qualification perceivable by the senses, such as its outward appearance, size, weight, measure, pleasantness, unpleasantness, usefulness, right and wrong, animate, inanimate, are experienced in the community in a specific way."[28] And it is why for an African "the inanimate world is also a source of symbolism of astounding wealth."[29] Put differently, the media of all earthly symbols, inanimate or animate, reveals to our moral

---

(Bantu) world, religion and culture. Ndombe, *Verständnis der negro-afrikanischen (Bantu-) Weltanschauung*, 7.

25. Mulago, "Bantu Strukturprinzipien der Gemeinschaft," 66; This is visible in Mulago's unpublished doctoral dissertation in Rome at the Urban Pontifical University: "L'union vitale chez les Bashi, les Banyarwanda et les Barundi."

26. Why are symbols weighty? "Für Mulago stellt das Symbol das Mittel dar, das den *Muntu* mit der Welt des Übernatürlichen in Verbindung setzen kann. In seiner Studie über die Gebräuche und Institutionen der Bashi, der Ruandesen und der Barundi hat Mulago die Funktion des Symbols folglich als Lebenszentrum charakterisiert, worin das Sein selbst, die unsichtbare Kraft sich im Menschlichen in seinen äußeren Erscheinungen bemerkbar macht." Ndombe, *Verständnis der negro-afrikanischen (Bantu-) Weltanschauung*, 160–61.

27. "Die afrikanische Zivilisation ist in gewisser Hinsicht eine Zivilisation des Symbols. In dem Maße, in dem die Beziehung von Mensch zu Mensch zur Natur sich durch das Unsichtbare ereignet, das der symbolische Ort ist, wo jede Realität zu einem kommen kann, ist das wirklich Reale unsichtbar und das Sichtbare nur der Schein: alles ist Symbol. Der Afrikaner lebt demnach in einem, 'Wald von Symbolen' als der bevorzugten Weise seiner Beziehung zum Universum." Éla, *Mein Glaube als Afrikaner*, 49. Éla's book, the German version, brings out the point home more clearly.

28. Merkies, *Ganda Classification*, 65.

29. Éla, *My Faith as an African*, 37. Éla's book, the English version, is clearer and brings out the point quickly.

senses what was obscure. Besides revealing what is in the visible reality, symbols bring out into the open what is concealed in the invisible reality. Through African symbolism natural raw materials turn into bonds that connect humans with other life forces. Together, they all cross the threshold to the invisible on their path to transcendence through the means of the transformed object. This is in agreement with the poetry of the first female writer (in the late 1960s) in the Belgian Congo.[30]

We note that both the invisible and visible community realities can be symbolized in a dialogue as concretized in a communicative palaver. But what is an African palaver? In discerning norms concretely, the African palaver is the place where norms that guide the community are elaborated. The palaver as the place for communicative ethics, for instance, dispenses justice in cases of conflict, completes matrimonial transactions, effects better medicinal therapies, and so on. It is within the communicative palaver, for example, that elders along the totality of the community interpret the notion of peace as much more important than the notion of equality, that the integrative aspect is more important than letting an individual be excluded from communal life, that dialogue, consultation, and deliberation have primacy before any decision-making[31] than purely focusing on individualistic pursuits. To symbolize the invisible reality as a crucial component of a communicative palaver, for instance, "a grave is where the presence of the invisible is concentrated."[32] This symbol of the grave reveals an invisible reality making it practically nearer in the African ethic given that its significance transcends the media of the visible earthly symbols. This explains why the African palaver of a communicative community emphasizes, on the one side, nonverbal forms of dialogue. It is the reason a *muntu* patient (sick or suffering descendant) will ask for forgiveness at the grave of her deceased parent or ancestor; whereby the grave symbolizes the presence of the deceased parent or ancestor, allowing not only a nonverbal dialogue between the relative of the visible community and the ancestral relative of the invisible community, but also an opportunity for the patient to fully unpack all the underlying

---

30. "Für die afrikanische Symbolik wird das Kunstwerk die Ausdrucksmöglichkeit schlechthin. Der Mensch verwandelt mit seinen Händen den natürlichen Rohstoff und macht daraus das Band, das ihn mit den Seinen verbindet. Gemeinsam überschreiten alle zusammen auf ihrem Weg zur Transzendenz durch das Mittel des verwandelten Gegenstandes die Schwelle zum Unsichtbaren." Faik-Nzuji, *Die Macht des Sakralen*, 125.

31. See Bujo, *African Ethic*.

32. Éla, *My Faith as an African*, 15.

burdens. Moreover, this is no different to the nonverbal encounter with nature (e.g., ancestral trees)—as it is the frequented place of the said communicative nonverbal dialogue between the earthly visible community and the invisible community of the living dead.[33] On the other side, there various forms of verbal dialogue. These are other communicative modes in Black Africa that also proceed in a contextual manner,[34] but not in generic terms like bioethical principlism.

The African participatory contextual verbal dialogue, which is not dichotomized, is synonymously understood as consensus that seeks a decision that is accommodative, where there are no losers and there is no attitude of winner takes it all.[35] The notions of dialogue and consensus awaken the conception of African moral thought of "live and let live," "be and let be," significant for verbally resolving interpersonal, inter- and trans-community disagreements and differences without excluding any life force or community of beings. This is no different from the image popularized by Julius Nyerere, the father of a communalistic sociopolitical ideology of *Ujamaa*. In Kiswahili, *ujamaa* means a family. The compound word *Ujamaa Vijijini* means not villagization of the world like in globalization, but rather a collective villagization—a strategy hinged on the African concept of familyhood—covering all the relatives of the extended family including the living, the dead, those not yet born, and those variedly attached to this three-dimensional community (visible and invisible).[36] The physical symbol of a *ujamaa* family enables a verbal palaver where African elders are comfortable sitting under a tree shade at some sunset, mutually dialogue uninterruptedly and discuss issues of community importance until they agree, find an undisputable solution, or reach a consensus. Everyone participates as a mutual comrade and as an inevitable dialogue partner with equal rights as some consensus is pursued. Coercing conversion to a certain category of thought isn't a priority but rather extending the spirit of familyhood, as Nyerere notes: "We, in Africa, have no more need of being 'converted'" to globalized principlistic constructs since they "are rooted in our past—in the traditional

---

33. See Bujo, *African Ethic*, 47; see also Bujo, *Ethical Dimension of Community*, 29.

34. See Bujo, *Ethical Dimension of Community*, 226.

35. See Eze, "What Is African Communitarianism?," 386–99.

36. Tangwa, "African Perception of a Person," 42; If we could borrow the German interpretation, *ujamaa* is an equivalent of a family or brotherhoodness. "'Ujamaa'— was so viel bedeutet wie Familie, Brüderlichkeit." Ndombe, *Verständnis der negro-afrikanischen (Bantu-) Weltanschauung*, 206; see also footnote 663; Bujo, *African Ethic*, 153–54.

society which produced us . . . Our recognition of the family to which
we all belong must be extended yet further—beyond the tribe, the community, the nation, or even the continent—to embrace the whole society
of mankind."[37] For Nyerere, embracing a universal conversion, it is why
for the "outside world the separate national states of Africa must cease
to exist. They must be replaced by Africa."[38] It is in the same spirit that
genuine dialogues between bioethical principlism and other outlooks
ought to pick a leaf since coerced networks of fellowships are outdated.
Each partner must fully participate as a dignified and dignifying crafts-
person in the building up of a collective future in the discourse of global
bioethics. In augmenting a totality of an African approach to bioethics,
varied forms of dialogue or consensus would be impossible without the
multifarious media of contextualized earthly symbols that cut across the
three-dimensional community inclusive of the natural environment and
the extended partnerships outside our own family and cultural zones.

Oral Tradition and the Word

An African approach to ethics relies on the media of *the word*, i.e., the
"incantatory power of the word."[39] According to Baganda wisdom, "ekitayogera tekirema ayogera," i.e., humans are endowed with the ability to
speak, express their feelings, thoughts, and opinions orally more than
other beings.[40] In short, humans can more exhaustively utilize the media
of orality. Beyond the word being orally passed on and heard by the ear,
according to Bujo, it is also *eaten, chewed, ruminated,* and *digested.* Apart
from the institutionalized councils of elders, there are also non-insti-
tutionalized corrective ones, in which elders indispensably give off life
whenever need be to the young generations through the word. This can
be illustrated in an elder's narration of the entire native wisdom as she
randomly eats with youngsters under a tree. The necessity of the eaten,
chewed, ruminated, and digested word here explains the respect youngsters owe the elderly.[41] It is the elderly that stand in the middle of the past

37. Nyerere, *Ujamaa,* 12; Nyerere, *Uhuru na Umoja,* 171.

38. Nyerere, *Uhuru na Umoja,* 338.

39. Éla, *My Faith as an African,* 2001, 46.

40. See Kasozi, *Ntu'ology of the Baganda,* 44–45, including footnote 4.

41. See Bujo, *Ethical Dimension of Community,* 70–71, 203–6; Bujo, *African Ethic,*
151–52, 122; Wiredu talks of a spoken word and techniques of communication har-
bouring virtues, passed on from one generation to another through maxims, logical

generation of the ancestors and the future generation of those not yet
born, meaning any opportunity of having a venerable communication
with the elder harmonizes the ancestral invisible world with the visible
young generation and gives off new life abundantly, even to the invisible
future generation. Moreover, Bantu parents bring up their children not
just to tell the truth to strangers, since they believe that a *miscalculated
oral word* could render the family vulnerable to any anti-social life like
witchcraft.[42] This simple calculation shows the basic rule elders employ
to pass on education by means of cultivating the right oral word.[43] And
in order to shape the media of the word so that it can fully give life, it
is why the Black African community palaver seeks *the best* and *correct
word*, tests its capacity to heal and increase community life; for instance,
in instances of approving a proverb as a constituent part of communal
knowledge. Citing a contextual example of the Baganda, they regard
proverbs as raw materials, textbooks, google engines, libraries, pedagogi-
cal instruments employed, brief summaries (compendia) of varied fields
of indigenous knowledge (native modes of knowledge production). For
them, the equivalent of a proverb is "olugero olusonge." Firstly, the noun
"olugero" is widely translated as "a measuring stick," deriving from the
verb "okugera," meaning to ascertain the measurements of something,
or simply to measure. Secondly, the adjective "olusonge" stems from the
verb "okusonga," which means to poke, prod, pierce (or being sharp or
even penetrating). Putting both together in a compound word, a proverb
("olugero olusonge") as a summary of wisdom is sharpened by diverse
experiences of life events and is summarized from accepted community palaver standards to which similar life realities can be eventually
measured so as to give direction like a pointed sharp arrow does.[44] Put

---

exercises, folk tales, poetry, drum texts, funeral dirges, myths, lyrics, art motifs, sometimes painted, other times carved out of wood or cast in other media, etc. Unfortunately, these elements, raw materials, resources of African moral thought were neglected by
colonial and missionary informants. Wiredu, "Oral Philosophy of Personhood," 8–10.

42. Since for Ganda, "nothing harmful happens by chance." Mbiti, *African Religions
and Philosophy*, 201. Witchcraft (*bulogo*) or (*busezi*) is regarded as a bad thing in Ganda
society and when witches (*balogo*) or (*basezi*) are detected they could be harmed or
killed. Roscoe, *Baganda*, 20. Witchcraft is understood to describe all sorts of evil em-
ployment of the magical force, generally in a secret fashion. Witches as *muntu* forces
use *kintu* forces to harm others. Sempebwa, *Reality of a Bantu*, 161–62; see also MacIn-
tyre, *After Virtue*, 192–93; Magesa, *African Religion*, 165–73.

43. See Bujo, *African Ethic*, 29–30.

44. See Kasozi, *Ntu'ology of the Baganda*, 75; Mbiti, *African Religions and Philosophy*, 86; *Lugero* means "measure," "measurement," or it refers to a story, fable, parable,

differently, African proverbs focus on protection (promotion) of life through *the word* as their sapiential character. The word of the proverb can formulate harsh truth or elderly moral wisdom and it is received by the young generation enthusiastically. In sharpening this word of wisdom, fairy tales, parables, songs, music, dances, pantomimes, allegories, and narratives are often employed in the palaver process as vehicles for summarizing proverbial wisdom. This means that the proverbial word of an individual must be tested before the entire community palaver—the council of elders acts as the community's ruminants—which can either approve or reject it, contradict or correct it, re-chew or re-digest it, based on whether it is well received as life-giving and not life-diminishing. But if a proverbial word fails in each circumstance to take in the whole truth of the previously experienced life events, including the non-literate community, or it has ambiguous solutions, it will be objected by the palaver community. In this sense, the spoken word is especially healthy and healing if willingly shared by the palaver community, but not monopolized by a community member who only possesses a small part of the word. All participants in the palaver must have broad ears to exceptionally listen before sharing the word[45] with the totality of the three-dimensional communitarian reality—the living, the dead, and those not yet born. The question of whether Africans are simply empiricists, since they value the oral experience of sharing the word in grounding wisdom, is absorbed in this three-dimensional communitarian life. An example can be cited from among Baganda. For them, their extensive wisdom as a corpus of knowledge consists of beings and phenomena not only in the visible physical world but also in the invisible worlds (spiritual and supraspiritual).[46] In reference to this account, the oral word would be too much for the visible community alone. It is the reason it must be shared with an even more comprehensive reality of the invisible community. Analogously, the word would be too much for one cultural thought; this is why it must be genuinely shared with the pluralistic cultural perspectives until

proverb, saying. Murphy, *Luganda-English Dictionary*, 287; For the Ganda, "there was a need to impress on men the moral truth . . . this need gave rise to pithy stories and proverbs which have gone on increasing in number until the language is rich in proverbs and folk tales." Roscoe, *Baganda*, 460. Proverbs are emphasized as morally educative for the young as well as the old about life facts (moral wisdom). Proverbs are used to settle disputes and quarrels. They are also used to effect proper communal relationships. See more in Nsimbi, *Olulimi Oluganda*, 17.

45. See Bujo, *African Ethic*, 51–53, 151–53, 155–56.

46. See Kasozi, *Ntuʼology of the Baganda*, 22–23, 107–8.

a consensus is reached in a genuine dialogue. This is the integral reality that is symbolized by the word itself as Éla puts it: "La parole elle-même est une symbolique de la réalité."[47] It is again the same word that the intercontinentally hailed Nigerian novelist Chinua Achebe narratively spoke in *Things Fall Apart*: "words are like yams, proverbs are the palm oil with which they are eaten."[48] This recurrent proverb among the Igbo is employed by the Umuofia clan as an agrarian community whose main crop harvests are yams and palm oil among others. Just as their palm oil quickens cooking while adding an aroma as a sauce for the cooked yams, proverbs easily facilitate passing on oral wisdom. And that may help us summarize at least two points about the oral word below.

Firstly, how the media of the spoken word can give off life (or lessen it), given that the word is so powerful in Black Africa. To put it plainly, the spoken word is the means of the *muntu*'s expression of the life force, and the expression of being in its wholeness.[49] Secondly, it signifies the authority of the *oral tradition*, as in "from mouth to ear." And even though writing has been improved over what it was in conventional Africa, the media of the spoken word still inevitably counts till today and we scarcely regard today's writing generation as completely liberated from orality. The thinker Bimwenyi-Kweshi likewise defends this thesis; namely, that the African generational experience is still handed down orally: "Die afrikanischen menschliche und religiöse Erfahrung wurde— und wird weitgehend auch heute noch—in der Struktur der Oralität erlebt; von Generation zu Generation wurde sie auf mündliche Wege weitergegeben."[50] In a like manner, not only Ndombe emphasizes "die Oralität als ein Kennzeichen der Kulturalität der afrikanischen Kulturen und Religionen,"[51] but also Éla narrows down on "L'oralité étant le trait dominant de la culture africaine."[52] Tangwa is then right to note that the word, orality and aurality, communalism and relatedness, still predominate even much more the complexity of health and practice of medicine inclusive of healing,[53] which all integrally contribute to the *muntu*'s daily

47. Éla, "Symbolique africaine," 93.

48. Achebe, *Things Fall Apart*; Orobator, *Religion and Faith in Africa*.

49. See Senghor, "Der Geist der negro-afrikanischen Kultur," 343–63.

50. Bimwenyi-Kweshi, *Afrikanischer Theologie*, 160; Wiredu, "Oral Philosophy of Personhood," 8.

51. Ndombe, *Verständnis der negro-afrikanischen (Bantu-) Weltanschauung*, 321.

52. Éla, "Symbolique africaine," 93.

53. See Tangwa, "African Thought," 104.

ethical living. Undeniably, the media of orality and the word remain alive in an African approach to bioethics.

Elders and Wisdom

In an African approach to ethics, there is also the media of the elderly and their native wisdom, which remain indispensable. The elderly includes one's parents and grandparents, chiefs, family heads, sages, religious leaders, and clan heads. It is typical of East African countries to use the term *mzee* (male or female elder, a grey-haired adult of a certain age). In Kiswahili, one refers to parliament as *baraza la wazee*, i.e., elders' consultative council, granting it collective wisdom. Also being a *mzee* or getting old is to become wiser.[54] Among Baganda, one is classified as *mzee* on presumption that she has *magezi* (wisdom) about many aspects of life, which may include tribal customs and laws, religious ceremonies and rites of passage. A *mzee*'s verbal testimony is unquestionable truth, granting her a moral prerogative to guide and pass on *magezi* to the young generation.[55] *Wazee* are consultants because "old age equals wisdom," as Baganda say: "*Obukadde magezi, takubuulira kyamukaddiya*," meaning, "Although old age is wisdom, an elderly person is not obliged to recite the details of her life-story." But they can be consulted since they are equipped with the ability to adjudicate and "to give advice"—*okuwa amagezi*, which literally means "to give wisdom or knowledge." Giving advice is transmitting knowledge, on the one hand, and an act of acquiring knowledge, on the other hand. This ability as a moral prerogative of the *wazee* qualifies them as having extensive self-consciousness and awareness of the other life forces around them, what Kasozi calls a "reality consciousness borne out of varied actual experience(s)."[56] The highest moral contributors to reality consciousness among the *wazee* are our mothers. They give life to African clan communities, making mothers the *highest incarnation of wisdom*, since the future and the fate of the clan community depends decidedly on their *elderly* contribution. *Firstly*, their evident educative role is highly indispensable; a factor that is specifically obvious in their

---

54. Assemblies within East African cultures, such as a *baraza* (plural, *mabaraza*), is effective in gathering critical cultural data and act as a forum to obtain a community's understanding of research, community consultation or consent. Vreeman et al., "Community Assemblies"; Bujo, *Ethical Dimension of Community*, 200–202.

55. See Kasozi, *Ntu'ology of the Baganda*, 73–75; see also footnotes 12, 13, 15.

56. Kasozi, *Ntu'ology of the Baganda*, 68–69; see also footnotes 4 and 5.

care for their infants by always carrying them on their back (not in a pram like one often sees on streets of industrialized nations) so that they don't miss out on participating *directly* in community life, even when the mothers are on fields digging. Among the Chagga and Nsungli of Tanzania, delivering mothers, akin to victorious soldiers or heroes, are greeted with songs of great esteem. *Secondly*, Africa doesn't not only have powerful queens and kingdom founders—the Kingdom of Nso' was established not by a man but remarkably by a woman, *Ngon Nso'*—but it has women religious priests in both *patrilineal* and *matrilineal* communities. Among the Bahema of the Democratic Republic of Congo, certain religious acts of a funeral ceremony can only be performed by women. *Thirdly*, they are life-givers in areas of art and economy.[57] *Fourthly*, African women are *custodians* of the cosmos,[58] making them to be in charge of the earth that sustains everything. In each of the above-mentioned reality consciousnesses, the young generation, as the hope of the invisible community of those not yet born, builds up bonds relying on the established unquestionable truthful experiences (knowledge or wisdom) of the genderless elderly, i.e., without discrimination, experiences originating from both women and men alike. And along this wisdom, the young are guarded from untruthfulness or lying, which are rebuked as destroying interrelational bonds. The evidence of the Swahili proverb in Congo-Kinshasa similarly cautions the young: "An old person's mouth smells badly, but it does not lie" (*Kinywa cha mzee kina harufu, lakini hakina uongo*).[59] And we can thus deduce that empowering socialization of women is one genuine guaranteed way through which we can equally empower men in their elderly dispensation of wisdom.

57. See Bujo, *Ethical Dimension of Community*, 122–27; Tangwa, *African Bioethics in a Western Frame*, 128–30; There is 16 percent of matrilineal social communities, i.e., the "matrilineal belt" in Africa, including but not limited to Malawi (75 percent matrilineal), Mozambique (74 percent), Namibia (67 percent), Tanzania (22 percent), and Zambia (41 percent). Also, in Burkina Faso (Lobi), Niger (Asben Taureg), Sierra Leone (Sherbro), and Zimbabwe (Chikunda, Nyanja, and Tonga), but all of these countries have less than 10 percent. Gottlieb and Amanda, "Effects of Matrilineality on Gender"; Berge et al., "Matrilineal and Patrilineal Landholding Systems," 61–69.

58. Tangwa asserts: "According to African cosmogonic myths and metaphysical ideas, God made woman the custodian of fire, water and earth. Custody of fire entailed responsibility over energy, principally, in the context of traditional Africa, fuel wood; custody of water involved responsibility over a most important substance, water, symbol of both cleanliness and survival; and custody of earth implied a doctrine of dual fertility of soil and of womb." *African Bioethics in a Western Frame*, 127.

59. Bujo, *Ethical Dimension of Community*, 200–202.

Dispensation of wisdom doesn't stop at the community of the living, for instance, among Baganda. The reality consciousness (*amagezi*) is too large for the living elders to dispense alone to the young generation. It extends to ancestral spiritual beings whose media in transmitting *amagezi* could be the elders, say the diviners, dream interpreters, medicine persons—who are powerfully revered religious persons.[60] Over and above, a message from a spiritual being is ranked higher than the verbal testimony of an elder. It is regarded not only as unquestionable truth, but also as true and certain knowledge. Baganda orality—as elderly people narrate stories—bears witness that the whole village has ever migrated in order to avoid a danger announced by a spiritual being through its medium, hence the proverb: "*Siroota kibula, asengusa ekyalo*," i.e., one whose dreams always come true, sets a whole village into migration. Related to that, a distant working relative may receive an instinctive bodily feeling (*ekyebikiro*) as a transmission of bad news (something has gone amiss) through the twitching of the eyelid (*ekisulo*) or a slight fever (*kitengo*).[61] Such substantive Baganda beliefs show the possibility of transmitting *amagezi* from one individual mind to another beyond our typically known channels of human senses. All in all, adhering to elderly wisdom as reality consciousness can promote life of the three-dimensional community or the same life can be lessened if the same wisdom is neglected.

In instances where the *muntu* neglects practicing or less practices this wisdom, it invites social sanctions. The Luganda word *emizizo* (singular *omuzizo*) refers to protective measures or prohibitions imposed by social customs and laws as instituted by the elderly wisdom to revitalize well-being. These instruments are aimed at hindering what interferes with community harmony or the *muntu*'s compliance with the communal

---

60. The powerful medicinal contribution of medicine persons has been hailed by Black Africans from centuries to date as Roscoe testifies: "Medicine men though not definitely connected with the temples and the Gods, were regarded as belonging to the religious class of the country, they formed a most powerful body and were greatly feared." *Baganda*, 277.

61. Also, a bee that flies several rounds over the head of a house woman preparing a meal, foretells a visitor who shall share that meal. Any visit that is not announced is welcome, but described as "Omugenyi atazunza njuki," that is, a visitor who does not "cause" a bee to announce her visit. On top of that, the cry of the jackal at night, the singing of a bird named *nasse-enswa* (literally, "I killed white ants"), which is strikingly "out of the ordinary singing" of the owl, and the reflex movement of the lower eyelid, are all exceptional fore-warnings of impending sad situations, say the death of a relative. Kasozi, *Ntu'ology of the Baganda*, 73–75 nn. 12, 13, 15.

binding codes.[62] For illustration, the protective character of social laws among Banyarwanda generates moral obligation by making them to be characteristically "formulated in the negative. They require from people not that they should do something, but rather that they should abstain from doing something. In a word, they are interdictions (*imiziro*)."[63] When the *muntu* adheres to *imiziro* or *emizizo* and abstains from doing what makes him less practice her wisdom, she promotes life by promoting continuity of social harmony. Promotion of a harmonious life relationally heightens one's social status in contrast to interference of it. For Bantu, reprimanding an individual who does not observe the moral codes is tantamount to presenting her as "having little or less knowledge or wisdom" than, for instance, a domestic fowl. Consider the common moral ridicule among Baganda: "*Enkoko ekusinza amagezi*"; that is, "The domestic fowl is wiser that you are." It is not that the *muntu* has less knowledge or wisdom than a domestic fowl per se, but actually behaving well is having knowledge or wisdom. Sarcastically, in doing what is expected of it, the domestic fowl has knowledge or wisdom, whereas a *muntu* who does what is not communally expected of her has less knowledge or wisdom than the domestic fowl.[64] In this regard, prohibitive sanctions arise due to less practice of wisdom, which implies guilt.

Guiltiness is serviced with reconciliation and forgiveness within the community palaver. Grave offenses, like those against the will of the ancestors, are sanctioned because they are repellent to the meaning of life of the entire community, which is beyond the world of principlism. Since African ethics does not split the totality of life dualistically, the community palaver essentially embraces both morals (concrete success of human existence) and ethics (universal formalistic principles) as a unified continuum in promoting growth in the quality of life for all members. In a different voice, the quilt of individuals effects all, and the sanctions must revitalize the quality of life for all community members.[65] In contrast, bioethical principlism doesn't see guilt, sanctions, and reconciliation as applied to the three-dimensional African community life within their upheld world of universal formalistic principles. Moreover, principlism does not pay attention to forgiveness with an intention of reconciliation, whereas for African moral thought, forgiveness is not necessarily based

62. See Kasozi, *Ntu'ology of the Baganda*, 89–90.

63. Kagabo, "Alexis Kagame," 239.

64. See Kasozi, *Ntu'ology of the Baganda*, 70 n. 7.

65. See Bujo, *African Ethic*, 57–58.

on mutuality, but can also come about "asymmetrically," i.e., a victim pardons a perpetrator who has not asked for forgiveness. For example, the Bahema of Congo-Kinshasha call someone "noble" when she does not "bear grudges," or when she refuses to seek revenge and instead grants pardon even before her enemy seeks for it.[66] Forgiveness and reconciliation extends further on to the ancestors, the victims of the historical past, as we discussed in the nonverbal palaver of a descendant at the decease's grave. Other examples include the asymmetrical reconciliations after the South Africa Apartheid and Rwanda genocide. Nonetheless, these approaches are inexistent in the bioethical principlism. As we conclude this section, it becomes clear that guilt, sanctions, and reconciliation as constitutive of morality in a communicative palaver sets the African ethic and principlism further apart as far as their relationship to the spiritual world is concerned. Sanctions as taboos express a spiritual (religious or even a spiritistic) reality. To comply with the spirits who are custodians and guardians of the moral and cultural codes of conduct, the Baganda fear being punished by the spirits in case of breaching such codes.[67] And in case one *muntu* consistently deviates from pursuance of the general good (life for all), one must heavily pay the price individually because the spirits do not discriminate in their attacks.[68] In that, the media of *wazee*, *magezi*, *emizizo*, *guilt*, and *reconciliation* as constitutive of morality in an African community ethic cannot be understood outside revitalizing the quality of life for the entire community of the living, the dead, and those to be born.

Homeomorphic Equivalents

Etymologically, the adjective "homeomorphic" originates from the Greek words ὅμοιος (*homoios*), which means, like, similar, or same; and μορφή (*morphē*), which means shape or form. The catchphrase "homeomorphic equivalents" has a connotation of similar forms. Just like Raimon Panikkar notes, I too seek a literal translation for a word or an activity, viz., the "homeomorphic equivalent," which describes the possible correlative moral activity in African localities or cultures. This disregards any

---

66. See Bujo, *African Ethic*, 62–63.

67. See Kasozi, *Ntuʼology of the Baganda*, 48–49. Kasozi borrows some examples from the works of Mbiti, *Concepts of God in Africa*, 135–57.

68. Merkies, *Ganda Classification*, 116 n. 1.

univocal word-to-word translation.[69] Panikkar's emphasis is no different to African scholarship. The idea of a genuinely inclusive global bioethics must neutralize the cultural domination of any approach to bioethics *in order to fit what good life means in different cultures*. Oceanian, African, or Asian approaches to bioethics cannot be accurately the same as European approaches. Just like an African approach to bioethics has existed since time immemorial—even when the practice of medicine has been either latently or orally passed on—a Western approach to bioethics remains likewise only *one among* numerous global approaches. It further means that cultural cries of non-Western populations should not be quieted but genuinely heard. The implicit moral imperative is to necessarily explore values beyond one's culture since no single culture can exhaustively capture reality or claim valid solutions to all complex global issues.[70] One tolerant way towards appropriated global bioethics is to mutually respect homeomorphic equivalents by hearing the crying voices from other traditions or approaches to bioethics. The reader should not forget that although we understood language as a cultural universal, the application of some unique language remains a cultural particular. This understanding is in dialogue with Tangwa's assertion that "every language captures certain aspects or elements of reality much better than others. So, there will always be meaningful expressions in every language (culture) which are not fully or completely translatable into other languages (cultures)," and as justification, "homeomorphic equivalents" come into play since "it is hardly possible to transfer corn or any other similar grain from one basket into another without losing some grains."[71] Accordingly, in several cultures, homeomorphic equivalents could be substantive beliefs. In retrospect, if I take the example of Aquinas's understanding of "moral knowledge," it is not only the first principles of *synderesis*, but also *beliefs* with general substantive moral content.[72] Before Aquinas, the emphasis on substantive beliefs as knowledge dates back to Socrates. Knowledge has been consistently broken down as true belief that has certain additional

69. Panikkar, "Homeomorphic Equivalents," 21, 43, 47.

70 Mbih, "African Perspectives in Global Bioethics," 208. It is "inappropriate, and . . . unethical to impose, either consciously or un-consciously, the dominant Western socio-cultural-moral construct to ethnic minorities in the West and the vast non-Western world." Chattopadhyay and De Vries, "Bioethics Is Western," 106–9.

71. Tangwa, "Bioethics: An African Perspective," 187.

72. See Hoffmann, "Conscience and Synderesis," 255–64; Herzberg, "Die thomanische Deutung des Gewissens," 189–208.

requirements. In spite of the fact that the accuracy of these requirements are not yet wholly fixed in our daily hunt for knowledge, "most extant accounts of knowledge start out from the assumption that knowledge is a species of belief—and hence that states of knowledge are states of belief."[73] If a Jehovah's witness authoritatively subscribes to moral beliefs derived from the authority of her institution and refuses a recommended blood transfusion or a Roman Catholic chooses against an abortion in deference to traditional Catholic belief,[74] thereby accepting these normative beliefs autonomously, why can't an African autonomously accept authoritative moral beliefs derived from a truly African lived social reality? In a nutshell, a few discussed preliminary elements of the media of African moral thought furnish the three-dimensional community with content. Besides, the African lived reality as discussed presents *life* as the supreme moral ideal to be pursued by all Bantu in their various relationships with other objects and diverse life forces. It is through this operative community that *continuity and vitality* can be increased or decreased, thereby enhancing the *solidarity* of the whole group and the *hierarchical existence* of the totality of life forces.

CHARACTERISTICS OF AFRICAN MORAL THOUGHT

Once the question of characteristics of African moral thought is answered, the direction to answering the question of how every entity increases or decreases life relationally will become clearer. Furthermore, if we grasp an African interpretation of bioethics, we may not only come to appreciate what the African approach can contribute to the general discourse of bioethics, but we will easily appreciate what can *bring together* rather than *take apart* an African approach to bioethics and a universally oriented principlism. I concentrate on five characteristics of African moral thought. They include promotion of life, community harmony, continued vitality, solidarity, and a hierarchical order of vital forces.

Promotion of Life Relationally: The Supreme Moral Ideal

The first characteristic of African moral thought is to do with the principle of life as the supreme moral ideal and the highest common interest.

73. Bermúdez, "Mindreading," 429.

74. See *PBE*, 103–4.

Primarily we wonder: what is the supreme moral ideal within African reality; namely, what is the chief end to be pursued, the highest common interest to be promoted, the main collective purpose to be fulfilled, or the fundamental ideal to be realized? Primarily, let us borrow two Malawian Bantu concepts of *moyo* and *munthu*, which have parallel etymological meanings—depicting Africanity or Africanness—in Kiswahili, Kikuyu, Zulu, Sipedi, Tsonga, Luganda and other Bantu languages. *Moyo* means life, "the foundation of all there is," while *munthu*—like we already defined, the *muntu* is an individual human being—is "a member of humankind or a particular community of people."[75] The *muntu* or *munthu* has *moyo*. "*Moyo* is from God. God is *MOYO*."[76] In line with the Black African moral thought, Sindima thinks life (*moyo*) can be achieved or decreased but within a *community*. An African "community is the heartbeat of African ways of life. A notion of power which does not pertain to building and maintaining community undercuts the concept of life and community."[77] Other Bantu, specifically the Baganda, refer to this community as *entababuvo*; the prefix *entaba* connotes "interaction."[78] Put differently, *moyo* is the life force dispensed by God, but while life is supreme, it cannot only be increased or decreased relationally; it must also be contextualized within an interactive relational community.

Although both bioethical principlism and African ethics articulate the ethical principle of life as the universal highest good, good life in the African sense must be concretized in the communicative community palaver process, which is never individualistic or particularly emphasizing the autonomous choices of an individual. Bioethical principlism fails in its application of the concrete of ethical conduct in a given relational community, but continuously develops numerous theories of principles about other theories of principles,[79] without practically translating them into a functional relationship between the individual and the communicative community. What we mean is a single person (Western thought) stands alone on one side, whereas on the other a single person stands in a close communicative ethic with a relational community (African thought). Argued differently, an African approach to bioethics doesn't only see life

---

75. Sindima, *Ethics in Africa*, 172.

76. Sindima, *Ethics in Africa*, 176.

77. Sindima, *Ethics in Africa*, 66.

78. *Entababuvo* (community): "Entaba y'abantu oba ebintu ebirina obuvo obw'awamu." *Wikipedia*, https://lg.wikipedia.org/wiki/Entaba-words.

79. See Bujo, *African Ethic*, 66–69.

as the highest good for an autonomous individual, but rather the highest collective good to be pursued by the triadic community of the unborn, the living, and the dead. It is to this fact that African people know by heart the inalienable fundamental principle of African communal ethics: I am because we are, and because we are, so I am too.[80] In explaining one's individual ethical conduct, since the framework of African communal ethics is not static, but relational, it continuously contextualizes the historical past, experiences of one's ancestors, the future community, and the life of the contemporaneous visible community. This relationality can be understood in Bujo's interpretation of the African conception of life, in which the *muntu* does not become a *muntu* simply by *cogito* (thinking) but by *relatio* (relationship, fellowship) and *cognatio* (kinship).[81] Accordingly, what is decisive for Black Africans is not relationships or fellowships between individuals, as it is in principlism, where each individual goes her own way in living out norms, but rather the reciprocal decisive relationships between the *muntu* and the triadic community, thereby upholding "we-ethics" that accompanies each *muntu* along the way.[82] So it is not the *muntu*'s accomplishments and creativity, as such are central in principlism, but it is rather the communal achievements that are granted a relational value. Thinking by *relatio* doesn't deduce moral principles *a priori*, akin to abstract bioethical principlism, but inclusively admits the contribution of that which cannot simply be justified in terms of reason-based classroom principles, a perspective that is not only disloyal to the

---

80. "Whatever happens to the individual happens to the whole group, and whatever happens to the whole group happens to the individual. The individual can only say: I am, because we are, therefore I am." Mbiti, *African Religions and Philosophy*, 4. "Der einzelne wird sich nur im Hinblick auf andere Menschen seiner Eigenart, seiner Pflichten, Vorrechte und Verantwortlichkeiten sich selbst und anderen gegenüber bewußt. Wenn er leidet, so leidet er nicht allein, sondern mit der Gruppe, der er angehört; wenn er sich freut, so freut er sich nicht allein, sondern mit seinen Artgenossen, Nachbarn und Verwandten, ob diese nun tot oder noch am Leben sind. Wenn er heiratet, so steht er nicht allein, und auch seine Frau 'gehört' nicht ihm allein. Im gleichen Sinne gehören seine Kinder der Gemeinschaft, mögen sie auch nur den Namen des Vaters tragen. Was immer dem einzelnen widerfährt, geht die ganze Gruppe an, und was der ganzen Gruppe widerfährt, ist ebenso Sache des einzelnen. Das Individuum kann nur sagen: *Ich bin, weil wir sind, und weil wir sind, bin ich*.'" Mbiti, *Afrikanische Religion und Weltanschauungen*, 136. Emphasis added. And so "Die Natur bringt das Kind zur Welt, aber die Gesellschaft macht aus ihm ein soziales Wesen, eine gemeinschaftsbezogene Person." Mbiti, *Afrikanische Religion und Weltanschauungen*, 137. See also Bujo, *African Ethic*, 5.

81. See Bujo, *African Ethic*, 28–29, 66; Bujo, *Ethical Dimension of Community*, 54.

82. See Bujo, *African Ethic*, 4–5, 71.

Western tradition of *cogito*, but also defines the *muntu* not solely through reason but in her full totality amidst all creation, given that the moral mystery that surrounds the *muntu* cannot be grasped by reason alone. Moreover, in defining the totality of the *muntu*'s life, an African approach counteracts the absoluteness or *imperialism of reason*[83] hailed in bioethical principlism as a one-size-fits-all approach.

Narrowing down, the basic prescription, on the one hand, is to do those actions that promote life relationally within the triadic community, but more-so those positive actions that promote life, and so even abundantly/fully/wholly (e.g., hospitality, respecting elders). On the other hand, one must resist from negative actions that do not promote life abundantly (e.g., suicide). We can now affirmatively answer that for Africans, "life is the highest principle of ethical conduct."[84] Godfrey Onah likewise summaries this moral end as the goal of African ethics: "At the centre of traditional African morality is human life. Africans have a sacred reverence for life . . . The promotion of life is therefore the determinant principle of African traditional morality and this promotion is guaranteed only in the community. Living harmoniously within a community is therefore a moral obligation ordained by God for the promotion of life."[85] The prescription can negatively influence or positively promote relational life. Some of the other classified communal examples of positive actions, which can also be negated, include: a) actions that promote "lineage or human solidarity," b) actions that promote "human vitality and continuity," and c) actions that promote "the hierarchical order of life," and d) all the above continuously interact within the triadic community to promote relational life. This means in fulfilling positive actions fully or refraining from negative actions, the *muntu* socializes along the continuity of the supreme value of life within a community setting. And if the *muntu* fails in increasing the continuity of life, then we can say she has disrupted the whole life, actually the welfare of the whole in the end, since each affects the network of forces of the whole. Conversationally, we can analogously think of the popular Latin proverb *Parvus error in initio, magnus in fine,*

83. It is "the imperialism of reason, if the person who is not able to argue counts as unreasonable. But if reason justifies everything, what justifies reason itself? So does reason reflect about itself, so that it criticizes itself as well? If reason begins with reason, does it itself have reason? It is difficult to see the rational motive that calls for the existence of reason! It grounds rationality on the base which itself is not rational." Bujo, *Ethical Dimension of Community*, 39–40, also 55; Bujo, *African Ethic*, 10.

84. Bujo, *African Ethic*, 3.

85. Onah is cited by Metz, "African Moral Theory," 329.

which literally means, "One's small error at the beginning leads in the end to major errors that affect not only her but actually all." To promote life fully, below, I illustrate the other four characteristics among Bantu people all closely linked to promotion of life, as understood by Baganda: community, solidarity (*bumu*), vitality (*bulamu*), and hierarachy (*bukulu*). In my discussions of these characteristics, it has become clear to me that from my own enriching local vernaculars and their approaches to bioethics, I could understand better other bioethical worldviews that I have encountered in the course of time. Moreover, I am not hindered from ascribing the Baganda, Swahili people, or the Nso' or Kom way of lived life as a Bantu paradigm or an African[86] moral thought, thus appealing to the wider Bantu family, and moreso to Africans south of the Sahara. This is partly because I cannot accurately describe myself as *Muganda* without being Ugandan and without enjoying an Africa origin. I am not saying—and it is unbeknownst to me—that the opposite is necessarily true in all cases.

Promotion of Community Harmony: Includes *All* Categories of Existence

We come to the second characteristic of African moral thought. One major group of Black Africans south of the Sahara is the Bantu people with their various forms *abantu, batho, ovantu, antu, watu,* etc. Among all, *ntu* is the basic irreducible static linguistic and/or grammatical root of

---

86. Varied African scholarships appeal to common strands and features (not every nook and cranny of Africa as Metz notes) that are shared as a way of life by many sub-Saharan communities. One argument why an idea that originates from the African continent can be called African has oftentimes hint me on the streets of Western countries, but also, it is similarly presented by Etieyibo: "When people wear a dress from an African country (say Ghana, South Africa or Nigeria), they say I'm wearing an African dress (even though they can also say I'm wearing a Ghanaian, South African or Nigerian dress). They do not say I'm wearing an American or European or Asian dress. Similarly, a restaurant may be called an African restaurant even though not all the cuisines on the African continent are represented in its menu list. If it is proper to call the dress African although it comes from one particular African country or the restaurant African although many cuisines from other African countries are not on the menu list, it is also proper, I think, to call an idea or view African even though it comes from one particular traditional African society or a specific region in Africa." See Etieyibo, "African Philosophy and Nonhuman Nature," 174–75; see more in Metz, "African Moral Theory," 321–41, and "Ubuntu as a Moral Theory," 375, and "African as Communitarian," 1176.

the word "Bantu."[87] Originally, in Alexis Kagame's maiden work, "La Philosophie Bantu comparée, Présence africaine, Paris, 1976," he explicitly gives these four categories of *ntu* based on the Kinyarwanda language: *Category 1. Muntu*, existents with human intelligence=Existierendes mit Intelligenz—("être qui a l'intelligence," includes God, spirits, the departed, human beings, and certain trees as substances). *Category 2. Kintu*, existents without human intelligence=Existierendes ohne Intelligenz—("être qui na pas l'intelligence," includes all beings which do not on their own but only under the command of *muntu*, such as plants, animals, stones and the like as substances). *Category 3. Hantu*, existents of time and space=lokalisierendes Existierendes—("être localisateur," includes time and space). *Category 4. Kuntu, existents of modality*=modale Existierendes—("être modal," includes items like beauty, laughter, etc. We can go further into quantity, quality, relation, action, passivity, posture, possession).[88] In quite a similar presentation, according to other scholars, the basic root *ntu* of Bantu worldview generally comprises of human and supra-human beings subdivided into "W*antu*," "K*intu*," "M*untu*" and "B*untu*" categories among Baganda.[89] Although some scholarship understands *ntu* not to be being itself, but rather very closely linked with being itself or some reality, given that *ntu* is a logical reality that has no existence of its own, independent of that given to it by the mind, in contrast, *ntu* is widely believed to be being itself, in the sense that it is a *common all-embracing universal principle,* a cosmic effective and constant force manifesting itself in the totality of all values and realities of Bantu.[90] Put differently, *ntu* as an all-embracing cosmic constant force is *moyo*; it is the supreme life force dispensed by God that the *muntu* contextualizes

87. Kasozi, *Ntu'ology of the Baganda*, 25–26 n. 2.

88. Kagabo explains Bantu ontology along the categories of being, *Mu, Ki, Ha, Ku*, in his study "Alexis Kagame," 235–36; see also Kagame, *Die Ontologie der Bantu Zentralafrikas*, 106–8; Kagame, *La Philosophie bântu-rwandaise*, 107–9. In Ganda (also known as Baganda) worldview, place and time are sometimes linked as synonyms. Sempebwa, *Reality of a Bantu*, 27, 54–55. Using Luganda (a Bantu Language), Kasozi formulates the four categories with some slight changes: category 1: MUNTU—the category of existents with human intelligence; category 2: KINTU—the category of existents without human intelligence; category 3: WANTU—the category of time and space; category 4: BUNTU—the modality category. Kasozi, *Ntu'ology of the Baganda*, 30. I maintain the original languages with the recent German and English translations for originality and clarity's sake.

89. Kasozi, *Ntu'ology of the Baganda*, 105–6, 108.

90. Jahn, *Umrisse der neoafrikanischen Kultur*, 101; Kasozi, *Ntu'ology of the Baganda*, 34–36.

within a communicative community of *Muntu, Kintu, Hantu* (Wantu), and *Kuntu* (Buntu) categories.

Muntu Category

Among Baganda, the *Muntu* category includes "human beings who are alive and those about to be born."[91] Besides, it includes God, spirits, the departed, and certain substances (living things). *Muntu* is the focus of this category. *Muntu*, the singular form of *Bantu*, means human being or human person, who was created by God and is part of the created contingent world.[92] Bantu are contingent existents not only with human intelligence (*amagezi*) and the capacity to speak (*okwogera*), but also endowed with, according to Kasozi, at least three outstanding unifying characteristics, i.e., blood, the tip of the sternum, and the heart. Whereas *omusaayi* ("blood") is what the *muntu* inherits from one's parent and it metaphysically qualifies one's link to her extended family or clan, and underlies the social interactions, *emmeeme* (the tip of the sternum) is the seat of the emotional life. Illustrative habitual expressions include: "*Em-meeme gy'esula n'ebigere gyebikeera*," literally, "The place at which the tip of the sternum spends the night, becomes the first destination of the feet on the following day." Its strict meaning is, "Where one's heart's desire lie, there is the base of one's emotional activity."[93] However, for Bantu, the heart (*omutima*) is the primary seat of the socio-psychological faculties and moral disposition, as well as the depth of the personality of *omuntu*. Illustrative day-to-day expressions include: a) "*omuntu ow'omutima omu-lungi*," which strictly means, "a good-willed person." b) "*omuntu eyafa omutima*," which literally means, "a person with a dead heart," whereas the strict meaning is "a person having a dead conscience."[94] These illustrative expressions of the heart (*omutima*) as the primary seat of the socio-psychological faculties in the *muntu*'s life have a unique prominence in African moral thought in contrast to principlism. Contrary to the Western principlistic thought of *scripta manet, verba volant* (i.e., writings

91. Mbiti, *African Religions and Philosophy*, 16.

92. Kasozi, *Ntu'ology of the Baganda*, 61–62; see also footnotes 4 and 5.

93. See Kasozi, *Ntu'ology of the Baganda*, 61–65 nn. 4–5. More examples on *emeeme*: a) The saying: "Emeeme eteebuuza, efubutula eggambo." b) Its literal meaning: "The tip of the sternum which does not consult or ask for advice, utters a foolish word." c) Its strict meaning: "An emotional person is liable to use slang or vulgar abuse."

94. Kasozi, *Ntu'ology of the Baganda*, 63–64.

remain, spoken words disappear), in the African moral world, words remain (*verba non volant*) through centuries, thanks to our heart, the primary seat of moral disposition. Bujo defends this thesis by appealing to the fact that the word comes from the heart of the speaking person, not the writing person, making the heart to be the central spot where ethical norms are concretely socialized among Africans. The Banyarwanda in Rwanda, the Barundi of Burundi, and the Bashi of Congo-Kishasa call the heart "the human person's little King," thereby expressing the essential dependence of human existence as a whole on the heart.[95] Having previously emphasized the authority of the oral word, which essentially comes from the *muntu*'s heart—as life-giving but not as death-causing—making the Black person to read books differently than they are read in bioethical principlism. With the *muntu*'s moral disposition (heart), she can increase or decrease life vitality through extended family or clan interactions for centuries, meaning relational disorders can be traced back as springing from *muntu*'s heart disorders, whereas the *muntu*'s well-disposed conscience likewise increases the relational clan vitality through epochs.

Kintu Category

In the *Kintu* category, the Luganda word *Kintu* (plural: *Bintu*) means thing, object. They are existents (*bintu*) without human intelligence but can be superior or inferior to humans; they include physical (e.g., inanimate things, plants, animals) and spiritual beings. The spiritual beings hold existence between the Absolute Being and human beings. We note a relationship therein, viz., the ultimate end of the *Muntu* category existents is in the *Kintu* category of spiritual *bintu*, which include the divinities, gods (*balubaale*), the spirits (*mizimu*). Though the spiritual *bintu* comprise the ontological destiny of the *muntu*, the physical *bintu* succumb to the *muntu*'s lead.[96] The *muntu* intermediates and plays a leadership moral role of bridging the spiritual and the physical beings. The *muntu*'s leadership signals that the *bintu* forces cannot act on their

---

95. See Bujo, *Ethical Dimension of Community*, 70–72; Bujo, *African Ethic*, 151–53, 122.

96. After *muntu*'s death, she joins the four spiritual beings, singular/plural: *muzimu/ mizimu*; *jjembe/mayembe*; *musambwa/misambwa*; *lubaale /balubaale*. Kasozi, *Ntu'ology of the Baganda*, 36–38. Spiritual beings originate from *muntu*. For instance, *balubaale* were "at one time human beings, noted for their skills and bravery, who were afterwards defied by the people and invested with supernatural powers." Roscoe, *Baganda*, 271.

own unless commanded by a *muntu*, since the *bintu* have no will of their own, as Jahn puts it: "Die zweite Kategorie *Kintu* umfasst jene Kräfte, die nicht aus sich selbst heraus handeln können und nur auf Befehl eines *Muntu*, sei er lebendiger Mensch, Verstobener, . . . die Planzen, die Tiere, die Mineralien, die Werkzeuge, Gebrauchsgegenstände usw. sie alle sind *bintu*, wie der Plural von *Bintu* lautet. Alle diese haben keinen eigenen Willen, . . . Die *bintu* sind 'geronnene' Kräfte, die den Befehl eines *muntu* erwarten. Sie stehen dem *muntu* zur Verfügung sind ihm 'zuhanden.'"[97]

Even-though in the hierarchy of beings the spiritual beings possess a more elevated position than the *bantu*, like we have intoned above, the physical *bintu* yield to the *muntu*'s authority. The *muntu*'s centrality—Baganda wisdom says, "*Ekitayogera tekirema ayogera*," i.e., the human being, who is endowed with the ability to speak, cannot fail to subdue a being that cannot express thoughts, feelings, or opinions orally—claims her capacity for consciousness and self-reflection as expressed in her language of the spoken oral word. The challenge I find with this moral prescription would be if the *muntu* presents herself as the sole knower, the subject, while the other beings are the known, simply the objects. If the *muntu*'s moral leadership, intermediary role, and centrality simply mean superiority and subduing other beings, such perspective would then insinuate a *strong anthropocentrism* given that the *muntu* plainly experiences herself as a being that knows the inabilities of the physical beings of the *Kintu* category.[98] Contrarily, I would interpret this moral perspective to situate the *muntu* in the center, purposely not as a superior, but as a lead between the physical and spiritual beings, mediating morally and harmoniously the two worlds, remaining contingently as a part of the whole without claiming superiority or to be the sole whole.

97. Jahn, *Umrisse der neoafrikanischen Kultur*, 106. See also ". . . gehören hierher Krankheiten, ferner Rauch, Feuer, Ströme, Berge und der Mond als Naturkräfte . . . auch Gifte, Medikamente etc." Meinhof, *Bantu-sprachen*, 32–33.

98. *Balubaale* enjoy more respect hierarchically than the ancestors (*bajjajja*) and the living persons, in the same way the heads of clan enjoy more respect than the heads of family. *Balubaale* are not regarded as the authors of moral order but as upholders or supporters of moral order. Mbiti, *African Religion*, 212; Kasozi, *Ntu'ology of the Baganda*, 44–45; see also footnote 4.

Wantu or Hantu Category

In the *Wantu* category of *where* and *when*, the prefix *wa-* expresses the ideas of down, below, above, there, or here.[99] The Luganda word *Wantu* means a certain place, somewhere, or space in general, indicating characteristics of whereness, somewhereness, whenness, occupiedness, or emptiness in relation to space and time. For instance, God is *somewhere*, in his abode *ggulu* (the sky or heavens). Besides, time cannot be empty, just as space cannot be a vacuum, since the universe is filled up with a certain dynamism through *events*.[100] Examples include the naming of persons: Zaavuga, i.e., there was gun fire, so one born during gun fire is named Zaavuga; Ziriddamu, i.e., fighting will resume, so one born during times of war is named Ziriddamu; Musisi, i.e., earthquake, one born around an event of earthquake is named Musisi; Kawonawo, i.e., survivor, one who survives in war, shipwreck, landslide, famine, etc, is named Kawonawo. For Baganda, even the future is not empty. It is also filled up with forces and events. One of the common persons' names expressing hope in the coming times is Tusuubira ("we hope").[101] In relationship to events, time and space are intimately linked, and often the very same word is used for both. For instance, *"Talinna bbanga kukoka kunonyereza"*—"She has no time to do research" or "She has no space to do research." When the Baganda say *"talinna bbanga,"* one should have in mind the third meaning, that is: "She has neither space nor time."[102] In this regard, if we are to grasp the Bantu moral thinking, every aspect of their life, inclusive of the concept of time, is intimate with the entire life. Let me illustrate further with cases amongst the Baganda, Banyakore, and Swahili people.

Firstly, the intimacy of time with the totality of Bantu life can be proved in virtues since they relate to time ontologically. The *muntu's* vital influence on the *wantu* forces (time) is exemplary amongst Baganda various virtue notions; a case in point is the virtue of *kugumikiriza* (patience), which is related to *gumiikiriza* (be patient, hold out, endure). The antithesis of the virtue of *kugumikiriza* is *kupapa* (rashness). To interdict the vice of *kupapa*, it is assumed that patience intimately includes the notion of *kwerowooza* (self-reflection) with time; whereby the *muntu* does not only have good time to consult with oneself, the *muntu* is also exposed

---

99. Sempebwa, *Reality of a Bantu*, 73–75.

100. See Kasozi, *Ntu'ology of the Baganda*, 36–39 n. 12.

101. Kasozi, *Ntu'ology of the Baganda*, 39–40.

102. See Mbiti, *African Religion*, 27–28.

to time and has the vital influence on time, consequently, defeating the menacing vice of *kupapa*. And to encourage patience, Baganda common-ly say: "*Bbugubugu si muliro*" ("Mere spluttering is not fire" or "Haste makes waste"); "*Akwata empola atuuka wala*" ("He who starts slowly or goes easy goes far)"; "*Alidde ggi, newesubya omuwula*" ("You have eaten the egg; but deprived yourself of the meat," i.e., if you had waited for the egg to be hatched, you might have had a good meal from the fowl when fully grown).[103] To put it briefly, being patient (virtuous) is a moral rendi-tion of the *muntu*'s sympathy with the totality of life, which intimately includes the ontological relationality with time.

Secondly, the intimacy of time with the totality of Bantu life is ap-parent in events. Even though the concrete idea of time is something that happens, time is generally a constitution of events that have occurred (time is only experienced), those that are taking place now (time is being experienced), and those that are expected (time is to be experienced). In the totality of Bantu life, time is a living life force that influences their own being and which they themselves influence.[104] Time is intimate with the *muntu* in its influence of events. Though time is concretely and en-tirely connected with events, it is not just for mathematics' sake like in principlism. This was obvious in our discussion of rationing, specifically with the principle-based fair innings argument that is regularly advanced as a justification for denying aged patients treatment if they compete with youthful patients on the basis of their over-lived time period. This mathematics of time fails in an African setting. In line with Mbiti's inter-pretation of African moral thought, it does not matter whether the sun rises at 5 a.m. or 7 a.m., so long as it rises. Even when the *muntu* says that she will meet another at sunset, it does not matter whether the meeting will take place at 5 p.m. or 7 p.m.; what matters is for them to eventually meet during the known period of sunset. Also, it does not matter whether people go to bed at 9 p.m. or midnight; what is important is *the event* of going to bed. It is why everyday examples from different Bantu tribes, such as the Banyakore and the Baganda,[105] show that time is meaning-

103. Another case in point is the virtue of *bunyiikivu* (perseverance, endurance, dil-igence, earnestness). To prove perseverance, the Baganda say: *Okuwumula si kutuuka* (Resting is not arriving i.e., to sit down by the wayside doesn't mean that the journey is ended). Murphy, *Luganda-English Dictionary*, 16, 45, 113; Sempebwa, *Reality of a Bantu*, 172–73, 282–83; Mbiti, *African Religion*, 19; Nason, "Proverbs of the Baganda," 249.

104. See Sempebwa, *Reality of a Bantu*, 77–81.

105. How the day events are spread out amongst the Banyankore—a Bantu ethnic

ful at the point of the event but not at the mathematical moment like in rationing as espoused in bioethical principlism. This explains why the Bantu continuously expect the events in the year to come and go, in an endless rhythm just like those events of a day and night. With their agricultural background, they naturally expect the events of the rainy season, planting, weeding of crops, harvesting, the dry season, the rainy season again, planting again, weeding again, the rainy season again and again, planting again and again, and so on to continue perpetually, in such an interminable ontological rhythm.

Thirdly, the intimacy of time with the totality of Bantu life is visible in their unified concepts of time periods that differ from the linear concept of time in principlism—thought as far as the indefinite past, present, and infinite future is concerned.[106] According to Baganda, the past, long ago, in former times (*edda*) and the distant past (*edda n'edda*), and present (*kakaano, kakaati*) have specific words, but the future doesn't, though it applies tenses in relation to words used in the past. Examples include, "*Tulibiraba edda*" ("We shall see them in the distant future"); "*Tunaagenda edda*" ("We will be going later on"—*in the near future*). So, the totality of life—the unity between the past, the present, and the future—is seen in

---

group in Uganda—who are cattle keepers: 6 a.m. *AKASHESHE*: Milking time; 12 noon. *BARI OMUBIRAGO*: Time for cattle and people to rest, since after milking the cattle, the herdsmen drive them out to the pasture grounds and by noon when the sun is hot, both herdsmen and cattle need some rest; 1.p.m. *BAAZA ABAMAZIBA*: A moment to draw water from the wells or the rivers—so as to avoid any hinderance to those interested to draw or carry water later—just before cattle are driven there to drink, what may eventually pollute the water source; 2 p.m. *AMASYO NIGANYWA*: Time for cattle to be driven to the water sources to drink; *AMASYO NIGAKUKA*: Cattle leave the water sources and re-start grazing; 5 p.m. *ENTE NIITAHA*: Herdsmen drive cattle back home; 6 p.m. *ENTE ZAATAHA*: Cattles enter their kraals; 7 p.m. *AKASHESHE*: Closing of the day. It is milking time once again before cattle sleep. Mbiti, *African Religion*, 19–21; Ganda phenomenon calendar/events in the year: *GATONNYA* (January): Banana harvest; *MUKUTULASANJA* (February): Hot month, withering of banana leaves; *MUGULASINGO* (March): Sowing; *KAFUUMULAMPAWU* (April): A variety of drone male termites begin to fly (known as *Mpawu*). A delicacy among the Ganda; *MUZIGO* (May): Month of heavy rains; *SSEBOASEKA* (June): Month of sickness and several deaths resulting from mosquitoes; *KASAMBULA* (July): Month when land is made fallow; *MUWAKANYA* (August): Month of lightning and thunder; MUTUNDA (September): Ntunda edible termites fly, a delicacy for Ganda; *MUKULUKUSABI-TUNGOTUNGO* (October): Rain washes away simsim; *MUSEENENE* (November): Nseenene (grasshoppers) fly, a delicacy for Ganda, *NTEVU* (December): Many harmful insects on the uncultivated land. Nsimbi, *Olulimi Oluganda*, 12–13.

106. See Mbiti, *African Religion*, 17.

terms of the past events, or what has been experienced,[107] which is not the case with principlism. The Baganda like the Swahili people have a comparable understanding. In Kiswahili, the *sasa* (present) period is complete, in the sense that people are not only aware of their existence, their limited future, and a dynamic present, but also aware of an experienced past. In contrast, the *zamani* (past) period is the graveyard of time, i.e., the final repository of all phenomena and events, as Mbiti states:

> *Sasa* ist die Sphäre, in der Menschen sich ihrer Existenz bewusst sind und von der aus sie sich sowohl in die begrenzte Zukunft als vor allem in die Vergangenheit (*Zamani*) versetzen können. Das *sasa* ist eine in sich vollständige, abgerundete Zeitdimension, die ihre eigene, wenngleich begrenzte Zukunft, eine dynamische Gegenwart und eine erlebte Vergangenheit besitzt . . . *zamani* . . . ist der Friedhof der Zeit, die Endstation, die Dimension, in der alles seinen Ruhepunkt findet. Es ist der endgültige Lagerspeicher aller Erscheinungen und Ereignisse, das Meer der Zeit, in dem sich alles in eine Wirklichkeit auflöst, die weder noch vorher ist.[108]

Mbiti's additional clarification shows that *Zamani* and *Sasa* periods form a unity of life and are harmoniously interconnected: ". . . die *Zamani*—und die *Sasa*—Perioden [sind] miteinander verbunden . . . Sie bilden ja eine Einheit. Des Weiteren kann angenommen werden, dass Zeit und Raum in Harmonie stehen."[109] The harmony between *zamani* and *sasa* periods (miteinander verbunden), forming an intimate union (eine Einheit) corresponds to an ontological unity between the distant past (*edda n'edda*) and the past (*mulembe*, denoting *the modern past*). After the *muntu's* physical death, she continues to exist in the *mulembe* period. She is not only remembered by relatives and friends who survived him, but she also participates directly in their being. The *muntu's* personality, character, words, not forgetting the unforgiven mistakes of ancestors, vitally influence the modern past (ancestors) and the living mutually, negatively or positively, which is quite unfamiliar in bioethical principlism. When, however, the last person who knew the departed also dies, then the deceased transits from the horizon of the *mulembe* into *edda n'edda*—the immortal, eternal, or endless state (the period beyond which nothing can

---

107. See deeper explanations and examples in Sempebwa, *Reality of a Bantu*, 89–90; Murphy, *Luganda-English Dictionary*, 62; Mbiti, *African Religion*, 17.

108. Mbiti, *Afrikanische Religion und Weltanschauungen*, 29.

109. Ndombe, *Verständnis der negro-afrikanischen (Bantu-) Weltanschauung*, 280.

go) of spiritual beings and God. Whereas the present (*kakaati, kakaano*) and the modern past (*mulembe)* refers to the period that constitutes "actual time" (the period of memory and experience), the distant past (*edda n'edda*) constitutes "endless time" (the period of eternity and immortality, the period of eternal peace and safety (*emirembe n'emirembe*), while the future constitutes "potential time" (the period of hope and expectation).[110] In contrast to bioethical principlism, African moral thought upholds a total unity of life where even victimizers in the actual time necessarily need to ask for forgiveness or reconcile with (through a nonverbal palaver) their deceased victims in the endless time so as to maintain harmony given that time and harmony are siblings (Zeit und Raum in Harmonie stehen). Joshau Wantate Sempebwa succinctly describes it:

> *Emirembe n'emirembe* overlaps with *mulembe* and the two periods are inseparable. After the events have been realized or actualized in the *mulembe* period, they move as it were backwards into the past, into *emirembe n'emirembe*. This *emirembe n'emirembe* which lies within the time of "endless time," becomes the period beyond which nothing can go . . . The Ganda expect human history to continue forever, in the rhythm of moving from the *mulembe* period to the *emirembe n'emirembe* period, from experience to immortality, and there is nothing to suggest that this rhythm shall ever come to an end: the days, months, seasons, and years have no end, just as there is no end to the rhythm of birth, puberty, marriage, procreation, old age, death, entry into the community of the departed and finally entry into the immortal company of the spirits. It is an ontological rhythm . . .
>
> The *mulembe* period generally binds the people, the spirits and the Gods and their immediate environment of *bintu* (things) together. It is the period of conscious living. The distant past period, *emirembe n'emirembe* provides the foundation or security to the *mulembe* period binding together all persons and all things, i.e., all *muntu* forces and all *kintu* forces, so that all these forces are embraced within the time region of "actual time." Like time, it is experience which defines space. Just as the *mulembe* period embraces the life that the Ganda actively experience, so does the place where they are. For this reason, they are very much tied to the land. The land provides them with the roots of existence as well as binding them to their departed,

110. See a thorough explanation in Mbiti, *African Religion*, 22–27. Baganda look at *sasa* as an "era," "epoch," what they call *mulembe*. *Mulembe* can also be used to denote "modern" or "up-to-date": *eby'okulwanyisa eby'omuleme* ("modern weapons"). Sempebwa, *Reality of a Bantu*, 91–93, 301 n. 134.

who are buried in their land. Thus land, place expresses their *mulembe* as well as *emirembe n'emirembe*.[111]

Reliably, we can give credence to Bantu substantive moral belief, a belief in an intimately unified ontological relationality that is not only between time and space (land as the root of the *muntu's* existence) but also between the period of an expected future, dynamic present, and an experienced past that is also closely tied to the *muntu's* indelible burial right to her historical ancestral land.

Buntu Category

Baganda philologists express *bu* in *buntu* as a state of being, or a condition in which nature is. *Bu* answers the question of how (the modality), in what interactive relationship (think of solidary bonds) God, spirits, man, animal, plant, object and other modes of existence. *Buntu* includes classifications and concepts of reality, value and other modalities of being like manner, procedure, form, variety, expression, manifestation, arrangement, pattern, education, aesthetics, freedom, responsibility, virtues, vices, norms, knowledge, health, character traits, quantity, quality, etc.[112] The Baganda say, "*Tuli bumu*," meaning, "We are one"—in *unity* and *solidarity*. According to Sempebwa, this confirms the *muntu's* totality of life immanent in the world itself, where the *muntu* enjoys oneself in harmony and solidarity with nature, re-emphasizing an ontological relational living unity of the *muntu* and other types of creation.[113] This living unity and interactive relationality can be expressed at least in *bumu* (solidarity), *bulamu* (vitality), and *bukulu* (hierarchy). Within a *communicative community*, all entities are not only in *solidarity* but also *continuously vitalize* and *hierarchize* each other in *promotion of life.* Given this necessity of unity in Bantu morality, specifically, Placid Tempels never considers social order as order if it is in conflict with this ontological relational order of life:

111. Sempebwa, *Reality of a Bantu*, 95–97; compare with Mbiti, *African Religion*, 23, 142.

112. Specific examples include: *Bukadde* (old age), *bukiika* (direction, sideways), *bulungi* (beauty, goodness), *bunene* (size, bigness), *buzito* (heaviness), *buwulize* (obedience), *bulimba* (dishonesty), *butono* (smallness), *bukulu* (seniority, greatness). See Sempebwa, *Reality of a Bantu*, 97–98; Kasozi, *Ntu'ology of the Baganda*, 42, 76.

113. Sempebwa, *Reality of a Bantu*, 97–98, 108.

Alles Gewohnheitsrecht, das diesen Namen verdient, also echtes Gewohnheitsrecht und nicht Missbrauch dieses Rechtes darstellt, ist von der Bantu-Philosophie, der Philosophie der Lebenskraft, des Lebenswachstums und des Lebensranges. Jedes Gewohnheitsrecht gewinnt Stütze und Kraft aus dieser Philosophie. Die Moral und das menschliche Recht . . . basieren auf denselben Prinzipien und stellen ein geschlossenes Ganzes dar . . . Lebenskraft, Lebenswachstum, Lebensbeeinflussung und Lebensrang. Die gesellschaftliche Ordnung hat ihre Grundlegung in der ontologischen Ordnung. Die Bantu werden die gesellschaftliche Ordnung nie als Ordnung betrachten, wenn sie mit der ontologischen Ordnung in Widerstreit steht oder fremd ist.[114]

Since the social order of Bantu moral life is premised on this ontological relational order, which may sound pre-logical, pre-scientific, or irrational to bioethical principlism, Bantu approach remains focused at enhancing the practical wholeness of the life force in a non-Western discursive rationality. In Sempebwa's interpretation of African moral thought, *bulamu* (Lebenswachstum) constantly increases or decreases the continuity of power, strength, health, life, and well-being of the group; *bumu* (Lebenskraft) enhances the solidary participation in the total life force and awareness of the primacy of the group over the individual, whereas *bukulu* (Lebensrang) promotes the hierarchical mode of existence of all beings.[115] I will demonstrate how these three characteristics of African moral thought interact with the other two characteristics, community and life, to promote a relational order of life.

Promotion of Vitality

We come to the third characteristic of African moral thought. This *buntu* concept of *bulamu* (vitality and continuity) includes ideas of life, health, and well-being, which increase or decrease the life force, quantitatively or qualitatively. On the one hand, Baganda talk of a quantitative increase— e.g., *bunene* (bigness) or *bungi* (quantity, amount)—or qualitative increase—e.g., *bulungi* (beauty, goodness, good quality). On the other hand, they talk about quantitative decrease—e.g., *butono* (smallness)—or qualitative decrease—e.g., *bubi* (wickedness, badness, ugliness, excreta). Increasing or decreasing of life force does not mean that the *muntu's* life

114. Tempels, *Bantu-Philosophie: Ontologie und Ethik*, 76.

115. See Sempebwa, *Reality of a Bantu*, 126.

becomes more or less than the other, but one's being becomes more or less forcefully influential.[116] In Bantu morality, however, humanness is constant, and can neither be decreased nor increased. We do not become more human than before and so our human nature cannot become more or less, except the growth in our qualities and faculties (*bulamu*). And given that the life force is not distinguished from being, it is this being that is strengthened and increased in itself, analogous to the gradual growth of grace in us as understood in the Catholic Church.[117] To emphasize this growth in *bulamu*, Placid Tempels, a Belgian Franciscan missionary for three decades in the Democratic Republic of Congo, comparably asserts:

> Wir sagen vom Menschen, daß er wachsen und sich entwickeln, Kenntnisse erwerben, seinen Verstand und Willen üben und so erstarken kann. Aber mit all dieser Entwicklung, all diesen Erwerbungen ist er nicht mehr Mensch geworden als vorher. Seine menschliche Natur bleibt dabei unverändert. Man besitzt die menschliche Natur oder man besitzt sie nicht. Sie kann nicht mehr oder weniger werden. Die Entwicklung geht nur in den Eigenschaften und Vermögen (Fakultäten) des Menschen . . . Die Lebenskraft wird vom Sein nicht unterschieden. Wenn der Schwarze also sagt, daß eine Kraft zunimmt oder ein Sein stärker geworden ist, dann muß dies in unserer Sprache (westlicher Sprache) heißen: Dieses Sein ist in sich selbst verstärkt, gemehrt. Was die katholische Theologie über die geoffenbarte Wirklichkeit der Gnade und des Gnadenlebens lehrt, daß es nämlich in sich selbst wachsen und sich verstärken kann, nehmen ähnlich die Bantu in der natürlichen Ordnung jedes Seins, jeder Kraft, an.[118]

Bulamu can increase or decrease beyond the living community. For illustration, one of the most common shared human activities (of *bulamu*) is *kulya*, which means to eat or consume in Luganda. Contextually, it symbolically implies force, strength, unity, vital union, fraternal love. In sharing a meal, one is thanked with the words "*Ompadde amaanyi*," meaning you have provided me with strength. It is the reason kinship goes with eating ("*Luganda kulya*"). The noun from *kulya* is *ddiiro*, which is the place where people unite to take meals (think of the Eucharist). It is

---

116. See Sempebwa, *Reality of a Bantu*, 97–98, 108, 129–32; Murphy, *Luganda-English Dictionary*, 31, 41, 43, 44.

117. *Catechism of the Catholic Church*, no. 1996: "Grace is favor, the free and undeserved help that God gives us to respond to his call to become children of God, adoptive sons, partakers of the divine nature and of eternal life."

118. Tempels, *Bantu-Philosophie: Ontologie und Ethik*, 30.

the place where people acquire more and increase in strength, force, and energy.[119] This vital energy gets over the physically living to include the deceased relatives. Sometimes before drinking or eating, one performs a symbolic act of unity and fellowship by pouring out a libation or throwing out a piece of food through the window as a sacrifice to the living dead.[120] This is not "performing a cult" of beer or food to the living dead, but rather "reliving a kinship relationship with them, actualizing such a relationship once again in the living present,"[121] as Éla argues. The living present should be further understood in the broadest sense possible as Tangwa notes, since prayerful libations to God poured on the earth are addressed to the cosmic earth as a whole. For example, it is why "the dead depend on the actions, especially ritual sacrifices, of the living for their well-being and the living, in turn, depend on the solicitations and intermediacy of the dead for their health, progress and well-being."[122] The focus is the entire vitality and continuity that obliges us to promote the well-being of all, including the ancestors, in sharing a meal; so long as it bonds kinships and allows us to participate vitally in the continual flow of created life, then it is an action to be embraced as realizing the supreme value of life which makes it as well morally worth to be done.[123] As far as vitality is concerned, even when it were to be understood in the anthropocentric sense, we shouldn't forget that along the meal, the ethical ideal is for all to have the fullness of life in common,[124] as a Kenyan proverb says: "One who eats alone, dies alone" (i.e., lonely).[125] And dying lonely

---

119. Given that Buganda unites all its inhabitants to acquire national strength, the whole country of Buganda is called *ddiiro: Obuganda ddiiro likunganya*. In Buganda, *Akuwa okulya y'akutwala omuluka* = "One giving you to eat dominates you." Still, one eats power or authority or office. Thus *okulya obwami* = "To eat chieftainship is to be raised to the rank of a chief." Lugira, *Ganda Art*, 19. Lugira gives more Luganda/Ganda examples showing vitalism on pp. 17–19; for example, he uses the verb *kuba* (beat, strike, win, use energy and force): *Kuba-giza* = "Keep company with one who is in sorrow"—"Mitgefühl"; *kulya bulamu* = to enjoy life or have good time; *Omuwala yamulyamu omwoyo* = "The girl won/stole his heart"; *Mwana muwala, oyo andya omutwe!* = "I am completely enamoured of that girl," lit., "She eats my head"; *Amaanyi si galya* = "Force does not prevail"; "Physical strength alone does not accomplish things." Murphy, *Luganda-English Dictionary*, 312–13.

120. See *Ntuʼology of the Baganda*, 102–3; Kyewalyanga, *Religion, Customs and Christianity in Uganda*, 286–88.

121. Éla, *My Faith as an African*, 19.

122. Tangwa, "Bioethics: An African Perspective," 191.

123. See Sempebwa, *Reality of a Bantu*, 131.

124. Bujo, *African Ethic*, 129.

125. See Mbiti, *African Religion*, 209, proverb 38.

or all alone is the worst experience that can ever strike any Black African! Otherwise principlism won't easily be incarnated in the African sense if hospitality (vitality of life) depicted in meals plays no role.[126]

Citing another example, the qualitative increase of *bulamu*, for instance, goodness (*obulungi*), is attributable to any object that is actually or potentially good. When goodness interacts with ethicality, it is *obuntu*, which is expressed as moral "civility" or "good manners." The combined form of good health or vitality (*bulamu*) and moral civility (*obuntu*) can be communicated through the compound term *obuntu-bulamu*, literally being a human being with *healthy humanness* (vitality). To note here, it is not primarily the bodily or physical health referred to, but the condition of being morally healthy, the state of possessing moral goodness. Hence, the *muntu* is referred to as *omuntu-mulamu* or *ow'obuntu bulamu* primarily because of her healthy humanness (conduct) that abounds in a particular kind of *obulungi* or (moral) goodness.[127] Moreover, this moral goodness often rests on the *muntu*'s good customary breeding. But when bioethical principlism outrightly fails in granting customary moralities their rightful place in our global pursuit of the common morality; in the first place, it fails in *appropriating* traditions as moral vehicles in unraveling core habitual dispositions resident in pluralistic cultures, while in the second place, it fails in appreciating homeomorphic equivalences prevalent in regional moralities. In African regional moralities, the *ntu* (in *omuNTU-mulamu or ow'obuNTU-bulamu*), as healthy being itself, cuts across the African world as the supreme life force. The notion of healthy humanness, various homeomorphic equivalents in relation to *ntu* can be understood as *vumuntu* in Mozambique, *utu* in Kenya, *bumuntu/utu/obuntu* in Tanzania, *obuntu-bulamu* in Uganda, *muntu* (Tswane) in Botswana, *ubuntu* (Zulu/Xhosa) in South Africa, *umunthu* in Malawi, *umuntu* in Malawi, *omundu* (Herero) in Namibia, and *gira ubuntu* (Kinyarwanda) in Rwanda.[128] Even when the said common morality (bioethical principlism) were to sit well in African cultural moralities, it doesn't mean subservience or loss of each other's rich values like *obuntu-bulamu*.

126. See Bujo, *African Ethic*, 159–60.

127. See Kasozi, *Ntu'ology of the Baganda*, 76–77. *Buntubulamu* can broadly imply morally approved rules. Sempebwa, *Reality of a Bantu*, 144. *Obuntubulamu* can also broadly imply the possession of courtesy, compassion, culture, etc. Murphy, *Luganda-English Dictionary*, 44. *Ak'obuntu* means humaneness. Karlström, "Culture and Democratization in Buganda," 485–505.

128. Nchangwi et al., "African Communitarian Ethic," 377–88.

Common and cultural moral traditions can richly appropriate each other without being obstructive to the functioning, serviceable, or usable past.

Firstly, living out an appropriated functioning tradition doesn't mean sticking in the past even in aspects that could have outlived their importance, but rather it means *we do not pour out the baby with the bathwater*. In other words, we neither need to adopt modernity's principlism for its own sake nor abandon our rich traditional values simply because we live in a new age of globalization. Contrarily, we need to appropriately drop what is no longer helpful and uphold what is still valuable, without occasioning any moral brainwashing or valorizing what is foreign in the name of modern bioethical principles.[129] Here the catchphrase "throw the baby out with the bathwater" or more accurately, "Das Kind mit dem Bade ausschütten," is a German proverb that dates back to 1512 and was initially recorded by the satirist Thomas Murner in *Narrenbeschwörung (Appeal to Fools)*. In the English world, it can be traced back to

---

129. In the Buganda Kingdom, as advocated by the Queen (*Nnabagereka*), they are twelve examples of *Obuntubulamu* core values. Some are:

- sense of shame (*ensonyi*). No shame in littering public roads, and in other instances, corruption seems like some normalized business.

- empathy (*okufa ku mmunno*)

- responsibility (*obuvunanyizibwa*). The engine of a responsible person is *okwetuma/ okwekubiriza*, as in *assigning or urging oneself*. A responsible person is one you would assign a task, and since she is self-driven, you aren't worried if she will do it or not.

- integrity (*obwesimbu* or *okuba n'ensa*). Here integrity denotes completeness in character, one fallows moral values.

- pro-active leadership (*obukulembeze*)

- civility (*obugunjufu*). We always have to sieve what we let out of our mouths; that is why the Baganda say that *akamwa akangu, kassa nyiniko* ("A quick mouth kills its owner"). One should tame the tongue (*okufuga olulimi*), because it can make or destroy communal life; e.g., certain things are not said by elders in front of children—sex-related language is packaged.

- humility (*obwetoowaze*) means lowering oneself, and its antithesis is *okwegulumiza*

- cleanliness (*obuyonjo*). The Baganda say that *ensiba mbi, edibya mutere* ("Poor packaging/branding kills the market of dry cassava")—people will judge you by the way you present yourself, and cleanliness includes the surroundings.

- selflessness (*okwewaayo*)

- morality (*empisa*) and transparency (*obwerufu*)

Nnabagereka and Ssentongo, *Obuntubulamu*, 7–62; Murphy, *Luganda-English Dictionary*, 236, 561.

the Scottish historian and philosopher Thomas Carlyle's essay on slavery in 1849: "Occasional Discourse on the Negro Question." His figurative imagery encourages everyone to join in the struggle to end slavery, but being so mindful, to an extent, of not harming the slaves in the process: "And if true, it is important for us, in reference to this Negro Question and some others. The Germans say, 'you must empty-out the bathing-tub, but not the baby along with it.' Fling-out your dirty water with all zeal, and set it careering down the kennels; but try if you can keep the little child!"[130] Nonetheless, when bioethical principlism does not keep the child as it does the emptying out of the bathing tub, it distances itself from customary, cultural, regional moral ways of life, without being mindful that it is harming the other would-be-enriching approaches to the globalized approach to bioethics in the long run. Against the norm of appropriating our traditions, the enumerated examples of numerous bioethical interventions currently happening in the Global South are surely throwing the baby out with the bathwater. And in a number of moral applications, principlism consistently underestimates, as critics aver, the morality of traits approved by cultures or social groups, indicating that their approval is not sufficient to qualify these traits as moral virtues,[131] thereby excluding cultural and group moralities from the localization of common morality. Throwing the baby out with the bathwater is underestimating cultural moralities, which subsequently underestimates appropriating or dialoguing with *the usable past* from the treasuries of our traditions in the present modern epoch. The Congolese philosopher Valentin-Yves Mudimbe reminds us of the necessity to metamorphose our historical usable past within our cultural moralities along our contemporaneous dialogues. He calls this "retrodiction":

> Retrodiction seems to be the main technique that establishes both the new right to speech (and the power of spatializing indigenous localities) and the intellectual efficiency of its interpretation. Retrodiction—from Latin *retro* (on the back side, behind, in time back) and *dicere* (to speak)—denotes the idea of speaking (and thus synthesizing) from an illusory, invented moment back in time. In the process, the present invests its values in the past with its questions and hypotheses, and rediscovers in

130. Wilton, *Word Myths*, 67.
131. See *PBE*, 32.

> the invented, reorganized spaces, laws, paradigms, or the truth
> of its suppositions.[132]

Retrodiction brings out the usable past in the present African moral thinking. The ancestral life and the contemporaneous community intersect and influence, increase and decrease, each other's life.

Secondly, without appreciating homeomorphic equivalents in the local contexts, adapting bioethical principlism remains the hardest venture. The question is: how can we proficiently pass on novel globalized bioethical approaches to village comrades within their own familiar ordinary experiences of life? In 1999 Tangwa noted that for the globalized principlism to sit well in the Black African milieu, he suggested a correct attitude towards the contemporary era of Western globalization. Globalization of Western science and technology should not be accompanied by the globalization of Western cultural thinking and value systems. Even when bioethical principlism dialogues with non-Western thought, it is unethical that it takes along with it all its Western packaging.[133] The Western packaging may not sit in the African receiving value system unless it is depackaged from its Western casing. Analogously, cultivating a correct attitude towards the offered meal of Western science and technology doesn't necessitate Africans taking along with them the serving dishes after eating the food. Tangwa suggests respecting the moral sensibilities within African value systems that should go beyond each value system learning from and enriching each other: "Western culture could empower African culture while African culture humanize Western culture."[134] This is one way bioethical principlism can humanize the local approaches to bioethics without its Western conceptual systems accompanying it to the detriment of locally cherished values.

To not morally damage valued local systems, foreign problem-solving bioethical interventions (out there) must be coordinated before incorporation in the local contexts without being contrary to local ways of life (here within) or moral sensibilities. Some Africanists, like Tangwa, are of the view that "it should never be forgotten that other cultures have their own word to say and that alternative values, ways of thinking and

---

132. Mudimbe, *Tales of Faith*, 95.

133. Tangwa, "Globalization or Westernization," 105; Tangwa, "African Perception of a Person," 42–45.

134. Tangwa, "African Perception of a Person," 43.

practices exist, and attempt should always be made to bring these out."[135] In order to take cognizance of local moral sensibilities (the consuming culture) in efforts to wholesomely intervene in the bioethical interventions, distinctive recommendations are advanced, for instance, of homeomorphically appropriating "decision aids" is morally fitting. Decision aids are interventions that would support people in making their decisions explicit, provide information about options and associated benefits/harms, as well as help clarify congruence between decisions and personal values. According to growing evidence, people exposed to decision aids feel more knowledgeable, better informed, and clearer about their values, and they cultivate a more active role in decision-making and have more accurate risk perception.[136] The current failure is to limit decision aids to individual autonomy as enshrined in its Western packaging without encompassing a broader collective community that would likely be the highest moral bidder and beneficiary. Both shortcomings, ensuing from not appropriating tradition as well as not appreciating homeomorphic equivalents, indicate a lack of a sense of collectivity as approved by cultural moralities. Just like in the theme of *obuntu-bulamu*, the collectivity of life in customary moralities remains trodden down by principlism.

In the practice of moral rules amongst Bantu, just like in their collective life, moral life can increase or decrease. Citing the *first* example, *obuntu-bulamu* (healthy humanness) increases life both singly and collectively, whereas deficiency in *obuntu-bulamu* affects both the individual and the collective group. We can substantiate this with two examples: one regards taboos, and the other core values of *obuntu-bulamu*. If the *muntu* disregards taboos, it is believed her conduct disrupts the values of vitality and the unbroken continuity of the whole life, whereas if observed, *omuntu-mulamu* concomitantly interrelates with the other supreme ideals of mutual human solidarity and collective lineage solidarity (I'll illustrate concretely in the next section) as well as social hierarchy.[137] To exemplify this, on the one positive side, given that respect cuts across all

135. Tangwa, *African Bioethics in a Western Frame*, vi.

136. See Stacey et al., "Decision Aids"; Kleinlugtenbelt and Kim, "Cochrane," 1298; *PBE*, 133–34.

137. See Sempebwa, *Reality of a Bantu*, 163–69. To foster the norm of hierarchical respect for the older people and good manners in children, "women always sat with their legs placed together under the knees, to one side; if they wished to change their positions, they leant forward on their knees, and moved their feet to the other side and sat back gain." Roscoe, *Baganda*, 48–49, 65–67, 80–96, 128–34, 393, 449; Murphy, *Luganda-English Dictionary*, 397.

interpersonal Bantu relations, the oldest child respectfully controls the rest of the children.[138] This interpersonal respect is common among Bantu from childhood. Drawing from my own experience, my younger sibling cannot address me by my name without an additional title (*baaba*[139]), and neither can I address my elder siblings by their names without an additional respectful title. On the other side, taboos in Baganda social-moral customs prohibit any intimate friendships with the daughters or sons of one's father's or mother's sisters or brothers. That is, one limits any intimacy with one's father's (gender can be switched) sisters' daughters (*kizibwewe*) or one's mother's brothers' daughters (*bakizibwewe*) or one's sister's son's wife (*muka mwana*) or one's mother-in-law (*nyazaala*). Failure to respect these constraining prohibitions not only arouses fear or presupposes illness to ensue from interbreeding, but also infers harm from the living dead, who are the key guardians of moral codes. So, a taboo acts as a moral protective measure for social harmony and it goes past not only the *muntu*, but also the *muntu's* collective lineage.[140] Like in numerous Bantu groupings, social taboos protect values of vitality like amongst the Bahema of Congo-Kinshasa.[141] Taboos are intended to promote vitality of social life.

Citing a *second* example, among the core values of *obuntu-bulamu* is shame (*buswavu*) or sense of shame (*ensonyi*), which has two aspects: *kafansonyi* (shame of the agent) and *kuswaza* (bring shame to others, put others to shame or disgrace others). *Kafansonyi* is the idea of becoming ashamed after violating a moral rule. In bewilderment, one community member may ask: "This person *talina na kukafansonyi*?" ("Does he not anticipate shame before others?"). People generally refer to someone who has violated a moral rule as not having shame. Likewise, an individual's shame is commonly assessed as unpleasant by the community:

138. See Haydon, *Law and Justice in Buganda*, 116.

139. Elder brother or sister, an address to a friend or an intimate friend. Murphy, *Luganda-English Dictionary*, 5.

140. See Kasozi, *Ntu'ology of the Baganda*, 109–110; see also footnote 4; Sempebwa, *Reality of a Bantu*, 163–69; Murphy, *Luganda-English Dictionary*, 217, 438.

141. "It is forbidden among the Bahema of Congo-Kinshasa for a young man to sit on his father's chair during his lifetime . . . To sit on his father's chair while the father is still alive amounts to wishing his father's death; a son who would do so is capable of actually killing his father, in order to succeed him as quickly as possible. In order to prepare a boy for proper conduct, the tradition has therefore developed a brief formula: 'Your father will die if you sit on his chair.'" Bujo, *African Ethic*, 127.

"Yaswaluka n'adda ne mu kati" (He was put to an endurable shame).[142] In this regard, the community is actually assessing the *muntu's* healthy humanness by evaluating her civility (*obugunjufu*). *Obugunjufu* comes from the word *okugunjula*, which means to impart morals. Cooling hot soup on a spoon before sipping it, or cooling hot food before eating it in public, smoking in public, chewing noisily, belching loudly publicly, littering public roads, eating or throwing wrappings and peelings out through car/taxi/train windows, not greeting others on stairways, inconsiderate use of roads and car lights against other road users, disembowelling an animal publicly, and socializing with corrupt officials are marks of poor moral upbringing, which distance people and the whole environment from us and the fulfillment of our civil duties. Moreover, the *muntu-mulamu* surmounts shame by cultivating selflessness and being considerate in public places to the elderly, the sick, pregnant women, physically challenged persons, and children. Selflessness is also visible in meals: *"Ono alya n'ono alya, ye mmere ewooma"* (Food is delicious only when each one is considered and is eating).[143] The bearer of uncouthness or inconsiderateness is disapproved of as not exercising healthy humanness. When this uncouthness and inconsiderateness become a habit, the blame does not only bare personal responsibility but also collective burden. The family or clan is collectively liable for such a shame or wrongs because *"Omulya mmamba aba omu, n'avumaganya ekika"*—When one member of the lungfish clan eats the lungfish, and thereby violating the taboo of not eating one's *totem*, she brings the whole clan into shame and disrepute. Put differently, the evil of an individual is the downfall of the group; that is, it is not only the individual who is ashamed but also the entire family and clan (*kuswaza ekika*). Any habitual wrongdoing—such as eating one's totem, which is every clan's national flag and intimately connects it to the natural environment—leads to a collective rejection or expulsion from the clan (*kuboola*).[144] Instances justifying col-

---

142. See Sempebwa, *Reality of a Bantu*, 253; Murphy, *Luganda-English Dictionary*, 134, 526.

143. See Nnabagereka and Ssentongo, *Obuntubulamu*, 7–62. *Ono alya n'ono alya, ye mmere ewooma* ("He is eating, and the other is eating; that is when the food tastes good"—It is not "I" have, but "we" have that makes a good society). Sekadde and Semugoma, *Ndimugezi*, 10.

144. *Toteism* is when *muntu* symbolically associates herself with non-human animals or plants, striking no dichotomy between nature and human beings. No person is allowed to marry or be married to a member of one's clan totem. One doesn't eat her clan emblem or totem (*muziro*). My clan emblem/totem is Kasimba (Genet Cat). With

lective rejection among Baganda include violations of taboos, e.g., *ekivve* (an abominable act or abomination, such as beating one's parent).[145] It is because these *muntu* acts contravene the social moral code of the collective group. If a *muntu* is collectively rejected, it basically means rejecting *muntu's* habitual disposition of core moral values, ideals, and virtues that would precisely reflect the *muntu's* moral character before the group. Similarly, it is this constant wrongdoing that implies defect in healthy humanness or character (deficiency in *obuntu-bulamu*) as rightly understood in bioethical principlism. For principlism, although it can be presumed that every conscientious person can make "honest errors" or "good-faith errors," consequences of habitual moral errors or defects in moral character such as failures of conscientiousness[146] are appraised as exceptionally grave, though in regard to an autonomous individual. That is, whereas life decreases for an individual in principlism, the group is not collectively affected in an individual's defect in *obuntu-bulamu* as it is in African morality.

To safeguard against personal or collective defects of *obuntu-bu-lamu*, it is the *muntu's* imperative prescription to "do all those actions which promote human vitality and continuity" and never to "do anything that will disrupt human vitality and continuity."[147] This African outlook could be challenged since handicapped children and twins were killed among some ethnic groups. However, the wisdom by then saw these irregular cases as shortcomings in human creation and so a threat to the life of the whole community, viz., disrupted vitality and continuity. Killing them was not counted as evil, since the ancestral tradition considered killing them as increasing the general life force, viz., promoting vitality and continuity for all. Contemporaneously, an African approach to biomedical ethics promotes life somewhat differently,

---

all my family, we cannot eat the Kasimba. The *Muziro* (totem) is "a specific food which cannot be eaten because of the traditions of the clan." "Eating one's totem is like burning one's national flag. It is an act of flagrant disrespect of one's ancestry." Sempbwa, *Reality of a Bantu*, 163–69, 253–54. On the meaning of *muziro*, see Murphy, *Luganda-English Dictionary*, 29, 338, 397.

145. A *muntu* who hits her parent (an act of disrespect) commits an abomination (*kivve*) and has to be expelled from the clan (*kuboolwa*, *boola*), meaning she becomes a foreigner, wanderer, homeless/stateless person (*mmomboze*) since he is an outcast (*muboole*) by not belonging to any clan or social group. See Haydon, *Law and Justice in Buganda*, 116; and Murphy, *Luganda-English Dictionary*, 29, 213, 349, 340.

146. See PBE, 34, 42.

147. Sempebwa, *Reality of a Bantu*, 131–32.

whereby the "physically challenged" and twins aren't a threat to growth in life anymore.[148] With improved knowledge, the family community accommodatingly loves these twins, and so abundantly takes care and grants them immunity without abandoning them. This is in agreement with principlism since it appreciates non-abandonment as one of a physician's central ethical obligations. Just like in the ethics of Black Africans such as African communitarianism, non-abandonment in principlism is an extensive collective commitment both to care about patients and to cooperatively seek solutions to their problems. Here principlism embraces a communitarian collective sense by being exceptionally different in facing an uncertain lonely future in contrast to facing difficult decisions with a committed knowledgeable community,[149] particularly when the diagnosis is unclear, the medication is too expensive, the patient is in a nursing home but not with the family; and all put together making the future of the patient's life seem so indistinct. Nonetheless, even though bioethical principlism widely views killing (a casual action that brings about death) as morally wrong, commitments to non-abandonment occasionally implicate death—*inter alia*, letting die (an intentional avoidance of causal intervention so that disease, system failure, or injury causes death), assisted suicide/physician-assisted dying or physician-arranged dying, voluntary active and non-voluntary euthanasia, as well as autonomous suicide—and are sometimes defended as legal and morally acceptable.[150] Moreover, the justification for actively terminating ill infants (according to the Groningen Protocol, which relapsed into a Hitler-esque style of eugenics),[151] and those who are "tired of

148. See Bujo, *African Ethic*, 103.

149. See Quill and Cassel, "Nonabandonment," 368–74.

150. When clinicians justifiably withdraw life-sustaining treatment, it is argued they allow patients to die but do not cause, intend, or have moral responsibility for the patient's death. Miller et al., "Moral Fictions and Medical Ethics," 453–60. According to the "bare difference argument," there is no bare intrinsic moral difference between killing and letting die (no distinction between an act and an omission). Perrett, "Bare Difference Argument," 131–39; Gert et al.," Patient's Refusals and Requests," 13–15; Lillehammer, "Logical Slippery Slope Argument," 545–50; Smith, "Practical Slippery Slope," 17–44; Lewis, "Empirical Slippery Slope," 197–210; Beauchamp, "Physician-Assisted Suicide," 437–39; Dworkin et al., *Euthanasia and Physician-Assisted Suicide*; PBE, 181–83. Decisions to withdraw or withhold potentially life-sustaining treatment are common in intensive care and precede the majority of deaths. Wilkinson and Savulescu, "Futility in the ICU," 160–65. The concept of futility does not justify unilateral decisions to forego life-sustaining medical treatment over patient or legitimate surrogate objection. Nair-Collins, "Laying Futility to Rest," 554–83.

151. Manninen, "Groningen Protocol," 643–51.

living"[152] is that "death can be of overall benefit, in which case it should also be facilitated in those who cannot consent."[153] But although many mental health professionals believe that suicide generally results from maladaptive attitudes or illness needing therapeutic and social support, principlists understand autonomous refusal of treatment, requests for aid-in-dying/aid in dying, and autonomous suicide attempts as *protected moral rights*, to which states, health professionals, and others have no legitimate grounds to intervene.[154] They are not only firmly obligated under ordinary circumstances to honor such autonomous requests, but also valid requests render it morally permissible for physicians (or other persons) to lend the requests or refusals. It is on these grounds that a valid request frees a responder of moral culpability for the death, while a valid refusal precludes culpability.[155] The moral focus remains not on respect for the collective life force, but on the powerful common denominator of respect for autonomy as the presiding moral principle of the individual. African moral thought rather defends a collective promotion of the life force as inalienable, in contrast to pursuing any casual action that brings about death or any intentional avoidance of causal interventions cherished as *protected moral rights*. Just like other authorships, it is morally wrong to consider patients as *eligible for help in killing themselves*. This is because the right to the protective promotion of life is inalienable for us all and principles shielding legalizations based on the patients' autonomous consents wrongfully treat as alienable what is in truth inalienable.[156] No created entity can ever be legitimately qualified for any assistance in terminating its own life. It's for this reason the promotion of life vitality (*bulamu*) in African moral thought remains collectively inalienable for the well-being of all without designating some eligibility to some in their own autonomous request to die.

152. Pereira, "Legalizing Euthanasia or Assisted Suicide," e38–45.

153. Jones, "Logical Slippery Slope," 379–404. See more in Daskal, "No Logical Slippery Slope," 23–48.

154. D'Angelo et al., "Aid-in-Dying," 164–73; Battin, "Physician Aid in Dying," 774–76; *PBE*, 239–42. Harm avoidance is a priority rooted in the value of autonomy, not in the value of well-being. See Keating, "Priority of Avoiding Harm," 7–39.

155. *PBE*, 189.

156. See Avila, "Assisted Suicide," 111–41.

Promotion of Solidarity

Treat the cosmic earth well. It was not given to you by your parents.
It was generously loaned to you by your children. We did not inherit
the Earth from our Ancestors, but we borrowed it from our Children.

—SAGACITY

Paradoxically, principlism takes each individual as autonomous and complete (self-contained) since each person has a world in herself; yet Black Africans understand humans as incomplete without the *other*, i.e., the different modes of existence. Settling this paradox requires *bumu*. *Bumu* refers to the supreme value of solidarity of life. For Baganda, *bumu* means oneness, collective unity, harmony, and *solidarity of all existential forces of the universe*.[157] That is, in the social reality of Black Africans, moral persons are intrinsically socialized toward different modes of existence (other). This can be articulated in three moral bonds of solidarity towards interpersonal humanity, universal institutions, and the entire cosmos.

In the first bond of solidarity, which is also in triadic form, the *muntu* relates with her fellow *living Bantu community* to foster human solidarity and oneness within interpersonal relationships. The moral prescription is: "'Do all those actions which promote human solidarity and oneness' or 'don't do anything that will disrupt human solidarity and oneness.'"[158] Solidarity isn't limited to the *first living community* only, but also extends to the coming generations. It is the reason the established interpersonal networks expect ceaselessly newborns (*the second community*) since those not yet born owe their place in this network of solidarity as forthcoming descendants.[159] Furthermore, this continuity in solidarity within the lineage life already foretells an intimate social reality of a collective harmonious living that neither interacts with the living and expects the newborns only, but also unionizes with those gone before us as the *third community of ancestors*. It is this continued vital participation of the physical and spiritual worlds, the visible and invisible communities, in an unbroken wholesomeness of communion that Washington classifies as basic to all "modes of expression of black African culture."[160]

---

157. See Sempebwa, *Reality of a Bantu*, 99–103.
158. Sempebwa, *Reality of a Bantu*, 129.
159. See Maquet, *Africanity*, 60.
160. Washington, *Negritude*, 56, 60.

Amid unsuccessful earlier objections towards the great reverence Black Africans have for the third community of the deceased,[161] it remains a general universal knowledge that the dead are treated with utmost moral respect across all cultures. Just like in the ethics of Black African contexts, Beauchamp and Childress (the chief principlists), stress this memoria of the dead in their most recent *PBE*, though never resolutely consider its weight in their network of principles:

> We dedicate this edition, just as we have dedicated each of the previous seven editions, to Georgia, Ruth and Don. Georgia, Jim's beloved wife of thirty-five years, died in 1994, just after the fourth edition appeared. Our dedication honours her wonderful memory and her steadfast support for this project from its inception. Tom also acknowledges the love, devotion, and intellectual contribution to this book of his wife, Ruth Faden, who has been the deepest influence on his career in bioethics, and salutes Donald Seldin, a brilliant physician and an inspiration to Tom and to biomedical ethics since the early years of the field. Don passed away at the age ninety-seven in 2018, when we were in the midst of preparing this eighth edition. He will sorely be missed, and never forgotten.[162]

Even when bioethical principlism (utilitarianism) is limited to the human sphere of the living, the *other*, for Black Africans, in this first bond of solidarity; the *muntu's* fulfilled happiness is only possible if the *other* (including the spirits of the dead) is happy too.[163] In reference to Smith, the role which "these incarnate spirits play in African life is based upon a belief that the human personality survives the death and decay of the body. It is not so much a 'soul' that lives on as the person . . . It is an essential element in African belief that 'living' and 'dead' live in *symbiosis*, interdependent, capable of communication one with the other. No iron curtain separates them . . . 'living' and 'dead' are linked in *an indissoluble*

---

161. In reference to Willoughby's authorship of *The Soul of Bantu*, Merkies interprets Baganda ancestor worships as "the psychological root of the primitive religion." This is grounded on the fact that many African societies have a sacred custom of preserving the bodily remains of the dead. Surprisingly, Merkies eventually concludes with a twist from worship to reverence: "In any case it explains the great reverence the people have for their deceased." Willoughby, *Soul of Bantu*, cited from Merkies, *Ganda Classification*, 95.

162. *PBE*, xi.

163. Jahn, "Value Conceptions in Sub-Saharan Africa," 63.

network by their force vitale . . ."[164] The rationale for implicating the dead in the moral life of the living, indicating no iron curtain separating them, is that they remain actively alive in Bantu daily social activities. Equally, it is the reason why for Mbiti the physically departed person is not really dead; she is still alive not only in the memory of those who knew her (in her life time by name), but also in the world of spirits, and hence referred to as the "living-dead."[165] Even so, the reader should keep in mind that it is only when a living-dead has fulfilled certain upright preconditions while still in the community of the living that she can be referred to as an "ancestor."[166] Such conditions may include having led a virtuous life and having executed heroic deeds, such as having been able to apologize in a way that healed communal conflicts.[167] Likewise, the "symbiosis," as Smith puts it, between the "living" and "dead" isn't just a limited companionship of community life, but it is rather further qualified as a great *indissoluble* community, as he puts it in his own words: "Ja die Stämme sind religiöse, nicht bloß soziale Ordnungen. In ihnen sind die Lebenden und die Toten zu einer großen Gemeinschaft vereint."[168]

And since to be is *to be in relation to*, this indissoluble three-dimensional community life, in many Baganda homes one finds a special place that is reserved for the symbiosis of the living-dead with the living relatives and those not yet born. Other times the bond is strengthened through the last funeral rites by installing a heir or heiress for the deceased, who succeeds to a hereditary title, making the living-dead to be honorably remembered by the living[169] before passing on the mantle to those not yet born. The inevitability of this indissoluble community as envisioned in a network of solidarity can be demonstrated when the *muntu* is at a point of crisis. She may not seek help from the community of those not yet born, but rather seek help in the solidary reality of spiritual existence. For instance, when the *muntu's* garden yields fail, she finds help in her faith

---

164. Smith, *African Ideas of God*, 24. Emphasis added.

165. Mbiti, *African Religions and Philosophy*, 38.

166. Each ancestor remains the apex of a triangle with a corresponding base that broadens with each generation, i.e., constituting the whole lineage of descendants who feel a close solidarity to the departed. Maquet, *Africanity*, 38.

167. Bujo, *Ethical Dimension of Community*, 15.

168. Ohm, "Stammesreligionen im südlichen Tanganyika," 20.

169. See Kasozi, *The Ntu'ology of the Baganda*, 102–3; see also Kyewalyanga, *Religion, Customs and Christianity in Uganda*, 286.

in the hidden forces of the revered ancestral spirits.[170] When the *muntu* needs to ask for forgiveness or reconcile with the ancestors, it is within the spiritual realm of existence that a nonverbal palaver can be fruitful. This is the desired peace and unity as foretold by the great ancestors and it is what the symbiotic unity entails. And since in augmenting a morally indissoluble symbiotic community every action that prevents the fulfillment of this cause (life) is evil, accordingly, the ancestors are *keenly interested*[171] in reconciling themselves with their descendants so as not only to promote life abundantly but also to continuously see their offspring remaining morally strong, peaceful, and united. In this regard, first of all, African community life is believed to be an *ontological symbiotic single unit* of the awaited descendants, the living, and the living-dead. Similarly, it is why for Jahn this indissoluble vitality "als geistige Kraft steht der Verstorbene, der Ahn, mit seiner Nachkommenschaft in Verbindung."[172] Secondly, the *indissoluble network of forces* connects the living not only to the living-dead, but also to the future generations in a manner that augments an *indissoluble triad of community* life that includes *all forces (other)*. We shall return to the question of the *other* later, a question that indeed goes beyond anthropocentrism. But even when we continue discussing morality that is exclusively anthropocentric, there is a problematic too. The symbiosis of the *indissoluble network of forces* prevalent in the three-dimensional community of African moral thought may have no place in bioethical principlism given its frequency with *moral residues*. In principlism, a moral residue occurs because a *prima facie* obligation does not simply disappear when overridden amidst conflicting obligations in a given circumstance. It arises when the obligations we were unable to discharge create new obligations. Principlism remains indeterminate on one's right while alive to decide whether to donate her organs after death, contrasted to fixed priorities or rights of the relatives to make decisions about the donation of the deceased relative's organs.[173] Moral residues are not only morally accommodated within the symbiotic indissoluble triad of community life that even goes beyond anthropocentric restrictions, but they are well sheltered in *Bantu* timelessness (immortality).

170. See Sempebwa, *Reality of a Bantu*, 104–5; see also Mbiti, *African Religions and Philosophy*, 16; Smith, *African Ideas of God*, 29.

171. See Bujo, *Ethical Dimension of Community*, 27–28; Sempebwa, *Reality of a Bantu*, 258–59.

172. Jahn, *Umrisse der neoafrikanischen Kultur*, 113–14.

173. *PBE*, 16–17.

The indissoluble symbiosis of the unified lineage predicts a state of moral "immortality"[174] that is exemplified not only through continuous dynamic care of one's descendants in whom some children bear the deceased's traits, but also through carrying out instructions (verbal or written testaments) of the deceased, either while she lived or when she spiritually appears[175] to descendants or children. Viewed in this way, the centrality of children is strongly remarkable. The popular Baganda saying "*Alifa tazadde, talizuukira*" (One who passes away without giving birth to a child will never come back to life) supports this strong belief in partial reincarnation of some living-dead who remain permanent guardians of the moral code and so intervene time and again to settle moral contraventions. Such an anticipated moral immortality is what Kasozi calls "the foreverness of being in relation to."[176] *The foreverness of being in relation to* originates from the principle that *to be is to be in relation*. This is parallel to the African thought that to increase the vitality of life forces between the spiritual and physical worlds, the anticipated immortal solidarity necessarily demands the *muntu's* practical participatory solidarity while still living. When this balance is upset, the living community experiences misfortunes and sufferings (like garden yields failing).[177] This reminds us of Mulango's interpretation of African moral logic as an indispensable participatory solidarity in life (*lebensnotwendige Partizipation*) where moral residues are practically handled within an endless solidarity. We can safely argue, given the solidary fellowship between the spiritual and physical worlds, present and past, *mulembe* and *emirembe n'emirembe*, just like Baganda, the same thesis is in dialogue and at home with the ethics of Black Africans of Nso' origin. They don't find death as a tragically shocking reality, since it doesn't mean the annihilation of a being, the lineage, the community or the world. Moral obligations even amidst moral residues will be amicably resolved.[178] This is because Bantu are never separated from the endless fellowship of life. The endless solidary

174. Mbiti, *African Religions and Philosophy*, 38.

175. Mbiti, *African Religions and Philosophy*, 25–26; Kagame, *La Philosophie bântu-rwandaise*, 371.

176. See Kasozi, *Ntu'ology of the Baganda*, 103–4; see also Kyewalyanga, *Religion, Customs and Christianity*, 286.

177. See Molefe, *African Personhood*, 110–2; Sempebwa, *Reality of a Bantu*, 104–5; Mbiti, *African Religions and Philosophy*, 59.

178. Sempebwa, *Reality of a Bantu*, 105; Maquet, *Africanity*, 65; Mbiti, *African Religions and Philosophy*, 59; Tangwa, *African Bioethics in a Western Frame*, 145, 158.

fellowship of Bantu embraces the dying in the company of the living, the living-dead, and future offspring indissolubly.

Bioethical interventions isolate the dying from the entire *community* into nursing homes, which doesn't sit well in an African context of solidarity. In nursing homes, one witnesses not only decline in executional autonomy, for instance of the elderly, but also decline in decisional autonomy.[179] Even when bioethical principlism aims at guarding individual autonomous choices, nursing caregivers repeatedly override the same patients' autonomous decisions that principlism all along has intended to protect. In a broader perspective of bioethical principlism, surrogate decision-makers such as hospitals, physicians, courts, or families generate decisions to terminate or continue treatment for patients with diminished autonomy, viz., patients who cannot competently accept or refuse treatment. In these instances, principlism employs a surrogate decision-making authority to compel every party to respect the principle of respect for autonomy.[180] It is not only difficult to compel each party to honor the principle of respect for autonomy, but quite often respecting autonomy becomes even more difficult in cases of medical utility. Here some scholars define " medical futility"[181] as a power struggle for decisional authority between physicians and patients/surrogates,[182] whereas others conceptualize medical futility as providing inappropriate treatments that will not improve disease prognosis, alleviate physiological symptoms, or prolong survival. The so-called futile interventions[183] allow physicians to judge a treatment as futile and are entitled to withhold a procedure. Some cases show how

179. See *PBE*, 139. Conflicts over autonomy can begin at the level of conceptualization. Collopy, "Autonomy in Long Term Care," 10–17.

180. *PBE*, 139–41.

181. Schneiderman et al., "Medical Futility," 669–74. Medical futility has both a quantitative and qualitative component. Schneiderman, "Defining Medical Futility," 123–31.

182. There is often serious disagreement between physicians and families regarding the benefits to the patient of continued treatment. Bernat, "Medical futility," 198–205.

183. Futility as uselessness. Chwang "Futility Clarified," 487–95. Each member of the team involved perceives futility differently. Morata, "Futility in Health Care," 1289–300. On lack of support for the physicians' unilateral decision-making on futile care, see Bagheri et al., "Medical Futility"; Brody and Halevy, "Futility," 123–44. Utility and futility are complimentary concepts. Kopelman, "Futile and Useful Treatments," 109–21. End-of-life decisions should be based upon the physician's assessment of the effectiveness of the treatment and the patient's assessment of its benefits and burdens. So, conditions for medical futility could be met either if there were a judgment of ineffectiveness, or if the patient were in a state in which she was incapable of a subjective judgment of the benefits and burdens of the treatment. Sulmasy, "Futility," 63–78.

physicians should act in concert with other professionals, but need not obtain consent from patients or family members as they help to facilitate the process of dying, whereas other studies indicate how exploring the familial social value of treatments may be as important in determining futility as medical scientific criteria.[184] The disagreements and tussles for decisional autonomy among varied key parties (like parents, doctors, surrogate guardian, lawyers) in arresting medical futility proves how surrogacy can be morally controversial. One of the most controversially discussed cases in the bioethics of surrogate decision-making and medical utility was that of nine-month-old Charlie Gard, who was born on August 4, 2016 with a rare and incurable disease known as mitochondrial DNA depletion syndrome.[185] The contentious heated question in medical utility was to establish whether it was in Gard's best interest as a patient to die since he was unfairly over-treated. Given that Gard could not exercise his executional or decisional abilities, amid currently maintained claims that making treatment decisions has a negative emotional effect on at least one-third of the surrogates,[186] further emphasis of a finer surrogate decision-maker other than the *community* leaves many conflicting questions open within an African ethic. At the outset, even if a community member is incompetent or non-autonomous given that she lacks the capacity to exercise her executional or decisional capacities, the surrogate cannot predict her *best interests* since it is easier to say what largely harms a person (overall benefit) than what should be done for a person's best

184. Shan and Peter, "Medical futility," 292–302; Schneiderman et al., "Medical Futility," 949–54. In cases of intravenous opioid use, resource allocation should consider the sociocultural stigma of addiction. Goldberg, "Futile Use of Futility," 4–5; Mitchell et al., "Medical Futility," 71–76. On the difference between futility and rationing, see Close et al., "Futility in End-of-Life," 373–79; White et al., "What Does 'Futility' Mean?," 318.

185. The parents of Gard fought a prolonged and heated legal battle to allow him access experimental treatment that they hoped would prolong his life and to prevent his doctors from withdrawing life-sustaining care. Charlie's clinicians believed that the brain damage Charlie had suffered as a result of frequent epileptic seizures, along with many other severe disabilities, would render any innovative therapy futile, and they disagreed with his parents' wishes to use an experimental therapy. They felt it in Charlie's best interest that he be allowed to die. A battle ensued among Charlie's parents, his doctors, and a surrogate guardian. Caplan and McBride, "Charlie Gard," 15–16. On April 11, 2017, the High Court in London ruled that it is in Gard's best interest to stop treatment and die. Even if it were clearly in Gard's interest to receive treatment, it could still be withheld on the grounds of distributive justice. However, those grounds do not apply because Gard's parents had raised funds to cover his treatment outside the UK. He did not deny others a chance. Savulescu, "Gard's Best Interest to Die?," 1868–69.

186. Wendler and Rid, "Decisions for Others," 336–46.

interest.[187] Apart from that, what if the patient with diminished autonomy was never previously competent, giving no traces for her previous autonomous choices from which a surrogate could reliably base or choose from? And what if one was previously autonomous but is now incompetent to articulate her prioritized interests? Even if the surrogate were to act on her preceding autonomous determinations, the surrogate might select from her former interests that only harmonize with the surrogate's own individualistic interests.

What if we move a little bit further than the limited principle-based surrogacy to an overarching surrogacy based on the sociality and endless solidary fellowship of the community? An integral fellowship of solidarity that is fully versed with guardianship of its members (as a pro-paternalistic custodian), has not only an upright discretion in decision-making, but rightly envelopes both executional and decisional competencies in alignment with the well-known history of the clan and its members. And since for a Black African the decision-making process is always a single unity, an all-inclusive authority as the surrogate decision-maker is the entire family of the clan fellowship,[188] i.e., the endless solidary fellowship of the community. It is within this community fellowship (the community as the *gatekeeper* in the sense of guardian) that a dialogic palaver is mutually verbal and nonverbal to cater for all parties involved towards achieving a harmonized oneness of the corporate community without excluding the incompetent individual. This can explain, in the African sense, why it is common that elderly and sick people die at home, not in nursing homes, but in the company of their own family community. Dialogue with the African spirit of community solidarity shows itself clearly during the hour of death, where a dying member is given a deep sense of belonging. Until the last moment of physical life, the dying

187. The harm principle is proposed in cases such as those of Vincent Lambert, Charlie Gard, and Alfie Evans. Bellieni, "Pain Principle," 41–49. Parental choices for their children should prevail unless their decisions subject the child to avoidable harm. Truog, "Best Interests?" 16–17; Coulson-Smith et al., "Defense of Best Interests," 67–69. For *pro-paternalists*, it is permissible to intervene in a person's choices if there is no wronging of anyone else. Jason, *In Our Best Interest*.

188. Families as surrogate decision makers are argued to replace the lost decision-making capacity of patients. Hyun and Kjervik, "Deferred Decision Making," 493–506. Shared decision making with the family. Heyland et al., "Cardiopulmonary Resuscitation Decision Making," 419–28. Legislation on living wills should include family decisions. Areen, "Withhold or Withdraw Treatment," 229–35; King, "Medical Decisions for Incompetent Patients," 76–79.

person has the assurance of dying within the community[189] (of elders as the community gatekeepers) but certainly not unaccompanied. Even at the hour of death, the communicative palaver ritually incorporates the dying into the wider solidary fellowship of the living and the living dead. African scholarship looks at this solidary dialogue as a process of growth even as one's physical strength declines. Though it may take the form of a nonverbal palaver, each active participant does not only learn afresh how to face suffering and death with the bravest humility ever, but also significantly contributes to the growth of each present in healthy humanness. The reader should note that growth in healthy humanness (moral personhood) is an unending process. Not even entrance into the community of the deceased ancestors ends it, for even ancestors remain always dependent on the collectivity of the community of the living for their growth in personhood.[190] So the transition from being simply humans to the "basic sociality of personhood," Masolo writes, defies any principlistic "boundaries of metaphysics, epistemology, or even ethics in the restrictive Kantian sense. It proposes far-reaching dimensions for a communalistic view of the world in which the project of becoming a person is always incomplete."[191] Even amidst medical futility, there is moral hope in the solidary fellowship of life that never ends with the community of the living. First, medical futility confirms unambiguously that human beings are mortal, and medicine's powers are also limited.[192] Second, the ethical controversy still stands that the patient's autonomy is overridden in varied forms of surrogate decision-making even when principlism has labored to protect this autonomy. However, the African moral medicine to these controversies is in the unlimited solidary fellowship of the three-dimensional community, which is the endless surrogate decision-maker without necessarily overruling one's individuality. This iterates the logic that Black Africans are never parted from the endless communal solidarity of life, even in matters of death or increase in personhood.

189. See Bujo, *Ethical Dimension of Community*, 21; Visagie et al., "Informed Consent in Africa," 174–77; Mbiti, *African Religions and Philosophy*, 108; Horn and Mwaluko, "Public Health Research Ethics," 103–4; Wasunna et al., "Informed Consent in an African Context," 61.

190. See Bujo, *African Ethic*, 89; see also Tangwa, "Bioethics: An African Perspective," 195.

191. Masolo, *Self and Community*, 13.

192. Schneiderman et al., "Abuse of Futility," 295–313.

The second and third bonds of solidarity (the other two modes of existence of the *Other*) concern the universal institutions and the entire cosmos of nature respectively. The *muntu* is squarely oriented to these modes of existence since they are closely "part of and continuous with reality."[193] Since we defined bioethics as respect or reverence for life, a respectful treatment of all people is only morally successful if its continuous reality is inclusive of the total respectful treatment of the communal and cosmic created environment. This is the culture of *Ubu(ntu)*, which is as old as the human race, a culture of ethics that is about a harmonious *sharing*[194] with the cosmos. According to Leonard Tumaini Chuwa, it is this rich harmonious sharing that gives off some inner peace as its product in the African sense. In the mind of African wisdom, this harmonious sharing is only integral if it embraces cooperation in creation at individual, communal, and cosmic levels.[195] In trying to describe this integral relationship, Metz awards this social harmonious sharing the highest good for Black Africans, what he describes as a twin-parallel relationship of identity and solidarity, well captured in his work "An African Theory of Moral Status: A Relational Alternative to Individualism and Holism." To be an object or a subject of life in this twin-parallel relationship of sharing a way of life is dependent on the *muntu's* capacity to harmoniously relate to other Bantus, in the first instance. Characteristic of this Sub-Saharan ethical thought, the *muntu relationally communes as her way of life;*[196] i.e., the *muntu* gets lost in solidarity with others. If we would cite the example of the recent COVID-19 pandemic, which didn't originate or have its headquarters in Africa, Tangwa thinks "many African countries" did display "a sense of solidarity in the face of COVID-19," but also "realized that ethics dumping must be avoided."[197] Here our relationships transcend from interpersonal life to structural-universal cooperations and cosmic-ecological relationships, in which we can obstruct the spread of COVID-19 and the wide-spread exploitive ethics dumping.

193. Sempebwa, *Reality of a Bantu*, 104.

194. Sharing is a dignified African way of life; e.g., it is depicted in the Baganda proverbs such as: *Akatono okalya ne munno* ("Even the little you have, you must share it with another"). Sekadde and Semugoma, *Ndimugezi*, 6. "And because it is family property all members have an equal right to a share in its use, and all have a right to participate in the process of sharing . . ." Nyerere, *Uhuru na Umoja*, 9–10.

195. Chuwa, *African Indigenous Ethics*, 1.

196. See Metz, "African Theory of Moral Status," 399; Metz, "Ethics in Africa," 102.

197. Tangwa and Nchangwi, "COVID–19," 1; see also 8; Tangwa, "Bioethics and Ubuntu," 239–49.

Obstructing exploitive ethics dumping not only asserts the necessity of total decolonization and neocolonialism, but also reminds us of demonstration of a viable globalizable solidarity. If we are to understand an African approach to solidarity, it requires us deeply understanding the *muntu's* identity in terms of the social *we*. It is not only to *share a common way of life* with *others*, but also, if we are not to underestimate the African approach, here sharing a way of life (solidarity) entails promoting the welfare of others for their own sake,[198] viz., the *muntu* seeks living in harmony with all created nature by caring and sharing in the life and strength of all creation (*otherness*, with exception of God).[199] In this moral pursuit, the Chagga of Uru in Tanzania, in their conventional wisdom to protect the entire planet have been communicating throughout generations: *Oruka lu n'maseiyano*. The moral of the statement is for the *muntu* to treasure the planet earth as a good moral steward, just like our preceding ancestors did,[200] before we can eventually pass it on in a befitting shape to those not yet born, the future community, which loaned it to us and remains the true owner. This reminds us that harmony with cosmic nature is an antithesis of contradictions in the cosmos, implying the universe is not a battlefield of struggling beings since everything is in ontological solidarity with another. The whole universe is a unity of undestroyable solidary forces that must be befittingly reaped. Destroying any one of the modes of forces means destroying the whole existence, since one presupposes all the others.[201] Then, it is not merely the case that if humans were taken away from the world the rest would become nothing, but also that if the rest of nature were taken away, there would neither be any future or humanity itself. Meaning, less of this functioning relational solidarity—what Sempebwa originally refers to as "eine funktionale Einheit"[202]—between all beings knocks down the African understanding

---

198. At the heart of Metz's African ethics are the twin-parallel relationships of identity and solidarity; see Metz, "Ubuntu as a Moral Theory," 532–59; Metz, "Ethics in Africa," 99–117; Metz, "Two Conceptions of African Ethics," 141–62. How the twin social relationships are related to the ethics of *ubuntu*; see Metz, "African Moral Theory," 337. For more on the meaning of these community harmonious relationships, see Mokgoro, "*Ubuntu* and the Law in South Africa," 15–26; Tutu, *No Future without Forgiveness*; Molefe, *African Personhood*, 80.

199. Chuwa, *African Indigenous Ethics*, vii, 1; Maquet, *Africanity*, 64.

200. Chuwa, *African Indigenous Ethics*, 1.

201. Sempebwa, *Reality of a Bantu*, 104; see also Mbiti, *African Religions and Philosophy*, 16; Smith, *African Ideas of God*, 29.

202. "Es handelt sich hier, um eine funktionale Einheit, in der die Sonderungen 'selbst' und 'Welt' immer integral des Lebens zusammengefaßt bleiben." Sempebwa,

of a relational unity. It is this integral shared way of life understood as relational solidarity that gives African ethics a pronounced secure infrastructure even for global cooperations. Accordingly, Black Africans are *other-oriented*, meaning their existential reality is only complete when understood in the solidary existences of the complete *other*.

In summing up this normative structure of solidarity, we confirm the collectivity of African life, a cultural representation of a natural metaphysics that operates outside the generally accepted verifiable-empirical-logical-scientific premises as espoused in bioethical principlism. Principlism precludes the solidary fellowship of the community of the living, those not yet born, and the mystical-spiritual-religious influence of the community of the living-dead on each community's daily moral life. Besides, it is less emphatic on the solidarity of universal institutions and the entire cosmic systems of nature. On that account as it stands, it is difficult to sit well in a Global South moral milieu that treasures the moral logic of participatory and cooperative relational solidarity. One notices an anti-metaphysical bias, an ethnocentric prejudice, assuming that the perspectives from the Global North are logical and scientific while those in Global South are prelogical. The Global South approach still refutes *Prälogismus* (of Lévy-Bruhl) as a failure to notice that even other perspectives are post-scientific since they go beyond the data of science and yet remain compatible with scientific data. Analogously, we cannot unambiguously subject the concept of God to scientific verification, but its postulates remain compatible to scientific evidence. That being the case, it is an enormous misfire not to appreciate the nobility arising owing to the relational solidarity of life but rather simply focus on *logical premises* such as those in bioethical principlism.[203] To get rid of this would-be moral cost, the African approach to solidarity additionally appreciates the relational hierarchy of life forces below.

Reality of a Bantu, 107.

203. See Sempebwa, *Reality of a Bantu*, 100–103. For more on criticism, see Lévy-Bruhl, *How Natives Think*, 101. Jahn proves that Levy-Bruhl revoked his theory of prelogism: "Levy-Bruhl hat am Ende seines Lebens seine Theorie des Prälogismus widerrufen und damit ein seltenes Beispiel wissenschaftlicher Lauterkeit gegeben. In seinen nachgelassenen Notizen fragt er sich selbst, wie er sich eine so schlecht begründete Hypothese nur einbilden konnte, und er kommt zu dem Schluß: '*Die Struktur des Geistes ist bei allen Menschen gleich*." Jahn, *Umrisse der neoafrikanischen Kultur*, 101. Emphasis added.

Promotion of the Hierarchy of Life Forces

The fifth and last characteristic of African moral thought in my classification relates to the hierarchy of life forces. The Luganda word *bukulu* means age, importance, greatness, high position, high status, and seniority,[204] which all implicitly point to authority. Authority hierarchically expresses vitalism since it is liable to decrease or increase in relation to power, dignity, or *bukulu*. The hierarchy (*bukulu*) of life forces extends from God to objects by virtue of their ontological participation in life within *different* African cultures.[205] Mbiti stratifies the hierarchy, the categories, or modes of Bantu existence, as we discussed, into five classifications. The Black God hierarchally occupies the highest position as the ultimate explanation, genesis, and sustenance of everything. God is followed by the spirits as beings made up of superhuman beings and the spirits of ancestors. Then humans, including those who are alive and those about to be born, come next, while biological life,[206] *inter alia* flora and fauna, plus other phenomena such as entities without biological life are hierarchized last. God not only sustains and stands above all Bantu known forces hierarchically, but also gives them existence and strengthens them, inclusive of *muntu*, animals, plants, and mineral forces, as Tempels similarly states:

> Über allen Kräften steht Gott. Er gibt den anderen Kräften die Existenz, erhält und verstärkt sie. Gegenüber den anderen Kräften wird er dann auch als Kraft und Lebensverstärker bezeichnet. Nach ihm kommen die Erstgeborenen unter den Menschen, die Stammväter der verschiedenen Clane. Die Stammväter, die ersten, denen Gott Lebenskraft mitteilte—zugleich mit der Macht, auf ihre ganze Nachkommenschaft stärkenden Lebenseinfluß auszuüben—stehen nach Auffassung der Schwarzen Gott und der Menschheit so hoch, daß sie nicht mehr als gewöhnliche Verstorbene betrachtet werden. Man bezeichnet sie . . . als vergeistigte Wesen von höherem Lebensrang,

204. Murphy, *Luganda-English Dictionary*, 38.

205. Each is aware of her mutual hierarchy in a functioning of life forces. Washington, *Negritude*, 60; Sempebwa, *Reality of a Bantu*, 114–17, 304 n. 207; Nsimbi, *Olulimi Oluganda*, 10; Roscoe, *Baganda*, 271. See the meaning of the name *Katonda* (God) and authority as a major expression of vitalism in the Ganda *Weltanschauung* in Lugira, *Ganda Art*, 18–20. For African ontological transcendence, see Mbiti, *African Religions and Philosophy*, 81. Tangwa shows Nso' culture's hierarchical authority from the smallest nuclear family to the king, the chief custodian of all land; see his *African Bioethics in a Western Frame*, 3.

206. See Mbiti, *African Religion*, 16; Sempebwa, *Reality of a Bantu*, 55.

als die—in gewissem Maße natürlich nur—unmittelbaren Teil-
haber an der göttlichen Kraft . . . dann [kommen] die hier auf
Erden lebenden Menschen, die ihrerseits rangmäßig geordnet
sind . . . Der lebende Mensch hängt aber nicht im Leeren. Er
wohnt auf seinem Grund und Boden mit allem, was darauf lebt:
Tier und Pflanze . . . Unter der Kräfteklasse der lebenden Menschen stehen die anderen Kräfte: die Tiere, Pflanzen, Mineralien
. . . Die Hochachtung vor dem Lebensrang, das Bestreben, sich
auf angemessenem Lebensplatz zu halten . . . [so dass] also eine
Art ontologischer Hierarchie der Kräfte oder Lebenseinflüsse
nach festen Gesetzen sich vollzieht.[207]

The ontological hierarchical order (stratification) of forces as understood by Tempels is respected along the life rank and can be condensed
by other scholarship in three types of "worlds" amongst Baganda: the supra-spiritual world, the spiritual world, and the physical world. The first
two are invisible, while the other is visible. *First*, the supra-spiritual world
consists of *Katonda* (God). He is the Absolute Being, with no beginning
or end, who necessarily created everything.[208] The same is depicted in
African anthropomorphism. God is not only the greatest person, but the
great powerful rational being and the great life force as Tempels states:
God is "der 'Große Muntu,' 'die große Person,' das 'große mächtige,
vernünftige Sein oder die große Lebenskraft."[209] In this regard, Africans
describe God as "den überragenden Weisen, der Seinswesen kennt, der
die Art und Natur ihrer Kräfte im tiefsten ergründet. Er, die Kraft selbst,
der die Kraft aus sich selbst hat und alle anderen Wesen gemacht hat,
kennt alle Kräfte . . ."[210] And since Africans understand God as the supreme sage who created all sages and other beings, God is life (*bulamu*),
the giver (creator) of life (*Katonda awa obulamu*), who can increase or
decrease life as well as the vitality of other beings. God dispenses *bulamu*
hierarchically to all creatures through other creatures who also extend
this creative character till the last creature. This creative concept is an
ontological relationship between God, spirits, *muntu*, *kintu*, *wantu*, and
other forces as we maintained in the categories of existence. By way of example, some *muntu* forces are "co-creators" and "the created" at different

207. Tempels, *Bantu-Philosophie: Ontologie und Ethik*, 34–38.

208. See Kasozi, *Ntuʾology of the Baganda*, 44, 50–51.

209. Tempels, *Bantu-Philosophie: Ontologie und Ethik*, 29; Ndombe, *Verständnis der negro-afrikanischen (Bantu-) Weltanschauung*, 149.

210. Tempels, *Bantu-Philosophie: Ontologie und Ethik*, 41.

stages.[211] For this reason, what I call the *co-creative* ontological concept of *Bantu* entails alteast three assumptions from broader African scholarships. For instance, Sempebwa maintains that every *muntu* can directly strengthen or weaken every other *muntu*, every *muntu* force can directly influence any *kintu* or *wantu* force, and a *muntu* using magical force can influence another *muntu* indirectly through an intervening *kintu* force. This influence must necessarily succeed unless the other *muntu* force is stronger in its own right like in the case of a medicine person (anthropologist), or the other *muntu* force is strengthened by some other distinct stronger *muntu* force, or the other *muntu* force protects herself with stronger *kintu* forces than those evoked by the first.[212] Put differently, in the Bantu ontological creative view of being, not only are *created beings more or less like creator and creature* (the word creator goes beyond God), but there is a causality of one force to the inner nature of a being whereby *one force can strengthen or weaken the other internally*, as Tempels states:

> In der Seinsschau der Bantu stehen die geschaffenen Wesen mehr oder minder wie Schöpfer und Geschöpf in innerer Seinsverbindung miteinander. Nach den *Bantu* können von Sein zu Sein, von Kraft zu Kraft Einflüsse gehen, Seinseinflüsse, die mehr als rein mechanisch, noch chemisch oder psychisch sind. In den geschaffenen Kräften nehmen die *Bantu* eine Ursächlichkeit der einen Kraft auf die innere Wesensart, die Natur selbst einer anderen Kraft, eines Seins an. Die eine Kraft kann die andere innerlich stärken oder schwächen. Die Ursächlichkeit innerhalb der Seinsweisen ist keine übernatürliche Kausalität, die etwa die Kräfte des Geschöpfes übersteigt. Diese Einflüsse— Lebens-Einflüsse—sind ganz natürliche Einflüsse.[213]

When we get down hierarchically from the Creator—the supra-spiritual world—we get to what the Creator creates and what further aids the Creator in co-creation. It is the *second* world, the spiritual world of the ancestors. Among Baganda, it covers spiritual beings, excluding *Katonda*. Some of its membership are divinities, deities, or gods[214] (*balubaale*)

---

211. See Sempebwa, *Reality of a Bantu*, 109.

212. Sempebwa, *Reality of a Bantu*, 113–14; see also Jahn, "Value Conceptions in Sub-Saharan Africa," 58.

213. Tempels, *Bantu-Philosophie: Ontologie und Ethik*, 32.

214. Till today, for the *Baganda*, heavens are infiltrated by spiritual beings and activities: a) *Ggulu*, the heavens, is the abode of the Absolute Being; hence, the *Baganda* open their hands in prayer towards the heaven. b) Thunder and lightning are produced by the divinity of the sky called *Kiwanuka* as he "beats" the clouds with a big stick. c)

who are a class higher than the rest of the spiritual beings, the ordinary spirits or ghosts (*mizimu*), the fetishes (*mayembe*), and the nature spirits (*misambwa*).[215] This spiritual world gets its operating symbols from the next, third world. The *third* physical world envelops human beings and physical beings (*bintu*); it encompasses animals, plants, inanimates, and so on.[216] Examples in human families include African kingdoms with their kings or chiefs, who are not purely secular rulers, but also mediators (gatekeepers) between the ancestral world and the physical communities. They increase the life force or decrease it if they lack moral qualities. They transmit the vital force coming from God via the ancestors. It is why their nominations cannot be intervened by governments, since in them lies the vital link binding the living-dead and the physical communities. For illustration, among the Bahema and Lendu of Congo-Kishasa, the expression "king of human beings" (*nrupi*) implies that their king is faultless in everything and a proven benefactor of humankind.[217] This is simply because, on the one hand, the king cannot be deficient in morals while, on the other hand, so believe Black Africans, intermediating moral concerns between the physical and spiritual worlds. The king's lack of appropriate moral qualities cannot increase life between these two African worlds. Elsewhere, the current leader of Baganda or Buganda Kingdom is the *Kabaka* (king), assisted by the *Katikkiro* (the prime minister), whose figure not only unites human beings with beings in the spiritual world, but also guarantees and unites all Baganda clans into one tribe. On the one hand, the *Kabaka* is the messenger or the ambassador (*omubaka*) of God, a person with divine powers to unite Baganda.[218] On the other

---

The clouds originate from the activity of the deity of the River *Mayanja* as he blows water upwards from lakes and rivers. d) After spreading that water out into clouds, the deity *Musoke* sends it back to the earth as rain. e) An alternative rendering on the origin of rain considers it to be the spittle of the Absolute Being. f) At any rate, rain is a blessing from the heavens for the growth and health of human beings, animals, plants, insects, etc. Kasozi, *Ntuʼology of the Baganda*, 47–48. Kasozi borrows some examples from the works of Kyewalyanga, *Religion, Customs and Christianity in Uganda*, 99; and Mbiti, *Concepts of God in Africa*, 138.

215. See Kasozi, *Ntuʼology of the Baganda*, 51–52; see also Kyewalyanga, *Religion, Customs and Christianity in Uganda*, 99; Nsimbi, *Amaanya Amaganda n'Ennono zaago*, 40, 139; Mbiti, *Concepts of God in Africa*, 47.

216. See Kasozi, *Ntuʼology of the Baganda*, 44, 50–52.

217. Bujo, *Ethical Dimension of Community*, 160–61.

218. The word *baganda* connotes a binding force (*obuganda*) on the one hand, and on the other hand it means the Kingdom of Buganda. Kasozi, *Ntuʼology of the Baganda*, 87–88; see also footnote 9.

hand, he guarantees *Bantu* solidarity as a figure of unity of the clan (*ekika*, plural *ebika*) system. *Ekika* has its etymological roots in the basic root *ka*; implying a home or family outside which there is no existence.[219] This means the intermediary role figure of the *Kabaka* and the homely clan system establish *oluganda* (unity) as a bond of solidarity amongst tribemates, which bond of tribemates cuts across the physical and the spiritual worlds. Today, it is why "there is nothing in Buganda as socially degrading as being clanless."[220] If the *muntu* is clanless, she is cut off from the other hierarchal worlds, just because promotion of life begins with the clan fellowship before it extends to the spiritual world, where other *muntu* tribemates are. It is one reason clanmates amidst other creatures typically "do all those actions which promote a hierarchical order of life in society' or Don't do anything that will disrupt a hierarchical order of life in society."[221] When Bantu defend a hierarchal order of life as one of the characteristics of their moral thought, first and foremost, they are not privileging humans over some creatures since no creature's vital role should be subordinated subserviently. Here, a subordination without subservience implies that "no creature's existence is . . . merely instrumental to the well-being of those above it in the hierarchical order, for each benefits from the other."[222] Secondly, such a hierarchical order without subservience isn't necessarily parallel to power in relation to status of some creatures.[223] For instance, for Bantu, an older man or village sage may enjoy a higher reputable status than his local chief, who is authoritatively higher than him. Besides, hereditary positions of princes or princesses and clan heads enjoy higher status but less authoritative power, whereas appointed positions like of chiefs enjoy more authoritative power but lower status. Moreover, a hierarchical order without subservience means that our life achievements are not entirely dependent

219. *Ekika* is organized thus: the individual (*omuntu*) is attached by birth to a household (*enda*); the household is bound by blood relationship to an extended family (*enju*); the extended family is further attached by blood relationship to a lineage (*olunyiriri*); the lineage is part of a sub-clan (*omutuba*); and the sub-clan is a subdivision of the clan acme (*akasolya*). Thus, unity amongst Baganda, i.e., *oluganda* is visible by means of *obuganda*, a binding force. Kasozi, *Ntu'ology of the Baganda*, 88.

220. Kiwanuka, *History of Buganda*, 5.

221. Sempebwa, *Reality of a Bantu*, 134.

222. Sempebwa, *Reality of a Bantu*, 133.

223. Proverbs relating to status: *Omumpi w'akoma w'atuuka* ("A short man can only hold where he can reach." One should recognize one's status). Sempebwa, *Reality of a Bantu*, 234, 283–84.

on our chronological age. For instance, this thesis is at home with the Nso', whose achievements are not necessarily attached to age or being an elder *(nggaywir)*. One's younger brother can easily become one's "father" if he is chosen and installed as lineage head, while one's sister or even daughter can become one's "mother."[224] Here I generally have in mind examples of inheritance and succession that are part and parcel of Bantu. They are typical moments visible in the morality of funeral rites; think of the last funeral rites, where an heir or heiress to the deceased is chosen and installed.

In short, the tripartite form of the three worlds reaffirms the immanence of spiritual life in the material reality of the physical world. But even though spiritual beings are immanent in the material reality of physical world of the living community, they transcend material existence. Notwithstanding the fact that spiritual beings are ontologically transcendent, they find their operative function in the symbols of material beings in the physical world. On that account, the immanence of the spiritual beings (religiosity) in material reality grants a unique significance to even natural phenomena in African substantive beliefs, and to their subsequent normative associations.[225] Given that all systems and forces as well as all material reality in the physical world has spiritual life, there is no creature or creation (all that isn't God) that is lifeless in Bantu moral thought. In a nutshell, we cannot erect an African approach to bioethics outside an integral hierarchical social interaction between all systems and *all* existents of the three-dimensional community (including the visible and invisible worlds). Their continuity and vitality, solidarity and hierarchical order, has to be harmoniously and vitally maintained in a social balance if biomedical ethics is to sit well, or rather promote life, in an African milieu.

---

224. Tangwa, "Bioethics: An African Perspective," 195.

225. Kasozi, *Ntuology of the Baganda*, 49–51. Kasozi's stance relies partly on the works (Christian existentialism) of Ruch and Anyanwu, *African Philosophy*, 123.

5

The Criterion of the Normative Structure of the Black African Ethical System

RELATIONAL SOCIAL REALITY

The ensuing normative moral prescriptions and supreme values point to how entities relate to each other. This relatedness revealed in the African ethical system reflects on the interactive social reality springing from their substantive relational ontological beliefs.

Black African Relational Social Reality Entails Non-Dichotomous (Non-Dualistic) Thinking

The Black African ethical system entails a non-dualistic thinking that can be demonstrated within the mind of Bantu by understanding the significance of the body and the soul and their relationship to life. For Baganda, the *omuntu* has a *omubiri*, literally the "body." *Omubiri* can be a substance either of good health or of bad (ill) health. In this sense, the adjective "bad" connotes the dead physical substance of the *muntu*, known as *omulambo* (or *enjole*, i.e., a "corpse" of a king). At the *muntu*'s death, *omubiri* loses *omukka* (a "gas," "smoke," or air inhaled and exhaled in breathing), which is the faculty of breathing or the highest life-sustaining thread. The loss of *omukka* from *omubiri* can be concretized in current Baganda catchphrases: "*Okumiza omuntu omukka omusu*" literally means to coerce or cause someone to breathe in (swallow) fresh air; the actual meaning is to murder someone. *"Okussa omukka omuvanyuma"* literally means to take the last breath; the direct meaning is to die. So after death, the non-material element (*omuzimu*) of the *muntu* deserts the body.[1] In these senses, firstly, we can infer that *omukka* is the topmost dividing line between the *muntu's* body substance of good health and of bad health, between *omubiri* and *omulambo*. Secondly, we can deduce that from the moment *omukka* from the *muntu's* body (*omubiri*) is lost, *omubiri* becomes wholly lifeless (*omulabo*). Thirdly, at death *omuzimu* deserts the *muntu's* body.

Though every *muntu* must have a *(o)muzimu* ("spirit" or "ghost,"[2] soul), not every *muzimu* is a *muntu*. *Omuzimu* can be understood variedly as: the spiritual element of a *muntu*, the ordinary spirits or ghosts, the non-material dimension of a *muntu* as well as the spiritual beings of the *Kintu* category (the spiritual *bintu*). Baganda substantively believe that it is by virtue of *omuzimu* that the media of *amaanyi*—i.e., the all-reality-permeating dynamism, as Kasozi calls it—is dispensed hierarchically to *muntu* or other physical beings through the spiritual *bintu* by *Katonda* (God). It is the media of *amaanyi*, *omuzimu* that is the immediate form of the manifestation of the "immanence of the spiritual in the material" that survives the *muntu* after death. It is also the means to transcend from the *muntu's* earthly community life to unite with other beings in the community of the ancestors (the living dead) and God ,who is above the sky,[3] in the mode of being that can simply be described as a cold breeze or wind (*empewo*). In the *muntu* joining the spiritual community of ancestors, she loses her *muntu*ness (membership to the *Muntu* category), since she joins the *Kintu* category of existents. Nevertheless, according to Baganda wisdom, the change in membership doesn't ultimately separate *omubiri* and *omuzimu*. It is why even today the dead are warmly

---

1. *Omubiri* can also refer to the skin, the physical stature of an individual. According to the mind of the Baganda, the non-materrial element (*omuzimu*) of *muntu* deserts the body at *muntu's loss of omukka* (death), and so disappears the visible shadow manifestation (*ekisikirize*) of that non-material element of *muntu*. Given the disappearance of this visible manifestation, it's why the lifeless body cannot cast a shadow even under sparkling sun rays. See Kasozi, *Ntu'ology of the Baganda*, 61–63; see also 98–99 nn. 4–5.

2. Murphy, *Luganda-English Dictionary*, 396.

3. See Kasozi, *Ntu'ology of the Baganda*, 65–66. Death is related to God. For Baganda, when the twins die, they say *abalongo babuuse* ("the twins have flown"); they have flown to God who is believed to be above the sky, from where they came from. Roscoe, *Baganda*, 124.

wrapped in bark-cloth (*lubugo*) before burial with a belief to safeguard them against the cold, windy spiritual world of the ancestors ("*Atamanyi mpewo y'emagombe, y'amma omufu olubugo*").[4] In this sense, we cannot dichotomize the continuum of life between *omubiri* and *omuzimu*, the earthily community and ancestral community, the invisible world and the visible world, etc. Given this unbroken unity, it is the reason Baganda consider, like other Bantus, entry (promotion) into the next world of the ancestors or demotion (banishment) into the world of the evil spirits to depend on how the *muntu* has lived morally and harmoniously with the other beings in the previous earthly community.[5] In this sense, Tangwa is right to note that "the customs, laws, taboos, ethics and metaphysics form a single unbroken piece within African worldview."[6] In other words, we cannot dichotomize Bantu thought; it is not dualistic. Bantu life, like their moral outlook, is a relational unbroken continuum.

For Bantu people, social reality is not dichotomized, in contrast to Western thinking, which is not only notoriously dualistic, and tyrannically monistic (philosophies of monotheism, monogamy), but also embraces binary thinking, as Tangwa oftentimes puts it in his eco-bio-communitarian approach. In my previous discussions, the *muntu* is not conceived as a duality, but rather as a unity of life, not only in the anthropocentric sense; but the pluralistic metaphysical tone of a single unbroken continuum of life renders the ontological dualistic separation of the human person, the dichotomy between the body and the soul, cognitive and non-cognitive, feelings and reason, realism and idealism, autocracy and democracy, universal and particular, mind and matter, egoism and altruism, anthropocentricism and biocentrism, body and soul, patriarchy and matriarchy, capitalism and socialism, living and deceased, God and humans, secular and spiritual, rationalism and empiricism, theism and atheism, communalism and individualism, one and many, monotheism and pantheism, etc. ultimately meaningless in African moral thought. There is no room for irreducible divisions or dualistic thinking rather

---

4. See Kasozi, *Ntu'ology of the Baganda*, 98–99. *Mpewo*, wind, breeze, cold air, cold; *magombe*, the place of the dead, the other world, hell; *mufu*, dead person. Murphy, *Luganda-English Dictionary*, 284, 344, 351.

5. See Kasozi, *Ntu'ology of the Baganda*, 102. "Außer den Seelen der verstorbenen gibt es andere Wesen, die manchmal sichtbar den Menschen erscheinen, oft aber unsichtbar ihre Wirksamkeit entfalten. Es sind selbständige Erscheinungen, die—mindestens in ihrer jetzigen Form—keine Beziehung zu den Ahnen haben. Wir bezeichnen sie meistens als Gesture." Damman, *Die Religionen Afrikas*, 18.

6. Tangwa, *African Bioethics in a Western Frame*, 58.

than experiencing reality as practically lived out in life. This is what some African scholarship understands as a non-dichotomized "vitology."[7] African vitology could be synonymously understood in the previously discussed concepts of *amaanyi*, *ntu*, and *moyo*, all indicating a unified vital force. One reason for this unified vital force is that existence in the Bantu languages is conceived as locative, that is, to be is to be at some place. For illustration, in his study "The Empirical Acceptation of Time and the Conception of History in Bantu Thought," Kagame points out that the expression for existence always has an *adverb of place*.[8] For instance, in Kinyarwanda, to exist is to *liho* or *baho*, where the *ho* means "there," at some place. But from the standpoint of a language in which to exist is to be in space—e.g., in English the spiritual would be that which is not physical or is unseen—nothing spiritual in the sense of being non-extended or immaterial can exist. But the soul is supposed to be spiritual in exactly this sense.[9] And if that is so, then the soul cannot have a place in African relational ontology. This is in line with other scholarship. Tangwa hints on latent thinking (though undocumented it remains openly instinctively acceptable) and its antithesis, forensic thinking, as documented and proceeding scientifically. Although forensic thinkin has more to do with bioethical principlism, African moral thought combines both forensic and latent thinking. African belief and ethical systems, within their latent and forensic forms, maintain a non-dualistic accord between creation, i.e., a single unbroken continuum. For instance, "the belief that humans can transform into animals, plants such as trees, or forces such as the wind, is very rampant within this belief system and has very significant implications for the way and manner nature as a whole and in its various specific manifestations is approached."[10] So this metaphysical conception of a single unbroken spiritual unity within African ethical thinking grounds social relatedness, to an extent that when taboos or sanctions are dishonored by an individual, as we previously underlined, the individual perpetrator's blameworthiness severely spreads to the entire clan family, given this unbroken non-dualistic thinking, necessitating cleansing and more so an entire communal purgative or penitential ritual if the individual is to likewise regain her moral standing.

7. Nkemnkia, *African Vitalogy*, 9.

8 Kagame, "Empirical Acceptation of Time."

9. See Wiredu, "African Philosophy in Our Time," 2; Wiredu, "Oral Philosophy of Personhood," 14; Tangwa, *African Bioethics in a Western Frame*, 144.

10. Tangwa, *African Bioethics in a Western Frame*, 157.

African Relational Social Reality Entails Relatedness: From Clan (Blood) to Universal Fellowships

The Black African ethical system entails *relatedness*. *First of all*, in an expanded form of social reality, relatedness of people or things (community) in an African approach to bioethics connotes a relational rather than an individual account of autonomy. If beneficence in bioethical principlism means self-determination of the individual, within the Black African ethical system, emphasis is put rather on the individual capacity to participate relationally in the life principle as espoused within the social reality of the community. If for principlism informed consent is grounded in the strength of individual choices and autonomy, for Black Africans, it means social relational autonomy inclusive of one's clan family. Two accounts of autonomy are itemized: individualistic autonomy, which is consistent with bioethical principlism, and relational autonomy, which is embraced in African sociality. On one reading, principlism endorses "a criterion of partiality—acting as an advocate in the incompetent patient's best interests."[11] On the second reading of other principle-based studies, a strong focus on individualistic autonomy has not only rejected the image of patients as passive care recipients (advocacy), but also remained skeptical about paternalistic influence. As far as principles are concerned, the initial focus in Western principlism has consistently dwelt on minimizing professional coercion in the patient's exercise of autonomy as we already noted, on the one hand, while on the other hand, there has been unshakable concerns of the family's complicity in the patient's medical decision-making in relation to varying values and priorities, which have sometimes contradicted the patient's autonomy, and likewise threw suspicion to the family members involved.[12] We need a unifying if not an integrative approach of these two extremes. But if an integrative role of relatedness in cultural moralities remains underestimated, as Tangwa already charged of principlism,[13] we risk also faulting in some professional moralities. An example lies within biomedical ethics of genetic medicine and reproduction, where a patient's diagnosis can mean their family or social members are at risk too. This raises further controversy regarding principlistic notions of confidentiality and consent in clinical genetics,[14]

11. *PBE*, 195–97.

12. See Ho, "Relational Autonomy," 128–35

13. See Tangwa et al., "Global Health Inequalities," 242.

14. See Callahan, "Communitarian Approach to Bioethics," 496–507.

and even more problematic, if it is unclear whether a patient has shared information. The partial solution lies in heartening pertinent familial sharing as a moral antidote, a social reality that supports a relational social moral account.[15] We cannot coexist as simply social atoms, but rather coexist as relational beings. Our relational social reality is integral in promoting not only the patient's overall life but also the overall social reality of life.

For principlism to sit well in an African context, individual autonomy should be embraced in relational autonomy (relatedness of life), whereby bioethics should focus on the overall relational communal good rather than focusing on *a priori* privileges of autonomous and individualistic pursuits.[16] By achieving the relational good, the individual good is concurrently attained, while by solely pursuing the individual good without focusing on the relational good, further moral controversies yield up. This entails embracing "relational ethics" as an alternative adequate account that responds to contemporary bioethical issues (like the ongoing global pandemics), issues that require a collective approach. This reminds humanity of its common vulnerability besides it acknowledging the central role of others in decision-making that relates to them.[17] But by emphasizing an African approach to relationality, we are not denying prevalence of relational ethics in some globally held accounts in bioethics. Even though some principle-based studies defend relational ethics, they do so only if it fulfills conditions of autonomy, like intentionality, non-control, understanding, etc. However, for the African outlook, autonomy is socially enshrined in the community relational competence and capacity to fully act. This community relatedness is sufficiently capable of cognitively, psychologically, and legally making not only an adequate but also a sufficient decision. For instance, in any case of

15. See Dheensa et al., "Genetic Medicine," 174–79.

16. Bioethics having focused on autonomy and individual rights in the 60–70s, while liberal individualism dominated the 80s–90s, communitarianism focused on the common good and the public interest. Callahan, "Communitarian Approach to Bioethics," 496–507; Etzioni, "Communitarian Approach to Bioethics," 363–74; Callahan, "Principlism and Communitarianism," 287–91.

17. The importance of one's sociality and relational responsibility in decision-making goes beyond individual autonomy. See Friedman, "Autonomy and Social Relationships," 2003; Gómez-Vírseda and Usanos, "Relational Autonomy," 493–505; Walter and Friedman, "Relational Autonomy," S16–23; Rodriguez-Osorio and Dominguez-Cherit, "Paternalism versus Patient-Centered," 708–13; Stoljar, "Relational Conceptions of Autonomy," 375–84; Donchin, "Understanding Autonomy Relationally," 365–86; Jeffrey, "Relational Ethical Approaches," 495–98.

diminished autonomy, the necessity of the individual human capacity for agency (intention), rationality, and independence (liberty) is not simply assimilated but positively *integrated in the community* capacity to competently execute a valid consent or refusal that will stand difficult times. An African will not plainly judge any community member as incompetent, since the collectivity of members has the moral mandate to competently perform a range of legitimate tasks on the *muntu*'s behalf, even when there could be some minimal limitations. The *social security* exercised through *relational autonomy*, an African form of medical insurance—in contrast to Beauchamp and Childress' reliance on the limited financial insurances[18]—is invested in the social reality of a relational functional community. For example, "the Nso' also value the community so much that they subordinate the individual to the community"[19] without subordinating the individual's own creativity. This relational legitimate authority of the community overrides any possible individual incompetence since it is the approved competent guardian and surrogate for decision-making, whose decisions remain autonomous, legal, and binding. This African outlook serves to neutralize the principlistic emphasis that the "obligation to respect autonomy do not extend to persons who cannot act in a sufficiently autonomous manner and to those who cannot be rendered autonomous because they are immature, incapacitated, ignorant, coerced, exploited, or the like."[20] Even though community social security remains the competent guardian of all cases of incompetences and diminished autonomy, this African position does not rule out "conflicts of interest,"[21] arising from close family members.

To prevent the conflicts of interest in the foreseeable loss of the *muntu*'s abilities, instructional directives from the deceased in form of the will (*kiraamo*) are typical of Baganda,[22] as an example. In these instruc-

---

18. *PBE*, 291.

19. Tangwa, "Bioethics: An African Perspective," 193.

20. *PBE*, 105–6, see also 193.

21. Warner and Gluck, "Conflicts of Interest," 36–46.

22. Why *muntu* should write a will (*kiraamo*): "Ekiraamo kiba kiwandiiko ekikolebwa ng'omuntu akyali mulamu, ng'alagira engeri ebyobugagga bwe, oba ebimubanjibwa (bisobola okuba ensimbi, ebintu ebikalu, emigabo, amabanja ge yeewola oba ge yawola) gye birina okulabirirwamu ng'afudde." "A kiraamo is a document made while a person is still alive, directing how his/her assets, or debts (can be money, real estate, securities, loans or advances) are to be managed upon death" (*Personal translation from the above Luganda rending*). Deutsche Gesellschaft für Internationale Zusammenarbeit, "Guide to Peaceful Coexistence on Private Mailo Land," 50; but also see deeply more about

tional directives, the testator, for instance, stipulates how her property and wealth will be allocated and managed in case of her eventual death. These instructional directives must be respected by the community and any legal courts of law. It is through the *muntu's kiraamo* that she not only authorizes allocation of her property, but also allots her body organs for possible transplant, gives out part of her body for research purposes, stipulates out her legal husband and biological children, mentions a specific choice of a heiress, proxy, or decision-making surrogate in addition to the legitimate relational authority invested in the living community, which the community also fosters inheritance rites according to acceptable conventional guidelines. Allocations and inheritance rites stipulated in *kiraamo* correspond to modern scenarios where agreements between insurance and policyholders—think of wills or testaments—institute a given sum to be given to the living children, trustees, or policy holders upon the death of the insured, deliberately for the children or the future care of animals. Disrespecting the deceased's *kiraamo* or modern misappropriation of an agreement is the same as embezzling money from an account established for a family dog named as a beneficiary (such as Helmsley's Maltese dog, allocated $12 million in a will)."[23] To respect the intended future care of the policyholder or the intended allocations, the language of rights is implicated. Biomedical principlism distinctly translates the language of rights into the language of obligations, whereby a right entails an obligation and an obligation entails a right. A *positive right* presupposes receiving a particular good or service, say a right to public health protective services, whereas a *negative right* frees one from some intervention, say a right of privacy or confidentiality (as discussed in Andreas Lubitz's plane crash case). A positive right entail another's obligation (think of the principle of respect of autonomy) whereas a negative right entails another's obligation to refrain from doing something

---

kiraamo on pp. 61–69.

23. "Helmsley's Dog Gets $12 Million in a Will." "In a 2009 case, billionaire Leona Helmsley left $12 million (later reduced by a judge to $ 2 million) to her treasured Maltese and millions to her brother and grandchildren for their assigned roles in caretaking," cited from Beauchamp, "Rights Theory and Animal Rights," 218–19, see also footnote 44; analogously, the value (rights) attached to nonhuman animals, such as a dog, corresponds to other modern scenarios where humans earn severe punishments when their relationships with nonhuman animals fall out, for instance, "the West Ham defender Kurt Zouma has been handed a maximum fine by the club, understood to be £250,000, after widespread condemnation over viral footage of him dropping, kicking and slapping one of his cats. Steinberg and Davies, "Kurt Zouma Fined £250,000."

(think of the principles of beneficence and justice).[24] So when a group of surgeons agrees to separate conjoined Siamese twins, the group of surgeons earns an obligation to the twins, whereas the twins secure a correlative right to get the surgery. Not to honor the surgery or not to respect the calculated future beneficial care of the policyholder (trustees, dog, children) by the insured deceased, in our earlier example, violates the biomedical claim that *rights impose obligations* and vice versa.[25] Violation of such insurance policies, agreements, testaments, or wills (*biraamo*) is violation of the children or animal rights implied in them, and in the African sense it is a grave distortion of the moral relations (atleast with the living dead and the non-human animals) as exemplary in African inheritance rites. This is because *biraamo* create a sense of obligation with mutual correlative moral rights amongst concerned community entities of the living and of future descendants, which *biraamo* can even disallow future transactions, e.g., on the deceased's land. In this African outlook, obligation in its genuine sense is not understood in relation to norms of obligations of conscience, love, and charity as understood in principlism. No person can claim another person's charity (which is a good thing to do, but it is not wrong not to do it) as a matter of a right, given that norms of charity only obligate individuals who are committed

24. *PBE*, 404; Beauchamp, "Rights Theory and Animal Rights," 206. The language of rights is translatable into the language of obligations: Do not cause pain or suffering to others—the right to not be caused pain or suffering or others; rescue persons in danger—the right to be rescued when in danger; nurture the young and dependent—the right to be nurtured when young and dependent; keep your promises—the right to have promises kept; obey the law—the right to have others obey the law. On pp. 213–15, Beauchamp discusses rights to non-maleficent treatment. Correlative to obligations of non-maleficence are rights to not be caused serious harms. Rights correlative to obligations of non-maleficence are negative rights only. For example, do not cause pain or suffering to others, the right to not be caused pain or suffering by others. If animals are in fundamental ways like humans—for example, in the capacity to experience pain and suffering—we have obligations to them from the principle of non-maleficence, and these obligations are correlative to rights. Pain is pain and suffering is suffering in laboratories for humans as well as mammals alike. In response to the principle of non-maleficence, practical examples include different countries banning or limiting Chimpanzee (and Great Ape) research between 1997 and 2008; countries include Belgium, Spain, Balearic Islands, Austria, Japan, Australia, Sweden, Netherlands, New Zealand, United Kingdom.

25. The protection of both positive and negative rights imposes obligations on others. Bradley, "Positive Rights, Negative Rights," 838–41. The European Court of Human Rights has long abandoned the view that human rights merely impose obligations of restraint on state authorities (the so-called negative obligations). Lavrysen, *Human rights in a Positive State.*

to admirable moral ideals. Supposing the walls and floor of the hospital room are ablaze all around the child, and the collapse of the room will almost certainly occur at any second. The would-be obligation of beneficence turns into a risky mission that can be described as a moral requirement. In this circumstance, a rescuer would have no obligation to rescue, and the child has no correlative right to be rescued. As risks for the rescuer increase in circumstances of fires, epidemics, raging rivers, cliff climbing, and other exceedingly dangerous circumstances, it naturally becomes increasingly less likely that a genuine obligation exists, and a rescuer at some point becomes a hero or a moral saint if she succeeds. But bioethical principlism is clear that the obligation not to interfere with an individual's free exercise of her right follows from the right, rather than the right following from the obligation. Meaning, although the obligation is entailed by the right, there is the priority of an individual moral right over a rule of moral obligation.[26] More emphatically, even *trump metaphorists*, like Ronald Dworkin, through their suggestive language of rights, argue emphatically for powerful individual rights that cannot be balanced or traded off or overridden through any magnitude of a public interest, i.e., rights are stronger—much stronger—than the moral claims created by community goals and preferences.[27] In some instances, however, at a cliff climber's discretion, she may permissibly determine to what extent to act beneficently. In case of wild landslides, the rescuer can withdraw the necessary rescuing support if the process becomes unbearable with intensifying menacing landslides. However, this may not hinder this climber from culpability if she forgoes the ordinary obligation of alerting the community for better combined expertise. In this scenario, culpability is shown to extend not only to the individual but rather ultimately to a larger communal group, which community could be more relationally successful and more accommodative not only of the individual moral

26. For Singer, regarding charity, it can be interpreted as wrong for one to spend money on luxuries instead of using the money to prevent starvation. Singer, "Famine, Affluence, and Morality," 229–43. And if one is capacitated to donate aid without loss of goods of comparable moral significance, according to principlism, it would be unethical not to contribute towards alleviating such starvation. Singer's moral emphasis is that we are morally obligated until we reach a level at which, by giving more, we would cause as much suffering to ourselves as we would relieve through our donation of time and resources. *PBE*, 220; see also 406–7.

27. A *trump metaphor* claims a right that leaves one in a position to assert that governments cannot override rights for utilitarian or communitarian reasons, even if the action would maximally promote the public interest. See *PBE*, 402; Beauchamp, "Rights Theory and Animal Rights," 202.

rights but also of the broader obligation. For example, when a *muntu* agrees to write a *kiraamo* before her death, as the community members earn an obligation to the deceased *muntu* of implementing her personal wishes, so does the deceased *muntu* secure a correlative right to have her *kiraamo* honored when her decisional capabilities get extinguished. Not to implement the *muntu*'s wishes according to the *kiraamo* or not respecting the calculated future allocations or inheritance prospects as written by the deceased *muntu* would be disregarding the biomedical privilege that rights impose obligations while obligations equally impose rights. Without letting rights override obligations as stronger and without prioritization of individual moral rights (individual rights of some community members who may not implement the will) over moral obligations of the community (implementing *kiraamo*), Bantu remain careful and try not to create a dualistic approach, where an obligation does not interfere with an individual's free exercise of her rights. It is not the question of whether actions should follow from the exercise of our rights or the rights should follow from our obligations, but according to the wisdom of the African relational community, there are no sharp divisions between obligations and rights, and it is through fulfilling our *relational obligations* that we fully attain our individual rights. More concretely, an example from Bantu life shows that each is required to commiserate relationally with a villagemate who has lost a grandparent or a dear one. It is an obligation of mutual aid (as justice) beyond an individual right, suggesting that the right follows from the communal norm of *mutual aid* as a binding obligation for all neighbors (village-mates), as I will elaborate later. In both cases of *kiraamo* and *mutual aid*, they are genuine senses of obligation, understood as requirements to be fulfilled for the community harmony's sake other than the individual, implying that genuine obligations suggest correlative rights, not to an extent, but all way through.

In the *second* instance of an expanded form of social reality, relatedness is only possible under a homely clan system—since it is shameless to be clanless—that further expands from clan blood relationships to non-blood relationships, and further on to universalistic interrelatedness of vital laws. Initially, under the clan, the common denominator to one's household plus the extended family and lineage is a life force of blood relationship, as Washington intones: "Blood is truly lifeblood for the black African. His blood ties are his strength, his source, his final resource when all else falls . . . Sign of life force, blood is also the sign of nobility of lineage in much the same way that a family name can connote the quality of one's heritage."[28] But relatedness doesn't stop at blood fellowships but includes non-blood relationships as well. It is one reason I find no justification in stressing an Africanness or Europeanness of bioethics. If *ethnos* (ἔθνος) means people/nation/race/culture, it is impossible to point out any *ethno*-neuter life, content, author, problem, method, or audience, etc., as some authors have put it. In that regard, the *muntu* unconsciously avoids moral costs such as naturalistic, ethnocentric, and ontological fallacies, since she is not confined to the limited world of her African clan, but openly encounters every other *muntu*. With simple illustrations of vital energies, modern Bantus commonly address each other on the streets of Europe or corridors of universities, to be precise, as "brother" (bro) or "sister" (sis) without having any antecedent acquaintances. It is the reason a *muntu* wishes to know from a total stranger whether the totality of her family is well. The *muntu* won't stop at asking, "How are you?," but also asks, "How are your people?" Besides, she won't only bid farewell to a stranger, but (even) also extend cordial greetings to the stranger's family at home![29] Beyond such a universalistic reception of these vital laws or energies, relatedness cuts across the three-dimensional community of the living, the dead, and those descendants not yet born.

Black Africans don't view relatedness simply "as biological, since relatedness signifies merely an openness that goes beyond what is present and visible in a given situation."[30] Beyond the physical world and the biological spheres, relatedness extends further on than visible clan and familial bonds to include not only broader communal blood relationships,[31] but also the relational triad of the living, the living-dead, and those not yet born. In this triadic relationship, consanguinity is regarded as "the result of ancestral guidance" and "the Guardians of the blood are the ancestors who protect the life force of the clan."[32] Yet still, in the same triad, it is unthinkable to exclude the non-human world. In ensuring relational harmony, the African relational palaver includes everything— not only human beings, but the whole nature—in one social reality that doesn't exist statically once and for all, but rather comes into existence

28. Washington, *Negritude*, 49.

29. See Onwubiko, "Re-Encountering African Culture," 103, 37; Bujo, *African Ethic*, 86; see also Bujo, *Ethical Dimension of Community*, 182–83.

30. Bujo, *African Ethic*, 3.

31. See Ndombe, *Verständnis der negro-afrikanischen (Bantu-) Weltanschauung*, 230; see footnote 762.

32. Washington, *Negritude*, 49.

continually. Again and again, indeed unceasingly, newcomers join, since the whole of humanity and the whole of the cosmos are one community moving towards a more perfect community.[33] To put it briefly, there is daily conversion, and each human being continuously becomes a person only in a fellowship of life with *others*. Firstly, solidary fellowships raise the quality of the vital force not limitedly to oneself, but to the entire community and humanity. Secondly, this is beyond ethnocentric tribalism. Family solidarity develops into clan solidarity, while clan solidarity in turn develops into tribal interdependent solidarity and vitality. Thirdly, and broadly, this interdependent solidarity extends respect and esteem from one ethnic group to another, since one human being is related to all humans, giving substance to the universalistic interrelatedness of vital laws of friendship, solidarity, respectful love, dialogue, and hospitality.[34] Bantu do strongly believe that these vital laws cut across the totality of the relational triad of community life, not only in local settings, but also in the global community.

Black African Relational Social Reality Entails Socialization: A Community Palaver

The Black African ethical system entails a *communicative palaver of socialization*. Primarily, my interest is in a "communal" but not a "societal" setting. For an uncomplicated distinction between "community" (*Gemeinschaft*) and "society" (*Gesellschaft*), the work of Ferdinand Tönnies, *Gemeinschaft und Gesellschaft: Grundbegriffe der reinen Soziologie*, is still relevant to my present discussion. Various scholarship, including Bujo's investigation, suggest that Tönnies takes the community as a unity, where members are essentially linked to one another, but each retains her own identity. The unity within rests on a harmonious "essential will," which is *long-term* in nature. In contrast, for the society, the members are indeed linked as contractual partners, but remain "fundamentally" separate. Here exists an "optional will," which, however, can only be unified at one specific point but cannot last.[35] For the "community will" to be essentially linked, it necessitates an indispensable relationality of life

33. Bujo, *African Ethic*, 65–66.

34. See Sempebwa, *Reality of a Bantu*, 148–49; Roscoe, *Baganda*, 346–48; Bujo, *An African Ethic*, 5–6.

35. Tönnies, *Gemeinschaft und Gesellschaft*; about Gemeinschaft (community), see pp. 9–45 and about Gesellschaft (society), see pp. 46–48. Bujo, *African Ethic*, 64–65.

between the individual and the community that is not preconditioned on contractual obligations between separate parties. It is why for Black Africans "a community is not plainly the aggregate sum of individuals (think of the personalistic Western approaches), but rather the aggregated sum is intimately interconnected."[36] Our understanding of the community in this investigation has so far gone beyond limitedness on the aggregate sum of individuals or on simply being a human affair, to relationally include and interact with other community members. For this reason, a community should, on the one side, point to the micro (as organism—*microcosm*) and macro (integrated within community—life *macrocosm*) unity of the whole dynamic relational beings. The *micro* and *macro* communities of Bantu include the conglomerate of families, clans, as well as ethnic groups, say of Baganda. For illustration, the word for one human member, *muganda* (plural *baganda)*, stems from the word *omuganda*, meaning a bundle, signifying unity, as well as suggesting a common ancestry, a kinship, or certain degree of blood relationship (fellowship or comradeship) that binds the Baganda together through *(o) luganda*, namely solidarity. Furthermore, *luganda (oluganda)* does not only signify solidarity, but it is also the law of unity that overcomes the natural barrier of social cohesion, as well as the language spoken by the Baganda.[37] This kind of solidarity understood as the law of unity or kinship is akin to a scar (*nkovu*) that can never be permanently lost even for morally troublesome members of the community like juveniles,

---

36. Ikuenobe, *Communalism and Morality in African Traditions*, 84.

37. From the static root "ganda," we get "(o)muganda" (a, or one member of the Baganda ethnic group), "(a)baganda" (two, or numerous members of the Baganda ethnic group). "Baganda" is the plural form of the word "(o)muganda." Kasozi, *Ntu'ology of the Baganda*, 19–20, 84–86; see also footnotes 2–4, 6–7. *Ab'oluganda bita bikoonagana naye tebyatika* (Relatives are like calabashes, they knock together but do not break) —"Blood is thicker than water." "Omuganda"—as the saying "kamu kamu gwe muganda," i.e., (collecting) one (stick) by one (you gradually make) makes a bundle—meaning "a bundle," signifying unity. Little by little fills the measure. Murphy, *Luganda-English Dictionary*, 286, 352–53. Also, "*kamu kamu gwe muganda*" implies unity is strength. For emphasis, *Luganda* is the language of the *Ganda* or *Baganda* people. *Buganda* is their country; see Sempebwa, *Reality of a Bantu*, 1–7, 279. More on unity and cooperation: *abataka abaagalana, be balima akambugu* (co-operation among kinsmen entails strength"). Nsimbi, *Olulimi Oluganda*, 6. *Ogwa kasajja gundeebukanya nga yeyamwezaalira* (however much a child troubles its parents, they must never give up their obligation to care for him). Sekadde and Semugoma, *Ndimugezi*, 9. The proverb shows interdependence of human life in the Kom tribe: *awu àmi'a ka'kɨ bu'lœ kul ibu*, literally, "one hand cannot tie a bundle." Each is not self-sufficient and cannot achieve anything great without the support of others. Mbih, "Foundation of Kom Proverbs," 12.

teenagers, youngsters etc. Baganda proverbs prove this indelible care of one's kin: "*Oluganda nkovu terugwa kumubiri*" (Kinship or solidarity is like a scar you can never loose).[38] But if we are to relationally interact with the other community members, on the other side, we have to go beyond the natural barriers of identification and fully include non-human members in our community cycles of solidarity. Such a strong integral sociality empowers Mbiti to refer to an African community as a "kosmische Einheit,"[39] so to speak, a "corporate community" that includes living humans, the living-dead, and those not yet born (all living and non-living things). Like I indicated already, the African ethical system, built on a community (*entababuvo*), extends relatedness not only to people but also to systems of forces and entities that don't necessarily enter into rational contracts but remain members of the community in its relational goal of promoting life.

Inclusion of all members is typical of the African palaver. The African community palaver is not preconditioned on who can powerfully reason and fully engage in contracts. The communicative palaver relatedly engages the entire corporate community with an aim of morally finding some therapeutic medicine for an integral whole. Bujo differentiates between a "family," "suprafamiliar," and "administrative" palaver. I will talk about the first two palavers since they directly serve the purpose of my investigation.

Since some moral questions relate to the existence of the family, the head of a family invites all the representatives for a family palaver, which aims at socialization of the vital force without forgetting the living-dead, our ancestors. Whenever the *muntu* looks for a stabilized life, she is guarding against tension between life and death, between the living and the dead, which cannot be eliminated, and what Bujo punctiliously interprets as an "anamnetic *poiēsis*" (reestablishing), in which every *muntu* historically remembers the presence of her ancestors in order to transmit life or give a new vital dynamism of the earthly communicative fellowship, which keeps alive the memory of the unborn and ancestors through the children.[40] I make use of Bujo's reconstruction of the African moral palaver with interest in an *anamnetic dimension*, whereby "the ancestors' words and deeds, the norms they set, are made available to the current generation so that it has life and continues to look after the

38. Nason, "Proverbs of the Baganda," 251.
39. Mbiti, *Afrikanische Religion und Weltanschauungen*, 129–31.
40. Bujo, *African Ethic*, 34–36.

deceased, and so that it prepares the future of the not yet born."[41] In this sense, African moral thought as understood anamnetically goes beyond the would-be problem-solving account of Joseph Ratzinger's account of "anamnesis" (recollection as the new *we*). This is because the African palaver socializes a *"memoria"* by reflecting on the words and deeds of the ancestors, in order to interpret these anew in today's communal life and approach to bioethics. Thusly, the anamnetic dimension is a living *memoria*.[42] It is a living *memoria* because it socializes vitality across the three-dimensional community, not only to the individual and her immediate clan family but also to the extended family of the living and the dead, with a to-do list that may include reading out one's *kiraamo*, expediting an inheritance, planning for a foreseeable marriage covenant, the deposition or installation of a family chief, resolution of family land conflicts, settling burial grounds, and so on. With the corrective participation of the elders and sages, who have not only "unquestionable wisdom" or are exposed to an "extensive reality consciousness" (*magezi*), but are also well conversant with the entire family history, which, when all combined together in a palaver, are relevant for a new family solution inclusive of the ancestral world. This is an indication of an *anamnetic* dimension, that the contemporary and future communities necessarily depend on the experiences of the past community. Some may question whether this palaver approach of the *anamnetic* dimension isn't simply an ethics of speaking on behalf of the deceased. Bujo argues that we cannot refer to this *anamnetic* perspective as "advocatory ethics," i.e., speaking on behalf of the dead; rather, Black Africans substantively believe that the dead are truly present in their lives and newly "speak" out relevant moral life concerns to the new community, such concerns include the need for reconciliation of the members of the living community with members of the ancestral community. And this is the reason this family palaver concludes with an exemplary lesson of a celebratory reconciliation, e.g., convincing two cohabiting close relatives (or those sharing the same totem) that the forbidden or incestuous marriage union that they desire is prohibited and won't be approved by both

41. Bujo, *Differentiations in African Ethics*, 433. Reference can be made to "remembrance" or "memory" or "recollection." τοῦτο ποιεῖτε εἰς τὴν ἐμὴν ἀνάμνησιν ("Do this *in memory* of me," Luke 22:19, NABRE); "Do this, as often as you drink it, *in remembrance* of me" (1 Cor 11:25, NABRE). Emphasis added. All Bible quotations are taken from the NABRE.

42. Bujo, *African Ethic*, 112.

lineages, that is, the lineage of the would-be future husband and the lineage of the would-be future wife. This is due to the fact that up to today, the liberty of choice is regulated by Bantu norms of exogamy, which prohibit any sexual relations within one's parents' clans (father's or mother's clan). Even though sexual relations could be leniently handled within a family palaver, marriage amongst clanmates is certainly not permissible. The immoral example exhibited in this couple about to disrupt exogamy norms can be cleansed with a sumptuous meal (*ddiiro*) where the parties reconcile with the ancestors and the entire corporate community. In this manner, the whole palaver not only re-energizes healthy interrelationships, but it has a medicinal and therapeutic character, besides its function as an ethical authority—where the ethical norms (proverbial wisdom) can be either confirmed, turned down, changed, or newly established in the social reality of Black Africans.[43] This therapeutic and medicinal character of an *anamnetic* dimension is lacking in bioethical principlism. In different principle-based biomedical projects like cloning, we miss out this true *anamnetic* representation as evident in an African approach to bioethics. With the recent reproductive technologies, a cloned human being can never replace an ancestor who had her own experiences beyond the biologically determined potentialities. The *anamnetic* wisdom and experiences continuously shared between

---

43. See Bujo, *African Ethic*, 49–50. The Nso' do not tolerate marriage between cousins up to the fifth degree. For them, incest is mainly understood as sex between close cousins. Tangwa, "Bioethics: An African Perspective," 197–98. More moral errors may arise from different cultural understandings of life if the community palaver fails to guide well, for instance, in matters of incest and marriage. Firstly, in some African peoples, the boundary against incest is reached much more quickly than in in the West, think of central Europe. In Western culture, atleast in central Europe, sexual intercourse between cousins does not count as incest, but this is strictly prohibited by many African peoples at all degrees of relationships—for instance the Bahema in the eastern area of the Democratic Republic of the Congo (p. 14). Secondly, in some "African languages, 'cousins' (in European languages) are 'brother' and 'sister,' in the case of children of one's father's brother or one's mother's sister. Even the children of what Europeans would call two cousins can be brother and sister. Further, the brother of one's father is equally 'father,' just as the sister of one's mother is equally 'mother.' They are not 'uncle' and 'aunt'—'aunt' is only the sister of one's father, and 'uncle' only the brother of one's mother. Consequently, the children of this uncle are just as much 'uncle' as their father, or 'mothers' in the case of daughters" (p. 15). The argument is, before we agree about the so-called *intrinsece malum*, who determines what is to count as incest? See Bujo, *African Ethic*, 14–15. For the Baganda, till today, marriage within a person's father's or mother's clan is prohibited. Mair, *African People*, 78; about sex within clan members see also p. 79. On no sexual relations with a man or woman of one's clan, see Sempebwa, *Reality of a Bantu*, 160.

the ancestors and the living descendants, in order to enrich their identities and personalities, have no relatedness to the clone. For a Black African, a cloned human being has no roots in the living community, which would actually deteriorate the vitality of the entire community. A similar challenge can be traced with in vitro fertilization, embryo transfer, embryo donation, and insemination. In these examples, it gets even worse when the donors of the sperms are anonymous, which anonymity does not marry with the *anamnetic* thinking of African anthropology. For Bantu, it is essential that the child, father, and mother mutually know their lineage and so remain in a *living contact* with each *muntu* after death.[44] This explains why Bantu avoid any possible interbreeding in advance by making sure each *muntu* intending a marital union knows the other's lineage, clan name, and totem very well so that the *anamnetic* contact may not be chaotic in the future. Since I am from the Kasimba (Genet cat) clan, having my totem as the Genet cat, I cannot marry a lady with the same totem. And since my mum is from the Ngonge (otter) clan, having her totem as the otter, I cannot similarly marry a lady with the same totem like my mum because I would be directly marrying my own mother. So, instances in bioethical principlism where anonymous donors from frozen embryos may have died several years ago—think of cryopreservation[45] of gametes introduced since 1960s—make not only a living contact impossible, but also parenthood difficult for the purchasers (or those who hire wombs) since a child's biological parent or both parents may have died. Not only that, but they also even unendingly raise legal fights from different possible living donors over the rightful biological parent to be part of the child's parenthood and blood lineage. As if such moral chaos is not chaotic enough, and even if we were to rely on hospital documents to ascertain who exactly gave birth as the rightful mother, others could go an extra mile to openly practice incestuous adoption of embryos from their children. A mother may adopt an embryo from her daughter and consequently give birth to her grandchild with whom they are biologically and genetically related, chaotically bringing two mothers into the moral picture (surrogate and donor), given that biological motherhood may not necessarily imply the woman who has given birth but may simply be a surrogate. And still worse, in all

44. See Bujo, *African Ethic*, 16, 94; Hallich, "Embryo Donation or Embryo Adoption," 653–60.

45. See more in Maryam Hezayehei et al., "Sperm Cryopreservation," 327–39; Gómez-Torres et al., "Effectiveness of Human Spermatozoa Biomarkers," 90–94.

the moral chaotic examples above, we wonder who has a right to have her name affixed as the ancestral parent onto the newborn's birth certificate or travel documents. In one example, there is the long-ago dead ancestor who donated frozen embryo, and in another, the grandmother who adopted the embryo. It becomes even more complex if the child cannot claim any right of lawful inheritance within a family clan since the dead ancestral donor never included her in the *kiraamo* (will). All these principlistic mix-ups and moral confusions with motherhood or fatherhood cannot sit well in African settings, where a family palaver is decidedly authoritative, and so these fall outside the anamnetic thinking of African moral thought.

Among Bantu, beyond unsatisfactory family or clan palavers, concerns of a larger group are tried before a court as the supra-familial (public) palaver, which is hierarchically superior to lower palavers. The participants are members of the regional "council of elders"—found among the *Barundi* of Burundi, known as *Ubushingantahe*, and among Baganda of Buganda Kingdom[46]—the delegate counselors of the king or chief. However, in the topmost palaver as the public palaver, it is the king or chief or a member of the council of elders that presides over every element of wisdom in every person speaking, trying to discern every solution in symbolic languages used in proverbs, fairy tales, and parables.[47] Given the central place of reconciliation in the *muntu*'s totality of social life, like the family palaver, and the public palaver having handled all moral concerns from the lower palavers or those moral issues that were unsuccessfully solved in the lower hierarchical palaver, it concludes with a reconciliatory ceremony as a visible symbol of relationally revitalizing life of the totality of the three-dimensional community. In defending the discussed distinctive communal social reality of Black Africans within which their ethical approach to bioethics is hinged, one question that arises regards whether the community does not suppress the *muntu*'s individuality. For further illustration, we may equally interest ourselves with practical communitarian examples in which stronger bonds of

46. A single *Mushingantahe* with exemplary sense of maturity in virtues is appointed by the people as advisor and representative to the king in order to lay all their problems as a just referee in controversial matters, sort conflicting issues, ensure reconciliation, promote peace and justice. The *Mushingantahe*'s office is sealed religiously with a consecration as a lifelong duty of caring for everyone's well-being. Not living up to this ideal, he would be relieved of his duties. Bujo, *Ethical Dimension of Community*, 159–61; Kasozi, *Ntu'ology of the Baganda*, 87–88.

47. See Bujo, *African Ethic*, 51–53.

relatedness have recently been observed in SSA. Let me attempt responding to these interests in the following sections.

African Communitarianism: The Identity of the Individual (I) in the Community (We)

We have already claimed that the ethical approach to principlistic bioethics relies on the autonomous responsible thinking, individual whereas
an African approach to bioethics is constituted on the relational social
reality of the community, which supplies the appropriate normative content along which the individual interacts and establishes norms through
a communicative palaver. Critics question whether the individual's moral
autonomous identity, freedom, and conscience aren't subjugated by the
moral constraints of the community palaver. In a condensed form, these
moral challenges are moreover distinctly manifest in principlism, whereby utilitarians and communitarians "maintain that individual interests
said to be protected by rights are often at odds with communal and institutional interests,"[48] if not directly in conflict.

First and foremost, in the communal social interactions, although
the individual is not the ultimate authority, her distinctive creative ethical conduct remains uncompromisable owing to the relevant normative
content given by the community. If the *muntu*'s unique creativity was
compromisable, then this would be no different to the threat caused by
bioethical principlism since it absorbs the individual and over-stresses
the solitary capacity of each autonomous individual and her autonomous
choices, subsequently unrolling into absolutism that takes apart our collective pursuit to unwind the global challenges, e.g., the current pandemics: Ebola, monkeypox, COVID-19. In contrast, however, for Bantus the
muntu's creative conscience is not absorbed into the community. The
muntu's new experiences enrich the entire community either positively
or negatively, increase life or decrease it. Even when the community
palaver always provides the relevant culturally specific and contextual
normative content, in order to safeguard the *muntu* from running simply
along with the community crowd or the excessive emphasis of her own
individual conscience, it is only in the sense of the community profit-
ing from the *muntu*'s unique creativity.[49] The *muntu*'s unique creativity

48. *PBE*, 401.

49. See Bujo, *African Ethic*, 164–66.

or her individual conscience is not the last instance, as in principlism, but rather the community conscience has the last word. For the *muntu's* individual conscience to be dependent on the community conscience means that she fulfills her task creatively for the sake of the well-being of all, whereby she necessarily needs the harmonious support of the whole extended family, dialogue with the clan fellowship, and the appropriate community infrastructure. There is no instance where the individual (her role or creativity) is replaced by the community, since the community owes its life force to each individual member.[50] There is an intimate relational interaction between the *muntu's* conscience and the community conscience that cannot be underrated. Mention of one's individual conscience here calls to mind some unique examplary works, like those of Thomas Aquinas in *Summa Theologiae*, on whose authority principlists partly ground their framework of bioethical principlism.

In retrospect, the Thomistic emphasis of the subjective or individual or personal conscience (reason) in grounding principlism—"inasmuch as by means of those principles naturally known, we judge of those things which we have discovered by reasoning"[51]—easily leads to absolutizing one's own private rational opinion in such a way that it becomes a law unto itself in concrete situations.[52] For an autonomous individual, it necessarily creates an obligation (a must) to obey even one's erring reason (*ratione errante*) or erroneous conscience (*conscientia erronea*). Thomas confirms that "conscience is nothing else than the application of knowledge to some action. Now knowledge is in the reason. Therefore when the will is at variance with erring reason, it is against conscience."[53] To obey

50. See Bujo, *Ethical Dimension of Community*, 74–76.

51. Aquinas, *Summa Theologiae*, Prima Pars, q. 79, a. 12c: ". . . inquantum iudicamus per principia per se naturaliter nota, de his quae ratiocinando invenimus." See the edition of *Summa Theologiae* cited in the bibliography; the same Thomistic works are cited throughout unless indicated otherwise.

52. See Bujo, *African Ethic*, 119.

53. Aquinas, *Summa Theologiae*, Prima Secundae, q. 19, a. 5c: ". . . conscientia nihil aliud est quam applicatio scientiae ad aliquem actum. Scientia autem in ratione est. Voluntas ergo discordans a ratione errante, est contra conscientiam." Aquinas gives an example: "for instance, to refrain from fornication is good: yet the will does not tend to this good except in so far as it is proposed by the reason. If, therefore, the erring reason propose it as an evil, the will tends to it as to something evil. Consequently the will is evil, because it wills evil, not indeed that which is evil in itself, but that which is evil accidentally, through being apprehended as such by the reason . . . We must therefore conclude that, absolutely speaking, every will at variance with reason, whether right or erring, is always evil." ("Puta, abstinere a fornicatione bonum quoddam est, tamen in

one's conscience is to obey one's subjective reason as absolute. And if this traditional Thomistic sense of conscience (*conscientia*) as the knowledge with somebody is confined to one's subjective reason (*scientia autem in ratione est*), then the pursuit of Aquinas's approach to conscience partly misses out on "objective" truth, which truth Africans believe cannot be attained by an individual self in isolation from the interconnectedness of the relational *we*.[54] If we randomly pick a contemporary example that follows the Thomistic line of thought, today bioethical principlism speaks of "freedom of conscience" by globally emphasizing the "age of maturity" (many governments talk of eighteen years of age) as freeing young people from control of elders or community members through restriction of external interference into their individual or private life. To be free for the juveniles is basically to be independent of the elders, the community members, parents, government influences, and so on. So, the right freedom of conscience for the contemporary juveniles is exercised outside the relational communal (*we*) dimension since an individual ethic reigns in contrast to the *muntu* living in a relational sociality, a *we* dimension that determines what is wrong or right.[55] Such an application of the individual conscience has influenced not only principlistic biomedical ethics, but also global church approaches as can be observed in the Catholic Church's magisterial moral debates, if we are to pick another example randomly. Debates in theological ethics have centered too much on the authority or primacy of subjective or individual autonomous knowledge, where one has to follow the judgements of her individual conscience or reasoning.[56] Such autonomous dictates of individual conscience basically lack an appropriate moral ranking within the African understanding of conscience, which I have interpreted as relationally communal.

Foremost, it is relationally communal because, as the African community palaver pursues the highest good, i.e., promoting the abundance of life, it richly embraces both the macro and micro ethical levels of Bantu life. It transcends any moral accounts that emphasize self-realization of

---

hoc bonum non fertur voluntas, nisi secundum quod a ratione proponitur. Si ergo proponatur ut malum a ratione errante, feretur in hoc sub ratione mali. Unde voluntas erit mala, quia vult malum, non quidem id quod est malum per se, sed id quod est malum per accidens, propter apprehensionem rationis . . . Unde dicendum est simpliciter quod omnis voluntas discordans a ratione, sive recta sive errante, semper est mala").

54. See Bujo, *African Ethic*, 119.

55. See Tangwa, *African Bioethics in a Western Frame*, 31; Bujo, *Ethical Dimension of Community*, 68–69.

56. See Bujo, *African Ethic*, 109–10; Bujo, *Ethical Dimension of Community*, 83.

the individual conscience, given that life to Bantu is to be realized both individually and communally.[57] Put differently, the epistemological moral imperative is for an integral communal conscience to be concretized, along the process of the communicative community palaver as it embraces both the macro and the micro ethical levels, and as it discerns and validates norms, it bridges between individual conscience (bioethical principlism) and communal conscience (African moral thought). To say it in a broader way, in the palaver's efficient institutionalization of codes that are appropriate for an integral communal conscience, it concerns itself relationally with Bantus' existential interests of both the living and the dead, not only at the macro-ethical level, but also and with the same intensity at the micro-ethical level. Given it is concretely prescriptively practical, the communal conscience doesn't operate at the level of principles or discovers norms in the study room, where the resulting morality would appear as an intellectual academic exercise without efficacy at the micro-ethical level.[58] In other words, we cannot separate between the individual conscience and community conscience in their *single* goal of promoting life abundantly, not only for the individual *muntu* but also for the community of Bantus. Even when we achieve a communal conscience that is practical and existentially sapiential to broader ethical problems, in all lived experiences, the individual's creative conscience can be neither crushed nor absolutized. Understood differently, "communal conscience, above all, measures and determines the individual conscience,"[59] meaning the community palaver of *we* forms the individual's conscience and translates it into a communal conscience, an experience of individual creativity tested with time against community creativity. The norms established in the palaver reflect the voice of the experienced creative conscience of the community.[60] The palaver's dialogic role continuously interrelates the individual creative conscience both to the community of ancestors and to the community of one's contemporaries, as well as to those descendants not yet born. In contrast to bioethical principlism, this African approach to bioethics is exceptionally concerned with promotion of life abundantly, but only in a three-dimensional community anamnetically (*memoria*-conscience). So, the living community cannot

57. See Bujo, *African Ethic*, 70–71.

58. See Bujo, *Ethical Dimension of Community*, 36–37; Tangwa, *African Bioethics in a Western Frame*, 7.

59. Bujo, "African Ethics," 433.

60. See Bujo, *African Ethic*, 1–2, 125–26.

keep away from the past experiences of the ancestors—whose guardianship remains contemporarily necessary since it is the ancestral wisdom that established the prevailing ethical norms—before it can institutionalize futuristic ethical norms to guide the forthcoming descendants.

Secondly, the *muntu*'s creativity not only enriches community life, but it also equally contributes to the *muntu*'s personality growth, which cannot be achieved outside the community. Put differently, the *muntu*'s ethical action either increases life for the whole community (due to the good done) or reduces life (due to the evil done). This means growth cannot be only asymmetrical, but rather must be mutual. It is mutual because the *muntu* reciprocates to the same community what she has received from it.[61] When both mutually fulfill the task of increasing the life force, the *muntu* reciprocally shares in the life force of the unabridged community relationally. The *muntu* remains incomplete without this mutual interaction, and likewise the community would fade away without her (individual members). The *I* cannot exist all alone except corporately as part of the *we* since the *muntu* owes her existence to other Bantus, including those of the past and current generations. In this sense, the community creates, fashions, and reproduces the individual newly. When the *I* suffers or rejoices, it does not do so in isolation but with the corporate community of the living and dead clanmates, villagemates, and relatives. It is only in terms of the totality of this mutual corporate life that the *muntu* becomes conscious of her own self in relation to the totality of other beings.[62] Put differently, the *muntu*'s success is the community's success, making the community and individual interaction a necessity to an extent that the ever-intended harmonious relationship cannot deprive the *muntu* of her freedom and creativity at the expense of the community, as Ndombe states:

> Stirbt ein Familienmitglied in der Stadt, wird immer die ganze Sippe zur Sammlung für die Beerdigungskosten aufgerufen. Erkrankt ein Mensch, wird die Familie für dessen Behandlungskosten zur Verantwortung gezogen. Heiratet einer, ergeht der Appell an alle Stammesmitglieder, an der Feierlichkeit teilzunehmen und zu deren Gelingen beizutragen. Der Auffassung, dass der afrikanische Gemeinschaftsbegriff das Individuum seiner Freiheit beraubt, halten wir entgegen, dass in der afrikanischen Gesellschaft Gemeinschaft und Individuum in Interaktion

---

61. See Bujo, *Ethical Dimension of Community*, 24, 189.
62. See Bujo, *Ethical Dimension of Community*, 73; Sempebwa, *Reality of a Bantu*, 127.

wirken, in einer harmonischen und schöpferischen Beziehung zueinander stehen.[63]

It is only in the community's fulfillment that the *muntu* is mutually and wholly fulfilled. It is one reason why African moral thought takes the *muntu* not to belong freely to herself if relational cohesion is to be achieved, but rather she belongs to the entire community. This does not only mean that the freedom of the *muntu* includes the freedom of the community, but more so it means that the *muntu*'s freedom is only achieved in the community's fulfilled achievement of freedom.[64] The *muntu* believes that fulfilled freedom does not only exist for her, but it is also for everyone else, since the *muntu* herself can be free only when every other *muntu* is free. It is only the *muntu*'s genuine community relationship that will grant her self-realization as well as grant other Bantu a genuine human existence,[65] thereby creating no room for community pressure to crush the *muntu*'s creativity or stifling her as an individual member of the community. Given that the *muntu*'s genuine self-realization is only possible with the genuine existence of Bantus, the *muntu*'s freedom or rights are only achieved in the community's relational achievement of its freedom or rights. Relatedness in Black African daily social living further suggests that the *muntu*'s rights must be relationally defined and preserved along community rights. It is only by respecting the rights of the individual that the community preserves its own identity, and the corrolary is true of the individual herself.[66] Analogously, we are emphasizing how the creativity of each individuality is inevitable and uncrushable in a broader discourse of an African approach to ethics. This is relatable to what Tangwa says. For him, even though "the principle of autonomy accords very well with an individualistic perspective of life and may be overemphasized in discourse within individualistic cultures like Western culture," it remains "equally important in all cultures, including communalistic cultures like African culture, in which individuality as distinguished from

---

63. Ndombe, *Verständnis der negro-afrikanischen (Bantu-) Weltanschauung*, 155–56.

64. See Bujo, *Ethical Dimension of Community*, 74–75.

65. "I am related, therefore I am." Bujo, *African Ethic*, 129–30; Bujo, *Ethical Dimension of Community*, 28.

66. See Bujo, *African Ethic*, 125, 163–64. For Kasozi, the community recognizes and makes sense of the questions of its critical individual members, realizes the necessity and urgency for change, though the community shuns any occasion of giving the full measure of credit to the critical individual members as it publicly elevates a simple social rule to a social custom. See Kasozi, *Ntu'ology of the Baganda*, 110.

individualism is also highly respected."⁶⁷ This emphasis of "individuality" rather than "individualism" reiterates how the *muntu* can only realize herself, her freedom, and her rights only in and with the entire community of the living, the deceased, and the unborn. Without claiming the community as dictatorial over the *muntu's* identity, freedom, and rights within an African approach to bioethics, we suggest a continuous dialogical palaver. This is because the *muntu* (the individual) cannot succeed enjoying her freedom or rights autonomously without other Bantus, suggesting a continuous necessary dialogical palaver (conversion) between the local and the universal community, without the universality overruling in anyway the local individualities or relapsing into universal imperialism.

Thirdly, the contrast between individualistic and communalistic moral tones can be further understood in name giving, where the *muntu's* individuality is not absorbed in the community or any universality—even when fundamental principles demonstrate their universality by being important in every global culture without being unconditionally absolute. Tangwa's mark of any "genuine ethical judgement is its universalizability"⁶⁸ while remaining valid spatially or timelessly. It is only in this sense that I understand Tangwa to take universalizability (cultural universals) as a litmus test of dependable moral judgements given that he believes "that every genuinely valid and uncontaminated particular moral judgement is universalizable, although not every such judgement is necessarily absolutely exceptionless."⁶⁹ It becomes difficult to presume *absolute exceptionlessness* from particularity or individuality as plausible in bioethics, even when each remains *indispensable*. Inferentially, via *reductio ad absurdum*, we can confront bioethical principlism in its superimposition of Global North–based philosophies as binding frameworks in other universal cultures. And if we risk the uniqueness of other cultures, interhuman, cross-cultural, and interactive global cross-fertilizations of bioethical thought won't be likely. We can primarily clarify this uniqueness of individuality in *name giving*.

Unlike in the principlistic Western sense, to emphasize the indispensability of each *muntu's* individuality in many African ethnic groups, every child has its own name depending on the distinctive circumstances

---

67. Tangwa, "Ethical principles," 55.

68. Tangwa, *African Bioethics in a Western Frame*, 102; see also 108.

69. Tangwa, *African Bioethics in a Western Frame*, 94.

of its birth. In respect of individuality, according to Tangwa, it is one reason that for the Nso' "each child is believed to come into the world with its own name which the parents or family only try to guess (shu'yir wan)."[70] In bringing out each *muntu*'s individuality, the child's name does not originate from parents since the self of each *muntu* is a unique narrative of its own—it is a personal ontological relational reality given that the *muntu* is a historical being destined to express her individual ethical conviction and creatively develop virtues. The *muntu*'s name bears an intentioned practical message that contains a comprehensive program for her whole life. Each *muntu* has to realize this message individually but within a family community, while remaining in solidarity with the entire future, the actual time (*kakaati, kakaano*), the modern past (*mulembe*), the distant past (*edda n'edda*), and the eternal past (*emirembe n'emirembe*). Naming is not only the *muntu*'s significant locus of ethical conduct, but also the community as the programmer pronounces itself through each *muntu* in the manner that the *muntu* remains unexchangeable and fulfills uniquely irreplaceable tasks. In this relational interdependence, the *muntu* and the community are not in opposition to each other, but rather complement each other.[71] Just as the celebrated Tanzanian scholar Nyamiti rightly asserts, "the individual's relationship to the community is so intimate that he belongs to it more than the community belongs to him."[72] It is why the highest beneficiary of this relationality is not the community but rather the *muntu* herself, given that the community vindicates her creative individuality (the self). This is the reason why even when the child is an exceptional memorial of the ancestor, if given the name of an ancestor (e.g., Buyondo, meaning the combined strength of several small hammers), she must still exceptionally develop her own unique distinct personality within the community. This is not only in contrast to the Western sense of rebirth or reincarnation,[73] but also to various interpretations of African moral thought, just like Tangwa's eco-bio-communitarian bioethics, which supports the possibility of transmigration and reincarnation within and across species (at least within humans).[74] In contrast to transmigrations, individuality is so important,

---

70. Tangwa, "Bioethics: An African Perspective," 190.

71. See Bujo, *Ethical Dimension of Community*, 28,148–49; Bujo, *African Ethic*, 90–2.

72. Nyamiti, "Incarnation," 9.

73. See Masolo, "Western and African Communitarianism," 495–96; Bujo, *African Ethic*, 6, 90–91.

74. See Tangwa, "African Perception of a Person," 42.

and it is the community that remains the alternative as well as the sure way of wholly pursuing this individuality. In the mind of Black Africans, neither does the community absorb the individual nor does acquiring an ancestor's name destroy the *muntu's* self-identity, but rather, it is the self that is *promoted*. This is further backed by Baganda counsel that holds self-reliance (*okweyimirizaawo*) of the individual—a core value of *obuntubulamu*—in high esteem. The hereinafter counsel can be understood to mean that a man who drives his parents' car cannot be entitled to speak in the council of men who own bicycles since he is still dependent, a status that still qualifies him as a child. It is also why "You cannot quench your thirst with water for which you have gone begging" (*Mazzi masabe: tegamala nnyonta*); or by the same token, "You cannot drive off a leopard from your own door with the stick that belongs to your neighbor." To restrain anti-communal habits, such as individual laziness, the community puts in place tools to instill the value of self-reliance. Such tools include further counsel like: "*Omuko omwavu, y'ayasa enku mu lumbe*" (A poor son-in-law is the one who splits firewood at the funeral); "*Omwavu takwana*" (A poor person doesn't make friends—no one finds her worth befriending).[75] The various counsel is intended to encourage hard work (individuality) as well as deter the *muntu's* lazy habits of begging that don't increase the life of the community. In a nutshell, the *muntu's* identity, individuality, self, or creativity remains indispensably pivotal in the community. We can allude to the African moral reality that the overall community "does not dissolve the ethical identity of the individual but affirms it. It is not that the subject does not exist, or that the community decides in the name of the subject; but rather, a stance that takes the individual as being nothing without the community, and the community as being nothing without the individual. It is in this interactivity that the regulations of morality are often elaborated."[76] The open

75. Adages that promote individual trust, individual self-reliance and development of the individual self among Baganda: "What you yourself have sown is better than 'Give me a piece of yours'" (*Ke werimidde: kakira mbegeraako*); "When a hair tickles your nose, pull it out yourself" (*Olwoya lwo mu nnyindo, olweggya wekka*) (see Katongole, "Ethos Transmission," 103 and 107); *Akezimbira tekaba kato* ("The bird that makes itself a nest is not young"). Additional proverbial counsel to instill value of hard work and deter begging includes: *Omwavu takwana* ("A poor person doesn't make friends"—no one finds him/her worth befriending); "*Muntu* should work like a slave to eat like a chief" (*kola ng'omuddu, olye ng'omwami*). Nnabagereka and Ssentongo, *Obuntubulamu*, 29–34.

76. Ikuenobe, *Communalism and Morality in African Traditions*, 210.

question is whether the *muntu*'s identity, though not dissolved in the community, isn't simply passive, implying passive communitarianism.

Though this moral interactivity hails the *muntu*'s identity as central and increases the life force of the community, less of it bloods miscon-structions and unhealthiness towards the welfare of all. First, one may misconstrue the community to assimilate the *muntu*'s moral identity pas-sively. According to the Baganda adage, "If someone says: 'Let me leap where my brother or my friend leapt,' he risks falling deep into the mud" (*Kambuukire baaba [munnange] w'abuukidde: kwe kunywa mu ntubiro*).[77] This adage illustrates that despite the fact that African ethics does not favor individualism, it is also emphatically against passive communitari-anism. Given that each *muntu* is indispensably unexchangeable, she can only enhance the relational community only through personal creativity and individual responsibility, however, not simply by following the com-munity herd passively. Blindly following the community destroys *oneself* and the community. Moreover, the *muntu* remains accountable even for community would-be beneficial deeds undone and remaining hidden in her heart.[78] Second, less of this interdependent corporate community life would already create temptations, such as suicidal fears,[79] egoism, and laziness, which cannot maintain and strengthen the overall social unit because a lazy *muntu* cannot rightly fulfill her obligation without contributing her share towards the common pool, according to Nyerere and Mair. And if the *muntu*'s obligation to work towards a social unit lacks, the social unity itself suffers, either with lazy members or with those who are after amassing wealth.[80] It is from this Bantu perspective that for Baganda, laziness (*bugayaavu*) is a vice not because it threatens the welfare of the individual, but because it jeopardizes the social welfare of the corporate community life,[81] confirming how "A lazy person kills

77. Katongole, "Ethos Transmission," 308.

78. See Bujo, *Ethical Dimension of Community*, 148–50; Bujo, *African Ethic*, 90–93.

79. For Baganda, "suicide (*kwetuga*) has always been treated with abhorrence." Hay-don, *Law and Justice in Buganda*, 280. "For the Ganda, the act of suicide is thought to be particularly pernicious, for not only does it deprive the clan or community of one of its members, but the spirit of the dead agent is supposed to be even more dangerous than other spirits. Thus, a person who committed suicide was buried without public lamentation (*kukungubaga*) and 'there were no funeral rites . . . nor did he have a suc-cessor.'" Roscoe, *Baganda*, 20.

80. See Nyerere, *Uhuru na Umoja*, 10–11. Though an African approach to work doesn't mean accumulation of wealth in the hands of the privileged few. Mair, *African People*, 141.

81. See Mbiti, *African Religions and Philosophy*, 108.

the whole community" (*Ekitta obusenze buba bunaanya*).[82] It is also the reason some basic necessities of life have been always freely and out of generosity (*bugabi*) available to every *muntu* who, instead of choosing *bugayavu* that kills the individual silently and the entire community, industriously takes perseverance (*bunyiikivu*) seriously, thereby implicitly promoting the welfare of the whole group by explicitly promoting the *muntu*'s individual welfare. To allow access to welfare, taboos against the commercialization and ill-intended globalization of basic necessities such as staple foods, water, firewood, etc. are common among villagemates.[83] They freely give to each other, receive, and mutually enjoy fruits, such as avocado, guavas, nuts, oranges, mangoes, etc., at no cost. The aim is good health. Ill health concerns not the individual alone but the whole community. Conclusively, the social reality of the community palaver indeed supplies the relevant normative content, which the individual must not blindly or passively follow, but rather interact with responsibly, without her identity, creativity, freedom, and conscience being crushed. Having clarified the question of the individual-communal relationship and that of passive communitarianism, now I attempt to present some practical success in the practice of bioethical communitarianism in SSA.

Fruits of Black African Communitarianism: Examples of HIV/AIDs and Stigma

> From the African perspective, the Western approach to the AIDS pandemic, like many other things Western, is overly empirical, statistical and businesslike. It is a question whether all problems that face us, including HIV/AIDS, can be solved by a purely analytical method where the base-line approach is to try to reduce

82. Katongole, "Ethos Transmission," 248.

83. In recent globalization, genes and gametes are seen in the same social-cultural-economic contexts, because, it is not only biomedical technology, but also the social-cultural contexts that are unfortunately globalized in bioethical principlism. Rwiza, *Environmental Ethics*, 137–47; Sempebwa, *Reality of a Bantu*, 145. Drawing from my own experience in my natal district of Masaka, Uganda (1990s–2000s), Tangwa similarly cites a personal example of an eco-bio-communitarian life in his natal village of Shisong, Cameroon (1950s–1960s), where welfare dispensation was more concerned with the welfare of the individual than it is dreamt of in the social welfare of liberalism. Tangwa, "Biomedical and Environmental Ethics," 393–94. Tangwa describes the marriage between medicine and market, where medicine has purely become commercial and highly a lucrative activity; see his *African Bioethics in a Western Frame*, 5, 46–47, 76–79.

complex systems to constituent parts, and where treatment of the parts of necessity implies salvage for the whole. This business-like statistical analyticity may, from some perspectives, appear like the epitome of rationality, but it ignores other perspectives and other aspects of being alive and being human.[84]

The initial question is: would we take stigma—the stigma arising for example from HIV/AIDs—in biomedical ethics as an individual or communal problem, thereby demanding an individual analytic technique or a communal therapeutic palaver, that is, metaphoric, filled with narratives, parabolic, intuitive, mythic, etc.? To circumvent any stigma in an African context, we have established the group as the lasting competent, legal, and therapeutic body, which assumes a paternalistic form in order to safeguard identity and sociality in bioethical interventions. Meaning if the *muntu* is stigmatized, the relational community is likewise affected. Analogously, Tangwa thinks that on top of stigma spoiling a person's identity, it leads to the *muntu*'s social status loss. Stigma disrupts not only the tranquility of social life, but also the established hierarchical status, as well as the solidarity of the social categories of Bantu. Among these social categories, it contradicts the twin-parallel binary constructs of relational harmony, i.e., identity and solidarity.[85] This is succinctly put forward by Tangwa. For him, in African communities

> where the social socialization is communal, and a group's norm is given an overriding importance or priority in order to maintain the cohesion, solidarity and harmony of the group . . . stigma has the potential to pierce the veil of social identity and solidarity (harmony), in that stigma is constructed as a pollutant which has contaminated the bearer, making them unacceptable, contagious and injurious to the well-being of the group and its social order; this leads to exclusion and rejection, consequences that in themselves inveigh against *ubuntu*.[86]

In African logic, stigma therefore embraces a harmonious relational solidarity, i.e., a dialectic relationship that demands a balance between

84. Tangwa, "HIV/AIDS pandemic," 217.

85. Tangwa, "Ethics in Occupational Health," 7; Metz, "African and Western Moral Theories," 49–58.

86. Tangwa, "Ethics in Occupational Health," 8; see also Metz, "African and Western Moral Theories," 49–58.

individual autonomy and communal values. African wisdom deals with the moral problem (stigma) implicitly by dealing with the root cause (HIV/AIDs). It is from this wisdom that the communitarian approach has registered success with HIV/AIDs—not through the challenged efforts of medical male circumcision (which have demonstrated that men have an approximately 60 percent less chance of becoming infected with HIV through unprotected vaginal sex, while leaving the remaining 40 percent risk open?[87]) or any magical operations, be it that the promotion of condom use remains still limited—but rather through an integrative prevention through change of the individual behavioral attitude and a renewed sense of communal responsibility emerging from a shared understanding of the collective common good.

Practical examples include the HIV/AIDS prevention collective efforts in Uganda, which demonstrated a communal sense of patriotic duty in preventing the further spread of the HIV virus: hands-on communitarian bioethics, as well known worldwide. In Uganda, not only the source of information originated from but also prevention messages spread horizontally through interactive and open local social communication networks, which were masterminded by close friends and clan families (who had information of those who had HIV or had died of AIDS), local chiefs, church leaderships, home-grown musicians, and community meetings at village levels. This interactive communication and shared perception awakened Ugandans not only of how personal irresponsible behaviors raise burdensome effects (of ill-health) towards one's neighbor, the solidarity of the community, but also an appeal to fight a common enemy together through each *muntu*'s behavioral change.[88] This Ugandan

87. Tangwa, "Ethics in Occupational Health," 2; Gwandure, "Male Circumcision," 91; Safurdeen and Cheryl, "New Prevention Strategies," 268–82; Reiss et al., "When I Was Circumcised."

88. Orobator, "Ethics of HIV/AIDS," 150; Wathuta, "Uganda's Early HIV Prevention," 114–17. The clearest example of declines in HIV prevalence and changes in sexual behaviour comes from Uganda. Low-Beer and Stoneburner, "Behaviour and Communication Change," 9–21. There are distinctive patterns for communicating through social networks about AIDS and people with AIDS in Uganda. Communication programs must take root at the level of social networks working through local networks of meetings, chiefs, churches and health personnel as well as the media. Low-Beer and Stoneburner, "AIDS Communications through Social Networks," 1–13. The study shows the successes of developing a multilevel governance while integrating the national program with the community response. Low-Beer and Sempala, "Social Capital and Effective HIV Prevention," 1–18; Kirungi et al., "Preventive Sexual Behavior," 136–41. Uganda is one of the only two countries in the world that has successfully reversed the course of its HIV epidemic. Slutkin et al., "Uganda Reversed Its HIV Epidemic," 351–60.

success in the HIV/AIDS communitarian bioethics did not only further play a role in its development of an antiretroviral therapy (ART) to prevent mother-to-child transmission (MTCT) at Makerere University in Uganda,[89] but according to Daniel Low-Beer and Rand L. Stoneburner's studies, such a communitarian bioethics became equivalent to a vaccine of 80 percent effectiveness.[90] President Museveni, a foremost leader in the fight against HIV/AIDS, affirmed this position at an AIDS conference, maintaining that the best attack to the HIV/AIDS pandemic was to reassert openly and candidly how every *muntu* owed respectful accountability to her community member.[91] This resolute position in the fight against HIV/AIDS confirms the common denominator consistent in the Ugandan approach as a hands-on communitarian bioethics.

Communitarian bioethics here means persuasion through morally correct communication, consensus of Bantu (community members) through dialogue, and commitment to responsibilities towards other Bantu, which are the chief practices relied on to achieve appropriate effective social behavior and the *muntu's* indispensable contribution towards a healthy social reality. The revitalized community life is deemed more ethical as it relies on relational education in the virtues, moral persuasion, and informal social controls, without stifling individual identity, agency, and capacity for self-determination.[92] The ensuing communitarian ethical life of *none can be better than us all* emphasizes a relationally functioning life (community), a bioethics that ought not to be different for us all even amidst an HIV/AIDS insurgence affecting directly some of our members. For Bantu, HIV/AIDS manifests a poorly lived relational communitarian life, a life channeled erroneously since an individual's misdirected action kills the community too. But if comradeship among a wider network of relations is reconciled, the evil is turned down and life forces of the community are increased for all, both mystically/transcendentally and physically.[93] In contrast, debates in principlism have focused on the rights and autonomy of the individual person living with HIV,

---

89. See Namara-Lugolobi et al., "Twenty Years of Prevention," 22–33.

90. Uganda provides the clearest example that HIV is preventable if populations are mobilized to avoid risk. Uganda has shown a 70 percent decline in HIV prevalence since the early 1990s. Low-Beer andStoneburner, "HIV Declines," 714–18.

91. See Museveni, "AIDS and Its Impact," xi–xvi.

92. Wathuta, "Uganda's Early HIV Prevention," 109–18.

93. See Magesa, "Contextualizing HIV and AIDS," 20–21; Bujo, *Ethical Dimension of Community*, 189.

extending similar influential debates to other bioethical interventions such as in the African fight against the virus. In relation to such principlistic debates, "individual confidentiality has raised several unforeseen problems for persons living with HIV/AIDS, ranging from stigma and isolation to feelings of dejection as it drives them away from their families as a way of trying to keep information about their conditions confidential."[94] However, such an approach doesn't originally sit well in an African communitarian fight against HIV/AIDS. In this sense, promoting a healthier community life force as the only possible dignifying and dignified bioethical life, the communitarian bioethics goes beyond *individual confidentiality* to embrace *communal confidentiality*. Community confidentiality seeks to promote life by protecting all various forms of third parties, whereby not only the patient's life increases by disclosing HIV status to the clan-family (unless there is serious harm), but also the clan-family and corporate community's life is integrally enhanced, while less of this would be murdering the social life.

The deep significance of what I have called community confidentiality is an awareness of suffering beyond the individual that can only be described as "social death," without forgetting the ancestral aim of promoting an integral life. The effects of AIDS gets beyond the world of the living, the afflicted and inflicted, calling us to reimagine the context of HIV/AIDS by understanding the African context of social death, which connotes the centrality of awakening the ancestral spirit of promoting life amongst Bantu communal sufferings,[95] as Peter Knox relevantly notes. To highlight the centrality of ancestors in social death, Laurenti Magesa, one of the finest global scholars of African descent, also notes that "to die of AIDS is viewed as 'bad death' that may disqualify one from attaining ancestral status,"[96] or as Knox notes, "that people affected by AIDS be reconciled with their families, neighbors, ancestors and God."[97] Here re-energizing life or healing from the effects of HIV/AIDS cannot be limited to the individual sphere or the living community. Encouraging openness or if an individual confidential status is married with the community confidentiality, individual sexual behavior is encouraged to change in broader relationships desirable to ancestral

94. Ndebele et al., "HIV/AIDS," 331.

95. See Knox, *Aids, Ancestors and Salvation*, 24–25, 228. See ancestors as central to Bantu suffering on page 23.

96. Magesa, "Contextualizing HIV and AIDS," 22.

97. Knox, *Aids, Ancestors and Salvation*, 229.

ideals of promoting life for all. Among the popular protective measures in Uganda is the proverbial wisdom that "Every goat must be tethered on its peg" *(Buli mbuzi ku kunkondo yayo)*, literally promoting *zero grazing* amongst human romantic relationships in this wake of HIV/AIDS. The sapiential counsel of the maxim is to stress faithfulness to one's sexual partner where unfaithfulness will lead to community suffering too. So, understanding individual-community stigmatization is to concurrently conceptualize stigmatization by the whole global community. In the HIV/AIDS social crisis, the *muntu* turns to the accumulated ancestral wisdom. It becomes clear, then, that the apt normative content provided by the clan-family or social reality of the community palaver can only be newly revitalized if the *muntu* adjusts her behavioral attitude to living out a responsible healthy community life as prescribed by ancestors. It is only through directing her action well (individuality) that she can vitalize the community life (community confidentiality). Living out the *muntu's* community social life well, let us say cultural ancestral empowerment, is to abstain from misdirected individual sexual actions (e.g., HIV/AIDS) that would otherwise diminish community life (inclusive of no entrance to the ancestral world) in addition to individual life. So, in arresting the insurgence of HIV/AIDs, communitarian bioethics consequently avoids stigma, relationally increasing not only the *muntu's* life but also the group's integral sexual health at large. Yet spouse inheritance after death of one's spouse in order to maintain one's lineage, as earlier instituted by the ancestral wisdom of past generations and as still practiced in some cultures, no longer contemporarily promotes integral sexual life given the threat of HIV/AIDS virus. So, it would be largely simplistic if the *I's* satisfaction is not manifested in the relational *we* satisfaction. To this end, the open question is the nature and exact relationship between Bantus' understanding of relational life and their subsquent normative codes. Before we turn to this question, it is worth noting that in recent pursuits of Western principlism in Black Africa, the principle-based autonomy of non-communicative individuals (individual confidentiality) has some-what sparked off new HIV infections.

AFRICAN SOCIAL REALITY DEFINES RELATIONAL ONTOLOGY AND NORMATIVITY INSEPARABLY

The previously discussed Black African outlook to bioethics, in contrast to principlism, focused on building up a practical *wholeness of life* in a strictly non-Western principle-based rationality. Promoting the practical wholeness of life has been inductively premised on characteristics such as "vitality"/*bulamu* (Lebenswachstum), "solidarity"/*bumu* (Lebenskraft), "hierarchy"/*bukulu* (Lebensrang), and an interactive relational social reality within a vibrant three-dimensional "community"/*entababuvo* (Gemeinschaft). In this section, we argue that an African cannot think of a normative system outside her relational ontological social reality. Etymologically, the word "ontology" stems from *ontos* (the Greek genitive of ὄν is ὄντος, "being," "existence," "reality" or "that which is"), and *logia* (λογία, "word," "reason," "discourse," "logical discourse").[98] The characteristics of an African approach to bioethics are substantially informed by "convictions of their social reality of existence or being" (their relational ontological world), consequently permitting a harmony of life forces as one unified continuum. For instance, under vitality, I have discussed the normative content of *obuntubulamu* as "healthy humanness," or even defect in healthy humanness—"*Oweddemu ak'obuntu*" (You have lost humaneness)[99]—which are inseparable from Bantu relational ontological framework of life. Moreover, if we take "the Nguni word 'umu-ntu', or the equivalent Sotho word 'mo-tho', it can be translated as a 'person.'"[100] The word *muntu* (person) has both ontological and normative connotations.[101] In this sense, healthy humanness can integrally increase or decrease the vital energy depending on how the *muntu* participates in the life principle, i.e., the relational ontological existential reality. Let me explore this inseparability (relational ontology and normativity) with the idea of personhood.

98. See εἰμί in Liddell and Scott, *Greek-English Lexicon*, and λόγος, λέγω in their *Intermediate Greek-English Lexicon*.

99. Karlström, "Culture and Democratisation in Buganda," 485–505.

100. Molefe, *African Personhood*, 23.

101. Ramose, *African Philosophy*; and LenkaBula, "Beyond Anthropocentricity-Botho/Ubuntu," 375–94.

Relational Ontology and Normativity are Inseparable: Idea of Personhood in Black Africa

For an instantaneous illustration, I quickly indicate two accounts of principle-based conceptions of a person from the Western thought. One account is from Tristram Engelhardt's influential book *The Foundations of Bioethics*. He gives this description of persons:

> Persons, not humans, are special . . . Morally competent humans have a central moral standing not possessed by human fetuses or even young children . . . Only persons write or read books on philosophy. It is persons who are the constituents of the secular moral community. Only persons are concerned about moral arguments and can be convinced by them. Only persons can make agreements and convey authority to common projects through their concurrence. To choose, to make an agreement, is to be conscious of what one is doing. It requires the self-reflexivity of self-consciousness. Otherwise, there is a happening, not a doing.[102]

Engelhardt goes on to define persons as "entities who are self-conscious, rational, free to choose, and in possession of a sense of moral concern"[103] and concludes that human "fetuses, infants, the profoundly mentally retarded, and the hopelessly comatose" are nonpersons, having no "standing in the secular moral community" and falling "outside of the inner sanctum of secular morality."[104] The second account is of John Locke, which is no different, if I would briefly describe it. For him, a person is a thinking, intelligent being that has reason and reflection, in different times and places.[105] To this end, we immediately notice a relationship to Tangwa's interpretation of African moral thought:

> The African concept of a person is not and in fact cannot be any different from the Western concept, unless there is some linguistic problem of translation or interpretation. *Both understand a person to be a fully self-conscious, rational, free, and self-determining being.* If the African perception of a person differs from the Western perception, this is not because it does not socialize the various developmental stages of a human being or

102. Engelhardt, *Foundations of Bioethics*, 135–36.
103. Engelhardt, *Foundations of Bioethics*, 136.
104. Engelhardt, *Foundations of Bioethics*, 139.
105. Locke, *Human Understanding*, 188.

qualitative differences based on the degree of attainment of positive human attributes or capacities, but rather because it does
not draw from these facts the same conclusions as are drawn in
Western ethical theory.[106]

Both Locke and Tangwa insinuate that if one stopped thinking due
to some transient ischemic attack, she would no longer be a valid person!
Again, they seem to imply that human babies or patients in the state of
transient global amnesia, since they cannot reflect across different times
and places, are not valid persons! There is some danger of a strong sense
of anthropocentrism. Locke's theory assigns value and moral status only
to the lives of those with the capacity for reciprocity and understanding,
as Sarah Chan and John Harris note. Thus the principle of respect for
persons applies uniquely to just a particular class of entities.[107] Tangwa's
first account in and of itself does not and vaguely so entails an African
approach to personhood because it incorporates only *moral agency* without catering for *moral patienthood*. Moral patients are legitimate entities
of moral concern whose well-being and interests need to be taken into
moral consideration when decisions about them are made, while moral
agents are entities who can be held morally responsible and accountable
for their actions.[108] In this scenario, entities that are incapable of reflection would scarcely therefore qualify as moral agents. Tangwa's second
account, however, moves a step further from Locke's principle-based
perspective of confining personhood to the exercise of reason (moral responsibility, liability, and culpability), which is suitable for moral agents,
to appreciating personhood as moral worth acceptable within a relational
community, an account that necessarily incorporates both moral agents
and patients:

> The Western conception of a human person, as a category or
> subset of human being, is appropriate only for the ascription of
> moral responsibility, liability, and culpability rather than for the
> ascription of *moral worth*, desert, eligibility, or acceptability into
> the moral community made up, as it necessarily is, of *both moral
> agents and patients*. By contrast, the African or Nso' perception
> of a human being, which applies to the *elusive entity underlying
> all categories, stages, and modalities of a human being*, although

106. Tangwa, "African Perception of a Person," 42. Emphasis added.

107. Chan and Harris, "Human Animals and Nonhuman Persons," 307.

108. See more in DeGrazia, *Taking Animals Seriously*, 203; and Bradie, "Moral Life
of Animals," 564.

conceptually less neat and analytically less firm, seems to accord better with our ordinary moral intuitions and sensibilities
and is thus more appropriate for nondiscriminatory morality in
general.[109]

In the relational account, being a subject of moral worth within a
moral community envisions a moral state of personhood that presupposes either healthy humanness (being a person) or lack of it (being a
nonperson), meaning that leading a genuine practical life by the standard
norms is saliently inherent in the system of values; i.e., it is a normative
ethic that is inseparable from the entire sphere of ontological relatedness.[110] Some of the leading African ethicists—Ifeanyi Menkiti, Kwame
Gyekye, Motsamai Molefe, Anthony Oyowe, Kevin Behrens, D. A. Masolo—identify themselves with three ideas of personhood, in reference to
personal identity, moral status, and moral virtue.

Personhood qua Personal Identity

The ambiguity in different scholarship—as we already extensively discussed in relation to bioethical principlism—lies in the fact that they ada-
mantly decline to define personal identity outside the *muntu*'s relational
ontological community realities. At least as presented by Bujo, Tangwa,
Mbiti, Kwasi Wiredu, Gyekye Kwame, and Ifeanyi Menkiti, personhood
qua personal identity depends necessarily on ontological socialization of
the community, given that the *muntu* is cultural by nature;[111] viz., without
and outside of culture, socialization is beyond the bounds of possibility.
Along this culture of socialization, Menkiti defines the idea of personhood *qua* personal identity in terms of the "environing community,"
namely, the socio-cultural facts and resources necessary for human socialization.[112] For Gyekye, a person is a communitarian being by nature
because, before embracing some culture, she is necessarily born into
an existing ontological reality of a community without any (voluntary)
choice. This communitarian social reality implies that she should not
live in isolation because she is not contingently but necessarily oriented

109. Tangwa, "African Perception of a Person," 42–43. Emphasis added.

110. See Metz, "African Moral Theory," 83; Wiredu, "Oral Philosophy of Personhood," 15.

111. See Molefe, *African Personhood*, 62; Wiredu, *Cultural Universals and
Particulars*.

112. See Menkiti, "Person and Community," 170; Molefe, *African Personhood*, 43.

towards and constituted by her relatedness to this ontological reality of life.[113] But then, would we say that personhood is external to the moral agent? Thinkers like Oyowe, Molefe, and Yurkivska in some sense understand personhood to be external. When the *muntu* is born, she is in some sense not yet a person. The *muntu* being born is necessary but an insufficient condition for possessing or guaranteeing personhood. This creates a difference between being *human* and being a *person*. Hence, it makes sense to think of personhood as somewhat externally achievable by purely responding positively to the process of socialization offered by the social reality of the community.[114] Given that our being persons is substantially of our own making, the necessity of external socialization aids us transcend from moral neutrality as we were created, navigate through moral agency, until we responsibly achieve moral completeness.[115] In aiding human agency, however, socialization plays either a constitutive[116] or an instrumental role.[117] Constitutively, the relational socialization between the individual and community remains forever dialogical, given that it inevitably insulates the identity of the person while allowing the community to indispensably dependent on this constitutive relatedness. Like we already stressed, neither the individual nor the community is prior to the other.[118] Instrumentally, "the community plays a vital role as catalyst and as a prescriber of norms,"[119] as Menkiti notes. Illustratively, employing the chemistry analogy of a catalyst, the community serves as a means to enhance the whole process of becoming a person, but it is not part of the final product. In a different voice, the community prescribes norms as moral guides,[120] gives content, on which the *muntu* works. Whether constitutively or instrumentally, socialization cannot be understood

---

113. See Gyekye and Wiredu, *Person and Community*, 105.

114. See Oyowe and Yurkivska, "African Personhood," 88–89; Molefe, *African Personhood*, 5–6, 18, 57; Oyowe, "Personhood and Strong Normative Constraints," 784. On the difference between humans and persons in the African worldview, see Behrens, "Two 'Normative' Conceptions of Personhood," 103–4; and Metz, "Philosophy of D. A. Masolo," 12.

115. Humanity is impossible without socio-relationality. Munyaka and Motlhabi, "Socio-Moral Significance," 324–31; Menkiti, "Person and Community," 165–66; Molefe, *African Personhood*, 19, 25–26.

116. Metz, "Two Conceptions of African Ethics," 141–62.

117. Molefe, "Individual or Community."

118. See Eze, "What Is African Communitarianism?," 386–99.

119. Menkiti, "Person and Community," 172.

120. Molefe, *African Personhood*, 27.

outside the relational ontological existential realities of Africans, more so the normative idea of personhood *qua* personal identity. In trying to streamline further the normative notions of personhood, there are some pertinent community realities in relation to the *muntu*'s moral state and moral achievements. Examining the works of Motsamai Molefe (2020), Anthony Oyowe (2018), and Kevin Behrens (2013), we come to discern further two more normative notions of personhood in African moral thought: the patient- and agent-centered notions of personhood.

Personhood qua Moral Status (Dignity): The Patient-Centered Notion

This idea of personhood *qua* moral status (dignity) relates to being a member of the species *Homo sapiens*. To have human dignity is to have the relevant human ontological features/capacities that make the acquisition of virtues possible. According to Gyekye, "the human person is considered to possess an innate capacity for virtue, for performing morally right actions and therefore should be treated as a morally responsible agent," which is why "the pursuit or practice of moral virtue is held as intrinsic to the conception of a person."[121] This means the patient-centered notion attributes moral status to an entity because it has a valuable intrinsic value in its own right, whereby having moral status creates moral obligations to moral agents.[122] It is why a Black African will judge the *muntu*'s conduct while implicitly suggesting a conception of moral personhood, in the sense that a person is defined as having "moral sense."[123] In this regard, the relational ontological moral qualities (moral sense) convey the idea of moral status or dignity. It is again this moral sense, depicted in the patient-centered notion of personhood *qua* moral status or dignity, that Menkiti associates with the idea of justice.[124] He confines justice to human persons since they are the sort of entities that are owed the duties of justice, the possessors of the rights, which possession depends on a capacity for moral sense, a capacity made evident by a concrete exercise of duties of justice towards others in the ongoing daily life relationships.[125]

121. Gyekye and Wiredu, *Person and Community*, 109.
122. See Molefe, *African Personhood*, 3–6.
123. Gyekye and Wiredu, *Person and Community*, 110.
124. See Molefe, *African Personhood*, 45.
125. See Menkiti, "Person and Community," 177.

As possessors of rights and a capacity to exercise duties of justice, Menkiti grounds moral status or dignity on the human capacity for moral sense, i.e., the capacity to pursue (achieve) moral personhood or live a life of moral excellence, a capacity he *doesn't extend to non-human animals*.[126] Menkiti so observes that "the constitutive elements in the definition of human personhood have become blurred through *unwarranted extensions to non-human entities*."[127] This understanding is indistinguishable to Menkiti, Gyekye, and Ikuenobe. For Gyekye, however, though human infants are not (yet) considered as moral agents who are morally capable (in actuality) of exercising moral sense; they are morally capable of exercising it in potentiality.[128] More clearly, Gyekye argues that beings that will in future be able to pursue personhood (moral perfection) have moral status, e.g., children, but non-human animals have no potentiality for moral sense and will never be able to naturally pursue personhood in the future, so they do not have moral status.[129] It means beings that are not able to pursue personhood in the future cannot anticipate having moral achievements of their own. Comparably, in Ikuenobe's recent studies, he defends human dignity hinged on moral achievements *qua* personhood.[130] Ikuenobe defines dignity in terms of "the capacity for, and manifestation of, self-respect and respect for, and responsibility to, others," taking a person with dignity to be the "one who uses one's capacity properly to promote harmonious communal living."[131] For him, even though ontological capacities are indispensable, they are inadequate for human dignity: "Human capacities are only an instrumental good; they are only a means for a good life,"[132] suggesting that human dignity or status is achieved through a doer's quality of moral conduct. Ikuenobe's perspective does not tally well, for instance, with the Universal Declaration of Human Rights, which is *egalitarian* in nature, and in which, based on innate ontological properties, all human beings equally merit moral consideration. Indirectly, Ikuenobe is agitating for an *inegalitarian* (moral performance as the basis for dignity) ethical world, a perspective that is

126. See Molefe, *African Personhood*, 46.

127. Menkiti, "Person and Community," 177. Emphasis added.

128. See Gyekye and Wiredu, *Person and Community*, 111.

129. See Molefe, *African Personhood*, 13–14, 84–85.

130. Ikuenobe, "Communal Basis for Moral Dignity," 437–69; "Personhood, Dignity, and African Communalism," 589–604.

131. Ikuenobe, "Communal Basis for Moral Dignity," 438.

132. Ikuenobe, "Communal Basis for Moral Dignity," 461.

unlikely to succeed in true African realities. Ikuenobe's *inegalitarianism* is comparably no different to Gyekye's emphasis of the actual use (now and there and then) of one's ontological capacities in order to achieve personhood *qua* moral status or dignity. Even when the mere possession of these capacities is instrumentally good, it is the actual positive use that is relevant in securing dignity or status.[133] Gyekye's argument too fails in some moral communities. If a *muntu* has diminished autonomy or is in any way incompetent in entering any contracts in the community with other competent community members,—or think of infants who cannot actually exercise their human capacities—would she have neither dignity nor moral status? In trying to find a solution, Ikuenobe argues that the idea of respecting unconditionally those who are not capable of acting to earn respect is supported by the moral principle of "ought implies can," which indicates that you cannot hold people responsible for what is impossible for them.[134] Precisely, even though we find difficulties in separating the normative notion of personhood *qua* moral status or dignity from the actual positive use of ontological capacities, the principle of "ought implies can" is claimed to secure unconditional moral approval for incompetent members of our communities since it is impossible for them to exercise those qualities *in actuality*. But even if incompetent members would be catered for, beyond them, some humans cannot positively use their capacities in actuality, which still challenges their relevant acquisition of dignity or status. If we are to go with this special interpretation of African moral thought about personhood, which seems indifferent to bioethical principlism, it would remain unsuccessful amongst *moral patients* (e.g., severe dementia) whose quality of conduct and the capacity for actual use of virtue is diminished to secure human dignity or status. Specific examples traceable in our previous discussions in bioethical principlism include taking cognizance of a surrogate as the rightful or lawful decision-maker in instances where the patient cannot positively exercise her actual capacities, which actually "entails that the incompetent individual has lost some rights of decision-making, and in this respect the individual's moral status is lower than it previously

133. See Molefe, *African Personhood*, 42–43.

134. Further on the moral principle of "ought implies can," see Ikuenobe, "Communal Basis for Moral Dignity," 464; Blum, "Kantian versus Frankfurt," 287–88. The principle of "ought implies can" is largely attributed to Immanuel Kant's works *Religion within the Boundaries of Mere Reason* and *Critique of Pure Reason*. For Kant, if we ought to be better humans, we must be capable. The ought must be possible under natural conditions.

was."[135] Other cases of moral patients in our communities include pregnant women who are brain-dead[136] but whose biological properties are artificially supported as a means to put to use their positive ontological qualities until the fetus is born. Principlists wouldn't think of dead people as having actual moral status. Besides, supporting a brain-dead pregnant woman's body with or against her advance directives to stop all technology at the actual point of her death presupposes her having a lower or no moral status (dignity), while the fetus presumably has a higher or partial status, given that her body is subjected to extreme measures (activating it to positive actual use) only to benefit secondary parties like the fetus, the woman's partner, relatives, or the next of kin.[137] The moral question remains: is the fetus the only community member in this bioethical technological intervention with actual moral status and rights at the point of the pregnant woman's brain death? The distinction in the quality, actuality, and potentiality of the moral status between the fetus and the mother is further understood differently in African moral thought, as I will illustrate in the subsequent paragraphs, but firstly I address the technology of transgenic animals.

Some of the principle-based artificial technologies generate genetically modified animals as new life forms by transgenesis, which morally challenges assignment of moral status to species. Transgenesis is creating forms of transgenic animals, hybrids, and chimeras referred to as "genetically modified animals" (GMAs). Transgenic animals are created by transferring genes from one species to another, hybrids by mixing

---

135. *PBE*, 66.

136. A thirty-year-old woman suffered massive brain injuries after a motor vehicle accident at fifteen weeks' gestation. The patient was diagnosed as brain-dead on her tenth hospital day. She was supported with intensive care for 107 days after this diagnosis, and a normal 1555-g male infant was delivered at approximately thirty-two weeks' gestation by repeat cesarean section. Bernstein, et al., "Maternal Brain Death," 434–37. In October 1992 a young woman died in a car accident. She was pregnant and her fetus appeared to be unhurt (the "Erlanger Baby"), so a decision had to be made: should the mother's body be artificially supported in order to give the fetus a chance to live? Anstötz, "'Erlanger Baby,'" 340–50. For more on pregnant women who have been supported after brain death until successful delivery of their infants, see Powner and Bernstein, "Pregnant Women after Brain Death," 1241–49.

137. Sustaining the pregnancy is not always a clear gain to the born child and because it may impose a substantial burden on the benefactor. Lindemann, "Postmortem pregnancy," 247–67; *PBE*, 66. An action-centered theory of right action associated with personhood states that "an action is right just insofar as it positively relates to others and thereby realizes oneself; an act is wrong to the extent that it does not perfect one's valuable nature as a social being." Metz, "African Moral Theory," 330.

the sperm of one species with the ovum of another, while chimeras by mixing cells from the embryo of one animal with those of a different species.[138] Though we note how GMAs would not only be useful for biomedical research, invention of disease-resistant species, but also generation of species that produce products that are useful to humans and xenotransplantation (human genes as introduced into pig embryos to generate organs that will not elicit immune responses when transplanted into humans), one wonders whether the new forms of GMAs would not morally change the state of moral status since some cells are strikingly of one species and others of another species. Even if we have with less moral difficulty embraced hybrids in biotechnological artificial intelligence, such as hailing computers as chess grandmasters and agricultural crops as mixed species (as we saw their rapid infiltration in Africa), the fusion of human and non-human embryonic material (admixed embryos) remains a disturbing moral chaos. For instance, if a GMA chimera would be born from human and ape/chimp cells, it could possibly act in human-like ways since the chimp/ape's brain would be developed under the influence of the human cells.[139] We could initially argue, given such a substantial human biological contribution, the GMA (since some GMAs have humanized brains) might have, for instance, markedly human properties for communication, procreation, or moral performance. Moreover, we are continuously committing a naturalistic fallacy and so deepening the moral confusion since we actually eat animal meat without animals actually eating our flesh! Again, we shouldn't forget that we have already assented to *xenotransplantation*, with a subsequent likelihood of eating human parts (or potential persons) from these GMAs. The question is

138. See Savulescu, "Genetically Modified Animals," 641, 643–44. Engraftment of human neural stem cells into the brains of nonhuman primates might alter the cognitive capacities of great apes and monkeys, with potential significance for their moral status. Greene et al., "Human-Non-Human Primate Neural Grafting," 385–86, On March 9, 2009, President Barack H. Obama issued Executive Order 13505, "Removing Barriers to Responsible Scientific Research Involving Human Stem Cells." National Institutes of Health, "Guidelines for Human Stem Cell Research," 32170–75.

139. See *PBE*, 69; Savulescu, "Genetically Modified Animals," 643–49. In the wake of "human-animal hybrids," Irv Wiessman and his colleague Mike McCune successfully created the SCID-hu mouse, a mouse that lacked its own immune system but depended on the transplantation of human bone marrow and other integrated human tissues or organs to have a functioning human immune system. Wiessman's Human Neuron Mouse focused on humanizing the brains of human/nonhuman chimeras, with a possibility of creating a mouse with some cognitive abilities, which creates a moral confusion. See Greely, "Human/Nonhuman Chimeras," 671–75, 676–89; see also McCune, "SCID-hu Mouse," 74–76.

whether anthropophagy or cannibalism or human-flesh-eating won't soon be morally permissible! The actual moral status of such an actual transgenic chimera, such as a human-chimp/ape chimera, remains a challenge since we currently lack adequate ethics to evaluate such personhood or even their creation.

If we are to focus on actuality or potentiality in understanding personhood *qua* status or dignity, this would make us to ask further if aborted babies are not then potential persons! The challenge encountered by an African approach to bioethics along its relationship with Western principlism is a perennial obsession with the concept of a person, to an extent that there is clear segregation in bioethics as far as assigning of moral status and moral treatment between adult humans, fetuses, human infants, mentally defective people, brain-dead patients, animals, and plants. Abortion "on demand," for example, may thus be justified on the grounds that the fetus, unlike the pregnant woman, is not a person.[140] One of the majorly maintained characteristics on which personhood of the above is hinged is moral status. For Tangwa, "to have moral status or worth is to count/matter, to have value, *from the moral point of view or perspective. The moral point of view or perspective is that from which anything is viewed and considered in its own right as an end.*"[141] If we pick out an example of an embryo as considered from the moral point of view and perspective, it cannot "pretend to have greater moral value/worth than, say, a zygote or blastocyst," and even if we relied on its greater attributes, they "might be significant and important only from the point of view . . . not from that of moral worth/value."[142] Like Tangwa interprets African non-dualistic wisdom, "a *morally significant* line of demarcation cannot be drawn between embryos and other categories of humans."[143] As far as upholding the moral status of embryos is concerned, in re-emphasizing the difference between African moral thought and that of bioethical principlism, again for Tangwa, the "African conceptions and attitudes, which are not disposed to admitting any significant difference between the moral status of the embryo and that of any other human being *at any other stage of its maturation,* would be markedly different from those of the dominant conceptions and attitudes of the industrialized

---

140. Tangwa, "African Perception of a Person," 40.

141. Tangwa, "Moral Status of Embryonic Stem Cells," 451. Emphasis added.

142. Tangwa, "Moral Status of Embryonic Stem Cells," 451.

143. Tangwa, "Assisted Conception," 297. Emphasis original.

Western world."[144] This can be well demonstrated in Tangwa's article "Moral Status of Embryonic Stem Cells: Perspective of an African Villager," in which he finds a logical error in taking a fetus as a potential human being. Unquestionably principle-based scholarship has contended that a sperm is only a potential human being, whose potentiality can only be actualized if it fuses with an ovum/egg. Even if we were to equate potentiality and actuality, whereby embryos are viewed as potential human beings and only prescribed a full moral status on turning out to be actual human beings who can consciously express their actual interests, then we are most likely to take on an individual spermatozoon or an ovum as a potential human being. Tangwa's reaction is: until something happens that causes the ovum and sperm to join, the danger will forever only remain a danger, an actualizable but unactualized potentiality; but if fertilization ever takes place, to a Black African, a human being has been irreversibly created.[145] This created human being can never be a potential human being again because nothing can be both in potentiality and in actuality at the same time in the relational non-dichotomized ontological sense of an African community. Even if fertilization or cloned human cells were extra-corporeally achieved—irrespective of how they come about or the intention of their creator—through reproductive mechanisms that are still criticized as unethical, this has nothing to do with the metaphysical moral status of the entities created in an African approach to biomedical interventions.[146] In contrast, Tangwa holds: "In particular, the differences between, say, a mentally retarded individual or an infant and a fully self-conscious, mature, rational, and free individual do not entail, in the African perception, that such a being falls outside the 'inner sanctum of secular morality' and can or should thus be treated with less moral consideration."[147] Given the non-dualistic thinking of African moral thought, the *potential*-vs.-*actual problem* fails. Yes, the embryo is a *potential adult*, but Black Africans cannot make the reckless logical mistake of considering it as a *potential human being*, because it is already human. In the African non-dualistic emphasis of moral status and worth of humans, Tangwa approaches all humans with great and uncompromising moral reverence: "A human being is a human being,

144. Tangwa, "Assisted Conception," 304. Emphasis added.

145. Tangwa, "Leaders in Ethics Education," 99–100; Tangwa, "Moral Status of Embryonic Stem Cells," 455.

146. Tangwa, "Moral Status of Embryonic Stem Cells," 456.

147. Tangwa, "African Perception of a Person," 42.

purely and simply by being a human being."[148] The reverential relational respect has no room for dichotomizing or individuating attributes: "If any moral status or worth can be assigned to any human being, it is by virtue of the simple fact of being a human being and not by virtue of possessing any particular individual attributes or characteristics, attributes and characteristics that we gain or attain and lose at different points of our lives as humans."[149] Assigning lower or different statuses to mentally defective people or brain-dead patients given their less use of their actual human capacities is unethical since they can never be potential humans in the African relational moral sense. They are forever humans, and they must be unconditionally accepted in the relational community. And what they are does not hinder them from their interests being incorporated in the relatedness of others. Even if we were to categorize some of our fellow human beings as entities with lower status or less life or no life at all, then as moral patients their overall subsisting interests still need to be considered. Tangwa is quick to argue that "it needs pointing out that nonconscious, non-sentient living things and even inanimate things should not so easily be brushed aside as having no interests, once it is admitted that interests need not consciously be formulated, let alone expressed"; metaphorically, it can be argued that "every living thing has an interest, a kind of metaphysical interest, in continuing to live and even that every existent (being), living or non-living, has an interest in continuing to be (subsisting)," to an extent that even if it were an animate, namely, "matter . . . it can change from form to form, cannot completely be destroyed or annihilated."[150] In this regard, the distinction between animates as moral agents and inanimates as moral patients makes no impact. Put differently, our discourse goes beyond any anthropocentric hierarchical superiority since we have argued that any distinction between

148. In Lamnso', Tangwa explains his position with this saying: Wan dzë wan a dzë lim Nyuy ("A baby/child is a baby/child, a handiwork of God"). The saying signifies the unconditional acceptance of a neonate, irrespective of how it comes about, no matter how it is, no matter what its particularizing and individuating physical and mental attributes. Wan dzë wan leads directly to wir dzë wir at the level of the adult human being. Wir dzë wir can best be rendered into English as "A human being is a human being is a human being, purely and simply by being a human being." Logically, for Tangwa, "wan dzë wan and wir dzë wir may sound like contentless tautologies, but within their linguistic and cultural universe, they connote the reverential respect with which anything human is approached." Tangwa, "African Perception of a Person," 39; Tangwa, "African Sociocultural Practices," 58–59; Tangwa, "Moral Status of Embryonic Stem Cells," 453.

149. Tangwa, "Moral Status of Embryonic Stem Cells," 453.

150. Tangwa, "Moral Status of Embryonic Stem Cells," 454–55.

the moral agent and the moral patient is totally meaningless. It is the reason African scholarship focuses on *telos* in awarding moral status in relation to the African teleological understanding of existence. Being or existence is relationally interrelated with purpose since each entity has a vital force in its own right (vitology) and so granting moral status at least, on the basis of *telos* as realized differently in different beings hierarchically, is plausible. Whether living or non-living, all beings have purpose for existence (*telos*) and so pursue teleological ends, say, positively or negatively, well-being or death.[151] In the first instance, it is unthinkable to eliminate what is human from what is equally human, viz., from the entire relational human family. This background thesis frustrates efforts of artificial technologies that dualistically allot higher or lower rights, statuses, complicated transgenic statuses, or exceptional qualities to different fetuses or human beings. Moreover, it basically matters less whether the attributes or interests are developed or not developed or even lost, but what remains morally significant is that the moral worth or value remains intact throughout as we discussed in the case of the embryo. So, to stress moral value or moral worth of every entity as intrinsic is far beyond simply settling for any utilitarian instrumental value or economic pursuits that stress quality control of babies.

In critiquing commercialized artificial reproductive technologies (ART), African scholars partly rely on African communal *moral imperatives*—whose moral statements are characteristically evaluative, restraining, directive, and similar to a doctor's medicinal prescription[152]—which emphasize that all children (the hope of the future community life) come from God, as the *Lamnso'* language wisdom confirmed above: "*Wan dze wan a dze lim nyuy*" (A child is a child, the handiwork of God). That being so, it doesn't matter what type of children, conceived or born in whichever manner and state; all are to be unconditionally, lovingly accepted and

151. Chemhuru, "African Teleological," 44–45, 48–49; Oruka and Jump, "Ecophilosophy," 117.

152. Some examples of African moral imperatives are to be specifically found in Baganda. For them, the nearest equivalent to the English "ought'" is *wasaana* (it is desirable), *kirungi* (it is good) or *kibi* (it is bad). Examples of moral statements: *Kirungi okusaamu ekitiibwa abantu abakulu* = "It is good to respect the elders"; *Kibi okutta omuntu okujjako mu lutalo* = "It is bad to kill another man except in war"; *Osaana okuba omuzira* = "It is desirable that you should be courageous". Other moral imperatives are suffixed with the term *nga* to indicate a repetitive or habitual action. *Tobbanga* = "You shall *never* steal." *Toyendanga* = "You shall *never* commit adultery." *Oyambanga abali mu buzibu* = "You shall *always* help those in need." See Sempebwa, *Reality of a Bantu*, 141–44. Emphasis added.

welcomed in the community since they come from the hand of God.[153] African relational community is not concerned with the quality control or quality increment, or better-quality status of embryos or babies born, since this is not a deliberate work of human hands. It is the reason why communal moral imperatives bestow greater value not on some, but on all children—since marriage for mere companionship, without fulfilling the obligation of procreation, is rare—even on handicapped children, if they so arise. Within Nso' culture, for example, a handicapped child is considered *mbuhme*, a special gift (of God), with unusual psychic powers, and often thought of as a messenger of God or disguises of spirits. If any technologies likely to assist the birth of healthy babies were to be embraced in African contexts, they would be morally misguided if they were presented as a matter of seeking a child of a certain higher (better) quality than another (maybe with trade and commerce aims)! Pursuing children with economic marketable propaganda under the influence of utilitarian calculative approaches to morality doesn't sit well in an African context, since a Black African's interactive communal life can't claim that one human life is of a better or higher quality status than another.[154] Likewise, human life or babies cannot be considered the *handiwork of any human* where we could, at our privileged liberty, throw away surplus embryos after sorting out our desired qualities. Over-stressing quality control, higher statuses, or economic terms creates danger of losing the positive actual unique community component of integral development that Black Africans substantively believe their foreparents fostered as the greatest gift to the world: a more actual human face as humaneness itself.[155] In African wisdom, since life remains a non-dualistic unity and always an unbroken constant in the three-dimensional community, African scholarship confirms that a moral doer or agent can do moral good or evil irrespective of whether the patient of her action is a person, a non-human animal, a plant, or even an inanimate thing. It means reciprocity isn't a moral issue. The heavy burden of moral liability, culpability, and

---

153. See Tangwa, *African Bioethics in a Western Frame*, 72; Tangwa, "African Sociocultural Practices," 58–59.

154. See Tangwa, "Biomedical and Environmental Ethics," 391; Tangwa, "African Sociocultural Practices," 59; Other shortcomings in human creation are present in Ganda people, which include: *nnamagoya*, an unpigmented individual; *nnamagalo*, a six-fingered person; *kigalla*, an individual who is hard of hearing. Merkies, *Ganda Classification*, 85.

155. See Gyekye and Wiredu, *Person and Community*, 208; Biko, *I Write What I Like*, 74.

responsibility rests not on whose status is morally special, qualitative, or higher, but who is morally liable. The question of reciprocity is answered in the sense that every entity collectively shares in the hierarchized relational responsibility and also participates in the hierarchical order of forces, even when each has a unique *telos* achievable hierarchically and differently. Even though as human beings we carry the whole weight of moral responsibility,[156] as Tangwa insists, no bioethical intervention can be efficacious within the African context unless it respects the order of hierarchical forces in their integral collective participation in the life principle. In this wisdom, every individual has an inalienable dignity, but as contently prescribed by the relational community. For instance, a handicapped individual becomes a person thanks to other human beings, just as these become persons thanks to the handicapped.[157] The *muntu* can perish when she is deprived of an opportunity of becoming a person in the community (because of ART). Expressing quality control through economic terms would be certainly accentuating the *mechanized production* of babies instead of reproduction as Tangwa intones:

> Above all, the wider implications of ART—its mechanization of reproductive processes, its severance of the link between "love-making" and "baby-making," its dispelling of the mystery of procreation through human mastery, its de-mystification of motherhood, so central to the status of women in African cultures, its reduction of reproduction to production and the human control and obsession with quality implied, etc.—would be hard to harmonize with the most general of African metaphysical and religious conceptions and beliefs, value systems, ideational and ideological thrusts, customs and consolidated practices.[158]

Human life remains morally reverenced beyond any justifiable mechanization or production philosophies. In other words, Tangwa is trying to relax ethical issues connected with the mechanization of life and its natural processes, which processes threaten to turn human reproduction into mere production.[159] If bioethical principlism embraces mechanized production, it dichotomously assigns non-personhood to

156. Tangwa, "Biomedical and Environmental Ethics," 388.

157. See Tangwa, "African Perception of a Person," 40; Bujo, *African Ethic*, 90–91.

158. Tangwa, "African Sociocultural Practices," 58.

159. See Tangwa, "Biomedical and Environmental Ethics," 391–92; Tangwa, *African Bioethics in a Western Frame*, 161–62.

some while awarding personhood to others, making no parallel moral significance in a true African moral community. If bioethical principlism remains a framework that robs the community of its integral positive non-dichotomized (between potential and actual, agent and patient, higher status and lower status) use of all ontological properties to secure personhood *qua* dignity or status, even within specific cases of diminished autonomy (the brain-dead patient), an entity's status or dignity remains intact. Moreover, as far moral liability is concerned, the hierarchical order of life forces does not stress reciprocity of responsibility amongst created entities but rather stresses that each entity has a *telos*, which according to the logic of Black Africans can originally be realized in respect to the hierarchical order of life forces that incorporates every entity, while secondly, the *telos* can be realized differently by each entity without forgoing the relational interaction between these entities. Moral liability is relationally shared on top of having an individual dimension.

Personhood qua Virtue: The Agent-Centered Notion

The agent-centered notion of personhood (*qua* virtue) is performance-based, reflecting the quality of the agent's moral excellence (virtuous character). In Menkiti's chapters "On the Normative Conception of a Person" (2004) and "Person and Community in African Traditional Thought" (2009), together with Wiredu's "An Oral Philosophy of Personhood: Comments on Philosophy and Orality" (2009), the normative concept of personhood is emphasized as a moral achievement. The normativity is not property-based (capacities) as in the patient-centered notion, but rather performance-based (moral achievement of the *muntu*).[160] To ground a performance-based personhood of the *muntu*, for Menkiti, the "word '*muntu*' includes an idea of excellence, of plenitude of force

---

160. Normative conceptions of personhood rely on moral achievements; see Menkiti, "Normative Conception of a Person," 324–31; Menkiti "Person and Community," 162–67; Wiredu, "African Philosophy in our Time," 17; Behrens, "Two 'Normative' Conceptions of Personhood," 103–5; Metz, "African Moral Theory," 331. A moral agent is commended or highly praised for her quality conduct; see Tutu, *No Future without Forgiveness*, 35; and Wiredu, "Oral Philosophy of Personhood," 15. Moral achievements of personhood reflect a virtuous character and moral excellence. Eze, "Decolonisation of African Political Philosophy," 1–17. What it means to say that some being has moral status: its interests have some moral weight; see DeGrazia, "Moral Status," 181–98; Oyowe, "Personhood and Strong Normative Constraints," 783–801; Molefe, *African Personhood*, 3–6; Metz, "Philosophy of Masolo," 12–13.

at maturation."[161] Specifically, Menkiti's association of the term "excellence" with the personhood of the *muntu* further implies that becoming a person is essentially associated with developing (or maturing in) virtue.[162] Personhood *qua* virtue brings out the view that humans perfect their humanity as genuine bearers of moral virtue or human excellence (maturation).[163] For him, this processual transformation (actualizing raw human capacities) culminates into moral excellence or maturation that is equivalent to "a widened ethical maturity," displaying *other-regarding* virtues of generosity, sympathy, compassion, hospitality, friendliness, solidarity, etc., which are more familiar to Africans. The *muntu*, according to Gyekye, humanizes the "moral practice" of these virtues[164] by exhibiting a sense of ethical maturity as well as conforming to gender-neutral social rules, as Oyowe and Yurkivska note.[165] As a moral prerequisite for the *muntu*'s maturation, the emphasis is on the normative content provided by the social realities of the communicative relational palaver. These include "the rituals of incorporation and the overarching necessity of learning the social rules by which the community lives, so that what was initially biologically given can come to attain social self-hood, i.e., become a person with all the inbuilt excellencies implied by the term."[166] Here selfhood is the ultimate goal of a person, self, or human; when the *muntu* becomes a full person, a real self, or a genuine human being; i.e., when she socially exhibits a virtuous life in a way that not everyone ends up reproducing.[167] Attaining a genuine social selfhood is attaining the normative quality of healthy humanness—the normative content of *obuntubulamu* as we discussed.

This makes African scholars agree that the ideal of personhood (virtuous life) is in the *muntu* performing actions that conduce to the promotion of life generally, and particularly promoting the welfare of others, through demonstration of norms, ideals, and moral virtues. These virtues

161. Menkiti, "Person and Community," 171.

162. See Behrens, "Two 'Normative' Conceptions of Personhood," 111.

163. See Metz, "Philosophy of Masolo," 12–13.

164. See Menkiti, "Normative Conception of a Person," 325; Menkiti, "Person and Community," 165; Wiredu, "Oral Philosophy of Personhood," 8–18; Menkiti, "Person and Community," 171–72, 176. To be human is a characteristic of moral excellence; see Eze, "Decolonisation of African Political Philosophy," 1–17; Behrens, "Two 'Normative' Conceptions of Personhood," 103–19.

165. See Oyowe and Yurkivska, "African Personhood," 89–90.

166. Menkiti, "Person and Community," 173.

167. See Metz, "Human Dignity, Capital Punishment," 83.

throw the *muntu* into robust indispensable ontological social relationships whereby, in addition to promotion of life as supreme for others,[168] the *muntu* reciprocally develops her personhood. This twin relationship—where the *muntu's* endeavors build harmony through identity and solidarity as we discussed already—is largely defended by numerous African ethicists, like Archbishop Tutu, Menkiti, Tangwa, Bujo, Gyekye, and Metz, who subscribe to Mbiti's maxim that "a person is a person through other persons."[169] This African maxim emphasizes the need for others in anticipation of the *muntu* developing her personhood. In adding weight to this twin relationship deriving from the maxim, Menkiti notes that the *muntu* "recognizes the sources of his or her own humanity, and so realizes, with *internal assurance*, that in *the absence of others*, no grounds exist for a claim regarding the individual's own *standing as a person*."[170] More to this relational twin relationship, for Tutu, "my humanity is caught up, (it) is inextricably bound up in yours,"[171] while for Gyekye, "the fundamentally relational character of the person (human being) and the interdependence of human individuals arising out of their natural sociality are thus clear."[172] Thus, the *muntu* cannot escape the indispensable relationships of other Bantus in her promotion of life as she reciprocally matures in her personhood.

In this twin relationship, the centrality of the self[173] can be *firstly* understood in the sense that the self-realization of the *muntu* is the proper moral end posited by the moral theory of personhood.[174] In this regard, realizing the *muntu's* self is only possible within a true community in which the *muntu* succeeds becoming a person only through other persons. And if selfhood (full personhood with a widened maturity of ethical sense) is only socially achieved, then social selfhood already points to a *self-realization ethic* in relation to the *otherness* within genuine *humane* relationships, signaling the moral goal of personhood. This is the view of Magobe Ramose: "To be a human being is to affirm one's humanity by recognizing the humanity of others and, on that basis, establish humane

168. See Gyekye and Wiredu, *Person and Community*, 110; Molefe, *African Personhood*, 66.

169. See Metz, "African Moral Theory," 331.

170. Menkiti, "Normative Conception of a Person," 324. Emphasis added.

171. Tutu, *No Future without Forgiveness*, 35.

172. Gyekye and Wiredu, *Person and Community*, 104.

173. Metz, "African Moral Theory," 330.

174. Molefe, *African Personhood*, 110.

relations with them . . . One is enjoined, yes, commanded as it were, to actually become a human being."[175] Here self-realization within humane relations as the moral goal of personhood binary translates into the *muntu* fully affirming the otherness of the *other* in her limitlessly boundless form, besides becoming a person. This further means, "in a true community, the individual does not pursue the common good instead of his or her own good, but rather *pursues his or her own good through pursuing the common good*. The ethics of a true community does not ask persons to sacrifice their own good in order to promote the good of others, but instead invites them to recognize that they *can attain their own true good only by promoting the good of others*."[176] It is by endorsing life of the other community members that I attain selfhood. I can realize my selfhood or promote my life only through enabling the well-being of the other members.

In humane relations, it is only through the *muntu* limitlessly promoting the common good of the community that she can affirm the otherness of the *other* while reciprocally realizing herself. The idea of common good, premised on human relational ontological endowments, can be well captured by the African moral imagery of the Siamese crocodile with two heads and one stomach. The motif's moral thought is that "at least the basic interests of all the members of the community are identical,"[177] that is, the shared stomach. Put differently, for the *muntu* to be able to realize herself, she requires not to be handicapped with "human goods," understood as the basic transcultural and commonly shared goods,[178] or be hindered from the requisite mutual collaboration as the moral wisdom of the Bashi of Congo-Kinshasa guides: "Two ants are able to carry a locust" (*Orhunyegere rhubiri rhurhayabirwa n'ihanzi*),[179] and "One bone cannot put up any resistance to two dogs" (*Ebibwa bibiri birhayabirwa*

---

175. Ramose, *African Philosophy*, 52; In most African languages, there is a saying synonymous with the *Sotho, motho ke motho ka batho*; i.e., to be human is to affirm one's humanness by recognizing the same quality in others; consequently, working towards establishing functional humane relations. See Ramose, "Emancipative Politics in Africa," 107–29.

176. Lutz, "African Ubuntu," 314. Emphasis added. Lutz argues for a needed theory of global management, consistent with our common human nature, that should originate from our cultural communities which promote the common good.

177. Gyekye, "African Ethics."

178. Gyekye, *African Experience*, 67; Gyekye, *Common Humanity*.

179. Ntabaza, *Emigani Bali Bantu*, 340, no. 2729.

n'ivuha).[180] So we note that in the mind of Africans, mutual collaboration (the shared stomach) in pursuing the common good gives off mutual life for all,[181] while reciprocally translating into the *muntu* realizing herself.

If we refer to the motif's moral thought, in the understanding that the common stomach is identical to every *muntu's* self-realization process, what then binds Bantu together is closer to the common good pursued than even *truth*. An instance is where the authority of the revered elder—whose authority binds the family towards the common good—is challenged by means of a message of truth delivered by an inferior. *Descriptively*, since realizing the self or becoming a person is only through other persons, without intending to comprise the truth, priority is *prescriptively* given to the role of authority in binding the clan family together, and not to truth when it unrelates the clan family in its pursuit of the common good. It is through a healthy relatedness in line with the common good that we can prescriptively foster self-realization of every *muntu*. The Baganda catchphrase *"Omukulu tasobya"* (An elder does not era or stray) isn't intended to claim the infallibility of elders in the community pursuit of the common good, but rather to warn inferiors who try to challenge elders unnecessarily while causing relational disharmony of persons.[182] Since the identical stomach as a pointer to self-realization ethics attracts everyone, pursuance of the common good isn't only central to the elders but also to the children. It is for this reason that Baganda espouse flexibility on the side of the elders in this proverbial counsel: *"Ne gwozadde akukubira engoma n'ozina"* (Your own child may set for you the tune for the dance). Even when the moral rules are largely sharpened by the experiences of the elders—because when an old African dies, it is believed a library has run to the ground—on the other hand, the children can fix, prescribe, or set the rules of life for the elderly to realize themselves.[183] Here the moral thought emphasizes the need for flexibility

---

180. Ntabaza, *Emigani Bali Bantu*, 63, no. 490.

181. See Bujo, *African Ethic*, 88.

182. Another authority-truth-related saying has similar meaning: *Nyinni mwana akubuulira: nti omwana yambuulira* ("The parent is informing you; and you say, 'The child told me!'"). Although truth is one, and it serves to maintain a harmonious universe; truth can be misrepresented, or its accuracy could be variedly understood. Two proverbs prove this: *Tekiwoomera matama abiri* ("Something can hardly have the same taste to two different mouths"); *N'omugezi awubwa: amatu tegawulira vvumbe* ("An intelligent person, too, can be mistaken; moreover, ears are not capable of smelling"). See more in Kasozi, *Ntu'ology of the Baganda*, 80; see also footnotes 8–10.

183. See Kasozi, *Ntu'ology of the Baganda*, 112.

amongst community members as they jointly pursue the common good that brings them closer. Realizing the genuine self is only possible when the *muntu* succeeds in becoming a person through pursuing the common good in company of the other community persons relationally.

In the *second* sense, self-realizing oneself is to let one's character shine. The normative facet of personhood revolves around the quality of conduct—since a Black African will judge the *muntu*'s conduct parallel to her having moral sense—within a relational ontological social reality. To say that one is a person is not only tantamount to moral praise for the quality of one's moral performance (moral sense), but also as good as moral approbation.[184] Akin to other interpretations of Bantu normativity,[185] Kwame Gyekye asserts how character, ethics, and morality are widely applied in practical social realities as synonyms. He writes: "Inquiries into the moral language of several African peoples or cultures indicate that in these languages the word or expression that means 'character' is used to refer to what others call 'ethics' or 'morality.' Discourses or statements about morality turn out to be discourses or statements essentially about character . . . The implication here is that ethics or morality is conceived in terms essentially of character."[186] Taking an example from the Kom people, to say the *muntu* has a good character (is morally upright) is roughly said as "*Wa kel nchini ijuŋ*," literally meaning, "You have morals," "You have a good character," or "You are good." The antithesis is that one does not have a good character (is not a good person): "*Wa kel wi nchini*," literally meaning, "You don't have morals/character" (you are not a [good] person, because your behavioral conduct has fallen short of the standard morality). Normatively, the Kom people would positively say "*Wa ghi wul*" (You are a person), or "*Wa kel nchini*" (You have morals), as opposed (negatively) to "*Wa ghi wi wul* or

184. See Molefe, *African Personhood*, 24, 58.

185. Desmond Tutu remarks normatively: "When we want to give high praise to someone we say, '*Yu, u nobuntu*'; 'Hey, so-and-so has *ubuntu*.' Then you are generous, you are hospitable, you are friendly and caring and compassionate. You share what you have. It is to say, 'My humanity is caught up, is inextricably bound up, in yours.' We belong in a bundle of life. We say, 'A person is a person through other persons.' It is not, 'I think therefore I am.' It says rather: 'I am human because I belong. I participate, I share.' A person with *ubuntu* is open and available to others, affirming of others, does not feel threatened that others are able and good, for he or she has a proper self-assurance that comes from knowing that he or she belongs in a greater whole and is diminished when others are humiliated or diminished, when others are tortured or oppressed, or treated as if they were less than who they are." *No Future without Forgiveness*, 35.

186. Gyekye, "African Ethics."

wa ghi nyam" (You are not a person, or you are a beast). In the normative sense, however, not every *muntu* is a moral/real person (*nkainti wul*).[187] Though all are born humans, one becomes special, a person, depending on how positively or negatively her behavioral conduct or character, not her being, affects others relationally. And if the goal of morality is to transform moral possibilities (potentialities) given at creation into actual moral realities by "decorating" one's humanity with excellences, then calling someone a person is to make a moral judgement (approbation) about the quality of her self-realized character (the actualized potential).[188] This would further insinuate that promoting the good of others (while attaining self-realization) within some particular sociality is to actually develop a morally excellent character (agent-centered personhood).[189] This not only affirms how character communicates culture, ethics, or morality, but also how the *muntu's* quality of her self-realized character reveals her personhood.

In all ways, the norms of the *muntu's* personhood, *obuntubulamu*, character, or self-realization remain largely synonymous, and it is out of question to neglect the ontological communal reality of the environs in establishing their normative content. The *muntu's* responsibility has a social nature; i.e., all community members take a share. In Burundi, "*Omuryambwá aba umwé agutukisha umuryango*," meaning, if one member of the family eats dog meat, all the members of the clan are disgraced since to eat the flesh of a dog is utterly disgraceful for the Burundi.[190] Blame does not only bare personal responsibility but also casts a social collective liability or burden (as earlier discussed under "promotion of vitality"): "*Omulya mmamba aba omu, n'avumaganya ekika*"—when one single member among the Baganda of the lungfish clan eats the prohibited lungfish, the action of the *muntu* not only violates the taboo (of not eating one's totem), but she brings herself as well as the whole clan into

187. Mbih, "Kom Ethics," 37–38.

188. See Molefe, *African Personhood*, 21. See more on character and personhood in Menkiti, "Person and Community," 172; Gyekye and Wiredu, *Person and Community*, 113.

189. Self-realization or achieving personhood (morally excellent character) are synonyms; see Molefe, "African Philosophy through Personhood."; Molefe, *African Philosophy of Personhood*; Matolino and Kwindingwi, "End of Ubuntu," 197–205. On retracing African humanism premised on our shared humanity, see Praeg, *Report on Ubuntu*. Given humane relations as a way of being-ness, every person is both an object and subject of duty and obligation. Etieyibo, "Ubuntu," 139–62.

190. Bujo, *African Ethic*, 87; see also 185 n. 61.

shame and disrepute.[191] What is significant is not an individualistic pursuit but rather the pursuance of the collective common good understood within an ontological social reality of a community.

In defense of Tangwa's emphasis of the fact that no one falls outside the sacredness of African moral life (as discussed in the preceding section of personhood *qua* moral dignity or status), a just community responds by providing common human goods for all to pursue personhood. In pursuing the common good jointly, we flourish jointly, which requires the creation of dignifying/enabling community/social structures/conditions for each *muntu* to realize herself. In contrast, an unjust community won't ensure the availability of the basic human goods. But still, a mere provision of these basic goods is still not dignifying or sufficient for securing moral personhood. Often, there are serious limits to a *muntu's* ability to contribute towards the realization of personhood by another *muntu*. This is why each *muntu* has an obligation to abundantly use the available maximal opportunities to achieve a dignified life of personhood *qua* virtue. The question is whether we could conclude that it is one thing for the *muntu*, as a matter of fact, to have dignity since she possesses the relevant ontological capacities, but yet quite another for her to be dignified. In an African response, for that matter, dignity (personhood *qua* moral status) within a just community entails the following: Firstly, not the mere possession of the relevant capacities (ontological properties), but all Bantu have the status of dignity, which renders them intrinsically and superlatively valuable, equal to that of every other *muntu*, i.e., *equal respect*, and every *muntu* has protection from the community, viz., *status dignity*. Secondly, providing the enabling human and common goods is dignifying. This provision grants a platform for the *muntu* to cultivate a dignified life or rather a decent life within the community in which she can access basic life amenities such as shelter, food, education, healthcare, stable political conditions, and so on, i.e., *community dignity*. Thirdly, a dignified life results not only from the *muntu* exercising her ontological properties—i.e., *performative dignity*—within the light of enabling conditions (*qua* human goods) to secure personhood per-formatively,[192] but also from the moral character of the community as the legitimate sympathetic surrogate for all inadequacies, i.e., *community sympathetic dignity*.

191. See Kasozi, *Ntu'ology of the Baganda*, 78; see also footnotes 4 and 5.

192. See Menkiti, "Person and Community," 165–66; Molefe, *African Personhood*, 25–26, 51–52, 104–7.

Just as sympathy reaches beyond normative principles and rules as a moral agent performs actions and the virtues that make her morally a worthy person,[193] so too does community sympathy override all previous conflicting inadequacies pertaining to the *muntu*, even in her circumstances of moral patienthood. But even the state of patienthood, as we have adequately argued in African moral thought, and more so not being able to act competently and responsibly, doesn't diminish the *muntu's* moral status and worth as a human being or even our moral consideration and treatment towards that patient.[194] Emphasizing equal respect, status dignity, community dignity, and performative dignity is good, but cannot sufficiently be dignifying or deeply efficacious without being deeply practical towards moral patients who lack the positive actual use of their ontological properties. To help fellow Africans grasp the fact that no human being falls outside the inner sanctum of community dignity, Tangwa finds it as one reason why "myths about, say, the superhuman psychic powers of physically or mentally handicapped persons or their liability to be used as disguises by God and other spirits no doubt arose in order to help people put prescription into practice."[195] Such symbolic myths positively demonstrate an actual collective sympathetic and empathetic receptivity towards moral patients. Though some scholars like Wiredu use the terms "sympathy" and "empathy" interchangeably, several African languages do not have a homeomorphic equivalent for "empathy," while those who do simply give a description rather than a definition of the term. Sympathy requires a *muntu* to be able to relate to, or even identify with or accept what another *muntu* is going through (e.g., pain), whereas empathy requires that a *muntu* also shares, understands, and takes on another *muntu's* feelings of suffering. Given that empathy is too demanding, an appropriate term that commonly features in African contexts is for sympathy, for instance, in *isiZulu uzwelo* and in *seSotho Kutlwelo bohloko*, meaning the ability to feel for or identify with another.[196] This can be illustrated in recent scholarship originating from Buganda Kingdom. Among the twelve core values of moral personhood, empathy (*okufa ku munno*) is explained with rather a befitting description of

193. See *PBE*, 31.

194. See Tangwa, "Moral Status of Embryonic Stem Cells," 453.

195. Tangwa, "African Perception of a Person," 42.

196. Wiredu uses both "sympathy" and "empathy" as the foundation of morality. He seems to take these two terms to be interchangeable, which they might not necessarily be. See Molefe, *African Personhood*, 73–74 n. 3.

sympathy. The moral description renders the *muntu* as caring toward the well-being of others. The description forbids exclusion of others simply because they are different from us. Imagine if a person crawling with crippled legs was your biological child and she was being laughed at by everyone. It is why the Baganda advise that "*Awali omulema, tewafuny-irwa lunwe*" (Where there is a lame person, one should be extra sensitive not to point with a finger),[197] whereby being sensitive doesn't necessarily mean empathy, but rather community sympathy. For prominent African scholars like Wiredu and Masolo, sympathy is "the root of all moral virtue."[198] On sympathy, for Masolo, the idea of personhood employs the moral wisdom of the "economy of affection," which demands every *muntu* to relate sympathetically by "creating humane conditions of life for everyone."[199] An since the economy of affection as sympathy is the root of all other virtues pertinent to the moral theory of personhood, to possess personhood is unambiguously to be sympathetic, which qualifies the *muntu* as inalienably endowed with status dignity. Besides, a decent life of healthy humanness, i.e., a performatively dignified life, is an outcome of living a sympathetic life. The *muntu*'s failure in attaining personhood is not synonymous with loss of status or dignity, but rather is a failure in securing healthy humanness, i.e., unsuccess in performative dignity.[200] In agreement with Molefe, we do not therefore achieve (status) dignity or equal respect or community dignity (since without our voluntary choice we are born into a particular community); it is a (performatively) dignified/decent life that we successfully attain or fail to pursue. And so, achieving healthy humanness (*obuntubulamu*), moral sense, social selfhood, or personhood is tantamount to living a performatively dignified or decent human life, but it is not the attainment of the human moral status or dignity or capacity for sympathy (community sympathetic dignity).

When we stress human moral status or dignity or capacity for sympathy, we sound quite strongly *speciesistic* and *anthropocentric*, which can be charged of some African scholarships. To show speciesism, when Tangwa and Henk propose contextualization of "genomics"—genomics as an important part of "techno-science," science mingled with technology—they emphasize contextual localization of scientific knowledge as "social constructivism," and "diversity" in relation to a concept of holism

197. *Munno*, literally "your friend." Murphy, *Luganda-English Dictionary*, 374.

198. Wiredu, *Cultural Universals and Particulars*, 71.

199. Masolo, "Western and African Communitarianism," 494.

200. See Molefe, *African Personhood*, 50–52, 71–72.

and communalism. Significantly, their discussion brackets living things, mostly human beings.[201]Their speciesistic emphasis on the living organisms becomes even clearer as they decisively argue "that diversity *exists* (among living organisms), whereas difference (between living organisms, peoples, human beings, sexes, etc) *is made*."[202] For Tangwa, it is the living things with a teleological purpose (inner reason), not the dried leaves used by a medicine person to cure the *muntu's* illness or the stones that guide riverbanks in the ecological system or the mountainous escarpments for therapeutic recreation: "Every plant growing in the wild, every creature crawling the earth, flying above or swimming beneath it, may have its own inner 'reasons' for being there, its hidden teleological and ecological purposes, or a value in the overall scheme of nature, unknown to human beings."[203] Moreover, Tangwa's emphasis on the questions addressed by the interdisciplinary field of bioethics (of biotechnology) remains focused on living things bracketing out non-living things, i.e., it involves "all living species on earth—plant, animal and human."[204] To prove anthropocentricism further, even when Tangwa tries to synopsize the traditional four principles, he captures them in a limited anthropocentric tone of two African adages: "A human being is a human being simply by being a human being," indicating respect for intrinsic value, autonomy and justice), and "The essence of a human being is having a good heart (will)," which indicates beneficence and non-maleficence. Both adages as summarizing bioethical principlism focus on humans and make humans to live in their human communities harmoniously.[205]One can likewise say that Tangwa forgets to emphasize his own interpretation of African moral thought as eco-bio-communitarian bioethics by restricting bioethics to controversial issues pertaining to living entities only. He even forgets that at least every non-living entity has a teleological interest and purpose in continuing to be (subsisting) since it shares in the vital life force hierarchically. And if *diversity* in Tangwa's eco-bio-communitarian bioethics was to mean the tolerant attitude of "live and let live," a moral attitude that doesn't mean assimilation or dominance of the *other*, it cannot be

201. Tangwa and Henk, "Colony of Genes," 1031, 1037.

202. Tangwa and Henk, "Colony of Genes," 1038. Emphasis original.

203. Tangwa, *African Bioethics in a Western Frame*, 93.

204. Tangwa, "GM Crops, Foods and Feeds," 69, 70.

205. See Tangwa, "Ethical Principles," s4–5. In Tangwa's natal language, *Lamnso'*, he claims we are infinitely perfectible, and this is briefly captured in the single idea of *shiliv she jung shi* (literally, "a good heart"). Tangwa, *African Bioethics in a Western Frame*, 95.

fully lived out without extending functioning relationships, towards not only living but also non-living things.

Even when Tangwa's ethical angle seems to go beyond the confinements of the living organisms, as when he believes that "ethics is limited to acts, actions and behavior that are free, purposive, intentional and liable to impact on other creatures,"[206] whether these creatures include only living organisms remains unclear. Tangwa's seeming limitedness towards the living indeed contrasts with other African scholarship. For example, in Mokgoro's thesis, "harmony is (to be) achieved through close and sympathetic social relations."[207] It is a question of renewed and functioning relations. Using an analogy of HIV/AIDS in relationship to *kairos*, just like Agbonkhianmeghe Emmanuel Orobator intones, "the body of Christ has AIDS'; 'Our church has AIDS,'" and certainly the church is "living positively with AIDS."[208] It is a social death requiring a relational healing. Our efforts towards this healing, i.e., "social *kairos*" according to Jacquineau Azetsop,[209] should be to further redefine these *malfunctioning relations* beyond humane spheres. This is the only way we can re-energize the communal vision of the social immune system. Moreover, this is clearly discussed in this work regarding the successful communitarian efforts towards the HIV/AIDS-pandemic's brutalization of our communities, with a further indication that good bioethical interventions must be integrative efforts and so must override the confinements of limited humane sympathy. I note here, however, that Tangwa's empathy (not sympathy) remains largely limited to the humane domain, which is a disservice to a charitable interpretation of African moral thought, which is sympathetically altruistic. Here true change or conversion begins with acknowledging that relational participatory solidarity is not only socially communal, but it necessarily encompasses an integral social-global approach to the vitality and continuity of life beyond humans, a life that is also interactively hierarchized.

Indeed, communities in SSA are not simply humanistic, but also largely practice both crop and animal husbandry. They are animal keepers just like the widely known Maasai or Karimojong (Nilotic pastoralists), Bantu cattle keepers of Ankole region, and Baganda farmers of Uganda. Among them, one gains personhood not only through one's relations

206. Tangwa, "Ethical Principles," s3.

207. Mokgoro, "*Ubuntu* and the Law in South Africa," 17.

208. Orobator, *From Crisis to Kairos*, 121.

209. Azetsop, "Time of AIDS," xviii.

to other persons but also one's relations to one's animals or good crop yields.[210] This points to the fact that if we take the example of personhood *qua* virtue as healthy humanness or social selfhood or moral sense, it cannot be attained or even understood outside the relational ontological social reality of a Black African. So, the notions would remain inexhaustive since the relational ontological world of an African goes beyond the living humane relations. We can therefore refine the African idea of personhood by defining a person not as self-realization or as an ontological act, but rather as a process of coming into existence in the reciprocal relatedness of the individual and community ontological social reality,[211] emphasizing a participatory and sympathetic solidarity towards all entities (*other*). Here participatory solidarity imbued in sympathy goes beyond humane empathetic relationships. More than often, life realities in the practice of bioethics demand renewed integrative values towards the holistic other. Indeed, Tangwa defends the African attitude and value of empathy if we are to understand a moderate altruistic form of communitarianism or communalism,[212] but his emphasis remains less relationally integrative of the whole altruistic value of the other as captured in the African moral system.

The moral contradistinctions posed earlier on by bioethical principlism between the moral status of the fetus and the brain-dead pregnant woman (personhood *qua* status) are handled in a novel and unusual way in a Black African approach to bioethics. Although different authors agree that patient-centered notions (some understand it as personhood-as-acquired or property-based) of personhood are more dominant in Western moral thought, like in bioethical principlism, whereas the agent-centered (others understand it as personhood-as-endowed or performance-based) notion reigns in African thought. However, the moral theory of personhood in African thought appreciates both perspectives as interdependent since there are ontological properties of being human (which inform the patient-centered notion) that are possibilities of pursuing the agent-centered notion of personhood; which, first of all, assigns respect to some agent relative to the quality of her conduct (appraisal

---

210. See Presbey, "Maasai Concepts of Personhood," 57–82.

211. See Bujo, *African Ethic*, 87.

212. See Tangwa, "HIV/AIDS Pandemic," 217–30, especially 220; Kopelman and Van Niekerk, "AIDS and Africa," 139–42.

respect); second, it assigns respect on the basis of the capacity for virtue (recognition respect).[213] We need to go beyond these classifications.

Firstly, we can all agree that the idea of personhood opens moral doors to a better understanding of integral humane conditions resident in sympathetic communities. The capacity for virtue (sympathy)—which appreciates a genderless status dignity inherent in all humans prior to their socialization—captivates the communal egalitarianism of African moral thought. Secondly, for a Black African logic does not necessarily imply that one becomes solely a person by means of achievements. More significantly, the person is not defined as an ontological act by means of self-realization, but by means of a functioning relationality, namely, a network of sympathetic relationships that not only secures an actual positive use of relevant ontological properties when the *muntu* gets incompetent, but necessarily constitutes the *muntu's inalienable dignity* or *status* as *a community dignity*. This is to suggest that the unborn child is already a person (*qua* status or dignity) as well the brain-dead pregnant woman. What principlism refers to as a "fetus," within an African moral thought is sheltered within an active sympathetic relatedness of the three-dimensional community of the living, the dead, and those to be born (the holistic *other*). Even if both the fetus and the brain-dead pregnant woman cannot competently handle an independent life of their own, they are embraced by the sympathetic and respectful love of the visible and the invisible community. There is a reciprocally shared hope of the future descendants, where the community of the living and the community of the dead both continue hoping for their continued life and ancestral-clan fellowship in the fetus and the dead pregnant woman (as an ancestor). This *reciprocal mutual interaction*, which is never unidirectional, but rather *shared between all members of the community* fellowship, *interdependently influences each community member positively or negatively*. The dead mutually depend on the living while the living mutually depend on the dead ancestors, who look toward those not yet born but mutually depend on the living and so on.[214] The bidirectional mutuality between

213. See Behrens, "Two 'Normative' Conceptions of Personhood," 105; Metz, "African Moral Theory," 331; Molefe, *African Personhood*, 5–6, 49, 69–70; Nyamnjoh et al., "Theories of Personhood."

214. See Bujo, *African Ethic*, 89. To understand what I emphasize as the *reciprocal interaction*, according to the African wisdom, in reference to Bujo's earlier writings, life coming from God flows in a hierarchical order. At the peak are the ancestors, who are followed by the elders of the community, and these include father and mother of the family, the clan head, and the chief or king. According to their function and task in the

the brain-dead pregnant woman, the living descendants, and the unborn child promotes the relational growth of life, given that the communicative nonverbal palaver (of the living community members, the brain-dead pregnant woman, and the unborn child) at the point of death contributes to the integral process of all becoming true persons.

Relational Ontology and Normativity Are Inseparable: Justice as Mutual Aid in Black Africa

On "partitioning of Africa," Tangwa notes: "And the most important single event in the colonial history of Africa is, arguably, the Berlin Conference of 1884 during which European imperial nations, like good hunters, stood over the map of Africa, like a game, and, with their imperial pens (penknives?) quartered her up amongst themselves, for its exploitable resources, with regard for neither the linguistic, cultural nor political state of affairs on the continent."[215] These inconsiderate divisions can be well captured in the historical slave trade. Argued by other Africanists, just like slave trade in the history of the African continent enormously

---

community, they form the link between the ancestors and the living, who only in this way can participate in the fullness of life. Here, it ought to be mentioned that this participation is *never unidirectional*, only from top to bottom. In the African context, it is a relationship of *mutual exchange between all members of the community*. Again this community is not restricted to the earthly dimension but it is a two-dimensional relationship; it includes the deceased members of the clan as well as all members of the family or clan *live in interdependence and can influence each other positively and negatively*. The good or evil done by the individual promotes or reduces the vital energy of the whole clan. Nobody is allowed to keep life for him/herself but has to share it with other members of the family or clan. Whoever behaves egoistically sins against God as the source of life. *If the dead and those who stayed behind form a single community, there is interaction between them.* The living know that they depend on the dead, especially the ancestors, for the increase of life. On the other hand, the dead, including the ancestors, cannot live happily without the support of the earthly community. The striving for the abundance of life and fear of reduction of vital energy is of extreme importance for this two-dimensional relationship. The admonitions, commandments and prohibitions of ancestors and of community elders are highly esteemed inasmuch as they reflect those experiences which have made community life possible up-to the present. Whoever despises the ancestors and elders and rejects the community laws and statues established by them, chooses death instead of life. And such death will not affect one person alone, but the entire community. In this solidary community, love and obligation are rooted, and an individual cannot become a fully human being outside the community group. Only in conjunction with other members of the clan community can one's identity be found. In the individual is the representation of the entire extended family. See Bujo, *Ethical Dimension of Community*, 197–99.

215. Tangwa, "Colonialism and Linguistic Dilemmas," 3.

contributed to the industrial-technological development of the Global North more than the Global North contributed to the collective development of the Global South, the disuniting and retarding consequences from the partition of Africa led ethnic communities with different languages to be clustered together or scattered without their voluntary wish or developmental benefit of their own. The deconstructed borders with superimposed foreign languages created more linguistic dilemmas, and stood against numerous administrative as well as social-economic realities like intercultural dialogic communication, pasturing of animals, fishing, trade, farming, cultural activities, etc. Contemporaneously, even though in Europe the nation largely precedes the state, Africans can narrowly speak of nations in the European sense, but of the artificially created controversial African states (by colonial masters). In today's Africa with over 2,000 living spoken languages, a single ethnic group isn't only scattered in different countries, but one country can have numerous local languages (in addition to the superimposed foreign languages): Nigeria has 522, Cameroon 275, the Democratic Republic of the Congo 214, Chad 129, Tanzania 126, along their several ethnic tribes![216] To be more exact, Bantu, who would have been a single nation today, are scattered in Kenya, Tanzania, Rwanda, Burundi, Democratic Republic of Congo, Somalia, Comoros, Uganda, Zambia, Malawi, Zimbabwe, Mozambique, Swaziland, Namibia, South Africa, Botswana, Angola, Lesotho, Equatorial Guinea, Gabon, and Cameroon. And still no country among the above has one language or a single tribe. Besides, as an established fact, this scattering was partly due to greed of wealthy natural resources. The colonial powers used the natural wealth (in Tangwa's language, "economic domination and exploitation"), artefacts of their colonies in Europe at their will, without consent of the colonized people. Examples include the museums of countries from the Global North (e.g., the British Museum), who are proud to show even mummies! One wonders if African

216. Tangwa claims that it was colonialism that imposed European languages and systems of education, creating several controversial linguistic dilemmas. For instance, English and French were superimposed on more than 240 indigenous languages in Cameroon. Tangwa, "Colonialism and Linguistic dilemmas," 4. See also Bujo, *Ethical Dimension of Community*, 171–72; Statista, "Number of Living Languages in Africa." Going against the conventional wisdom that colonialism brought modernity to Africa, Olúfémi Táíwò claims that Africa was already becoming modern and that colonialism was an unfinished project. Africans aspired to liberal democracy and the rule of law, but colonial officials aborted those efforts when they established indirect rule in the service of the European powers. See more in Táíwò, *Colonialism Preempted Modernity in Africa*, 42.

mummies are also venerated as the dead in Europe! And by the way, entry to museums is not free; hence Africans, who wish to admire the works of their past ancestors, must fly to Europe at a fee.[217] In the failure of principlism, which kind of normative structure can we collectively erect to heal and reconcile the present realities with the victims of such past injustices of exploitability and exploitation so that we can contemporaneously have an integrated life? To an extent, Africa paradoxically remains the richest as well as the poorest continent on earth.

Justice as Communitarian Mutual Aid/Cost/Assistance in Black Africa

African communitarian justice understands justice not literally as fairness but as *mutual aid* or *mutual assistance* or a *social cost* (social obligations).[218] Mutual aid qualifies as an imperative of sisterhood or brotherhood that we can understand as "the economics of affection or eating together" (*ddiiro*) that supersedes looking alike.[219] Mutual aid connotes supererogatory acts that meaningfully go beyond the moral domain of duty as it is standardly known. For example, for the Tonga, a Bantu tribe in the northeastern part of South Africa, the principle of reciprocity and mutuality in services is the basis of all their moral rules. Concrete scenarios include their mutual duties and obligations to one another in instances of marriage arrangements.[220] Similarly the principlistic understanding of moral supererogation goes beyond what moral duty requires,[221] though this is not justice (as fairness). For a supererogatory act in principlism, which would be an example of justice in the African sense, one performs beyond what is owed or, more generally, does more than what is required, but still with differences as I will show

217. See Bujo, *Ethical Dimension of Community*, 177–78; British Museum, "Egyptian Death and Afterlife: Mummies."

218. See Wiredu, "African Culture," 201.

219. See Magesa, *What Is Not Sacred*, 147–49.

220. See Macbeath, *Experiments in Living*, 180.

221. The concept of supererogation can meaningfully be applied beyond the moral domain. McElwee, "Supererogation across Normative Domains," 505–16. An adequate ethical theory must include *both a theory of right conduct and a theory of virtue*. Not just outward conduct but also inward vice and virtue can appropriately be subjected to public scrutiny. Deontic principles, deontic standards, and standards of vice are all conventional. Trianosky, "Supererogation, Wrongdoing, and Vice," 26–40. There is no good reason to include praiseworthiness in the analysis of supererogation and so the standard analysis should be rejected. Archer, "Acts of Supererogation," 238–55.

in the African sense. Though from the outset such acts in principlism fulfil four key conditions: "First, supererogatory acts are optional and neither required nor forbidden by common-morality standards of obligation. Second, supererogatory acts exceed what the common morality of obligation demands, but at least some moral ideals are endorsed by all persons committed to the common morality. Third, supererogatory acts are intentionally undertaken to promote the welfare interests of others. Fourth, supererogatory acts are morally good and praiseworthy in themselves and are not merely acts undertaken with good intentions."[222]

In contrast, although mutual aid promotes an integral harmonious life of the community, it is nonverbally prescribed, required, and approved by customary social realities without calling for any moral praiseworthiness like in principlism. To bring out a closer meaning of justice (as mutual aid) in African moral thought demands us to amalgamate the principlistic notion of supererogatory acts with their notion of moral heroism or moral saintliness. A moral hero is celebrated for an exceptional action or virtue. With an exceptional moral character (inclusive of moral and personal integrity) that disposes the heroine to perform supererogatory actions of high order and quality, she regularly realizes duties and moral ideals over time while foregoing luxuriating over certain advantages where most people would fail to do so.[223] Although mutual aid is an obliging imperative in the Africa sense, in principlism moral heroism or moral saintliness doesn't carry this weight—what actually makes our amalgamation unsuccessful. Moral heroes or saints "who act on moral ideals do not always consider their actions to be morally optional. Many heroes and saints describe their actions in the language of ought, duty, and necessity: 'I had to do it.' 'I had no choice.' 'It was my duty.' The point of this language is to express a personal sense of obligation, not to state a general obligation. The agent accepts, as a pledge or assignment of personal responsibility, a norm that lays down what ought to be done."[224] In

---

222. They rely partly on John Rawls's theory of justice to interpret supererogatory acts; see *PBE*, 46.

223. The four criteria of moral excellence: (1) constantly faithful to a worthy moral ideal; (2) motivational structure, prepared to forego certain advantages in the service of a moral ideal; (3) an exceptional moral character, moral virtues that disposes one to perform supererogatory actions of high order and quality; (4) a person of integrity—both moral integrity and personal integrity. Although a person can qualify as a moral hero only if she meets some combination of these four conditions, on the contrary, a person must satisfy all four to qualify as a moral saint. *PBE*, 52–54.

224. *PBE*, 46.

contrast, an African moral hero (exceptional moral excellence) won't say, "I was only doing my duty," given that he doesn't regard his success or failure as personally acquired,[225] but rather a community achievement, and an *aid* that can *mutually* be evaluated by the same community members. Since duties or supererogatory acts are not morally optional, but a mutual shared cost, the *muntu*'s sense of duty is the general communal sense of duty. Supererogatory acts are not only an interest of some but what every community member is supposed to mutually do, and so not necessarily requiring praiseworthy. Given the continuum of a non-dualistic mutual life, an African won't therefore separate the quality between a low-level supererogation, such as walking a patient lost in a hospital's corridors to a doctor's consultation room, and a high-level supererogation, such as heroic acts of self-sacrifices.[226] Both require the same mutual assistance channeled towards an integral life of all. It is the reason why moral excellence or heroism is both mortal and immortal since mutual life is a continuing unity, a unified continuum. Given that the process of becoming a person continues even after death like we stated before, Africans come to realize earthly moral heroes (moral personhood), while after the earthly life, through the personhood of heroic ancestors, Africans come to realize moral saints (immortality of personhood). I will briefly illustrate the meaning of mutual aid/cost/assistance in the African understanding of procedural, corrective, and distributive justice.

In the *first instance of justice*, among contemporary Baganda, is procedural justice, which is *bwenkanya*, understood as equality and fairness.[227] *Bwenkanya* as a positive social junction—a general rule of promoting the welfare of others, i.e., their life—is always strictly executed through care of the *muntu*'s neighbors (*balirwaana*) in numerous ways. *Firstly*, the *muntu* takes care of her neighbors or village-mates through financial help, aid, or assistance (*buyambi*). Besides, the *muntu*'s safety or reaction

---

225. See *PBE*, 46.

226. See *PBE*, 47–48.

227. Murphy, *Luganda-English Dictionary*, 57, 420. Some proverbs show *bwenkanya* (fairness): *Enkima tesala gwa (mu) kibira* ("A monkey does not decide a case involving the forest")—meaning: no one is a good (fair) judge of his own case (situation) (Nason, "Proverbs of the Baganda," 253); *Tosala gwa kawala nga tonnawulira gwa kalenzi* ("Do not decide the girl's case until you have heard the boy's")—i. e., *audi alteram partem* (listen to the other side) (Haydon, *Law and Justice in Buganda*, 333). "The right of appeal (*kujulira*) was well understood and utilized. An appeal from the decision of a chief and so through the various court even to the Kabaka himself." Haydon, *Law and Justice in Buganda*, 5.

to an alarm depends largely on one's sensitive neighbors, who are bound to offer assistance thus mutually assisting one another. Each village member does *bulungibwansi* periodically, a public duty/joint communal work for the common good of the community, a core value of personhood/ *obuntubulamu* notable as civic engagement. Here mutual aid insinuates civic engagement/responsibility that is commonly publicized through a drumbeat called *saagala agalamidde* (I don't want anyone seated) or *gwanga mujje* (come all). These calls attend to civic responsibilities like cleaning up a community source of water, any theft, or murder. Moreover, it is everyone's civic responsibility to report corruption: one's silence is conspiracy with the evildoer.[228] Here justice is to mutually promote the welfare of my neighbor through aid or assistance.

In the *second instance* of procedural justice, one ought to show respectful sympathy or compassion (*busaasizi*, literally *Mitgefühl* in German) to others, in case of sickness, a house burnt down, some misfortune of any kind, death, etc. The drumbeat *gwanga mujje* may be sounded. Specifically, relations, friends, and neighbors are obliged to attend the burial, or come to console as soon as possible after the burial. Failure in personhood is to show no sympathy to the bereaved family. Here *busaasizi* implies a moral imperative of duty (*buvunanyizibwa*); otherwise complete failure to console the grieving family would be interpreted as unkind (*ttima*)—that the *muntu* is a non-person. Given the weight attached to mutual aid/cost/ assistance, an employee will easily give up work if she is not released by the employer to attend a neighbor's burial.[229] She has to respond to this imperative call of the community before she fails in personhood.

The Black African moral thought of mutual aid as justice raises questions of obligations and ideals, obligatory beneficence, and non-obligatory moral ideals of beneficence. In principlism, the "principle of beneficence refers to a statement of a general moral obligation to act for the benefit of others."[230] Here mutual aid as a prescribed obligation is not only increasing life of the others (the bereaved family) but also increasing

228. See more in Nsimbi, *Amaanya Amaganda n'Ennono zaago*, 27–35; Sempebwa, *Reality of a Bantu*, 149–50. *Oluguudo lw'obulungibwansi*, public highway, road built by communal effort, for the good of the country. Originally, *bulungi* means beauty, goodness, good, quality; *Buvunanyizibwa* originates from the transitive verb *vunaana*, hold responsible, charge, blame, accuse, calm, demand, have a reason. Murphy, *Luganda-English Dictionary*, 41, 45, 53, 574; Nnabagereka and Ssentongo, *Obuntubulamu*, 25–37.

229. See Nsimbi, *Amaanya Amaganda n'Ennono zaago*, 27–35; Sempebwa, *Reality of a Bantu*, 150–51.

230. *PBE*, 219–20.

life for the village-mate (maturation in moral sense or personhood or self-realization), thereby promoting life mutually for all. Mutual aid is not just an act of charity, and so persons would violate this vital obligation if they failed to act beneficently. In this regard, even though both the Samaritan's (village-mate) motive and her actions are beneficent, they don't suggest that positive beneficence is here an ideal rather than an obligation in the African sense, given that the Samaritan's act does not exceed the ordinary morality of Bantu.[231] The moral question would be if the Samaritan (villager) coincidentally met a relative or non-relative. Here principlism distinguishes between what is specific from what is general beneficence, pointing to the distinction between non-obligatory moral ideals of beneficence and obligatory beneficence. Principlism takes specific beneficence to rest on moral relations, contracts, or special commitments—directed at particular parties such as children, friends, contractors, or patients; while general beneficence is directed beyond special relationships.[232] If we consider an instance of ethical obligation (obligatory beneficence), it can be fairly illustrated in instances of community health emergencies like the onset of HIV/AIDS in the 1980s, where the *compassionate use* of the *antiviral drug ganciclovir*[233] was a mutual duty of rescue independent of clinical trials. Such a compassionate use may be interpreted by some as having created an obligatory (prescribed) ethical obligation, not merely a moral ideal of rescue (nonobligatory moral ideal of beneficence). However, in the mind of *PBE*, to proceed with a *compassionate use* of drug azidothymidine (AZT) in the treatment of HIV/AIDS is not an ethical obligation since most investigational products do no survive clinical trials to achieve regulatory approval, and many turn out to have harmful side effects.[234] Nevertheless, as far as procedural justice from an African outlook is concerned, the distinction between the quality of mutual assistance doesn't dualistically discriminate between specific and general beneficence, obligations and ideals, obligatory beneficence and non-obligatory moral ideals of beneficence, rather the focus is not on the specific moral ideals of beneficence, but rather on the prescribed

231. Contrary to *PBE*, 218.

232. See *PBE*, 218.

233. Compassionate use is the use of unapproved drugs outside of clinical trials. It is primarily a kind of treatment rather than biomedical research. Borysowski et al., "Ethics Review in Compassionate Use," 136; Buhles, "Compassionate Use," 304–15; Cadman, "Ganciclovir Implants," 3–6.

234. See *PBE*, 226.

sympathetic ethical obligation of mutual aid (*busaasizi*) that relationally promotes life and abundantly promotes it, not only for every *muntu* but all Bantu.

Thirdly, the *muntu* offers hospitality generously as another form of procedural justice. "*Luganda kulya olugenda enjala terudda*," literally meaning, one who goes away hungry does not come back again. It means people should be hospitably received. This is well put by John Roscoe, a British anthropologist and missionary, through his impression of the Baganda: "Their extreme politeness was proverbial . . . Their manners were courteous, and they welcomed strangers and showed hospitality to guests." He concludes, "they were charitable and liberal . . . kind and generous."[235] Such a prescription of hospitality and generosity to others is obligatory and goes beyond the *muntu*'s neighbors, friendships, and relatives to include strangers who may share in a meal (*ddiiro*).[236] This is not to suggest a barter ethical system, given that even when Bantu are mutually interdependent, hospitality accorded to visitors through gifts like animals, which aren't expected to be returned, though when reciprocated, it is not a return but a new act of good will.[237] We should however not inaccurately misconstrue such hospitality as a passport to lazy around as the Swahili proverb testifies: "*Mgeni siku mbili; siku ya tatu mpe jembe*" (Treat your guest for two days, and give him a hoe on the third day!)[238] Otherwise laziness is dishonorable and should never abuse the

235. Roscoe, *Baganda*, 6, 267. There are various concrete examples to sustain Roscoe's impression, which moral examples in particular prove the magnitude of friendship and hospitality: *Akwana akira ayomba* ("One who makes friends is better/happier than one who finds fault"); *Awava munno tewadda munno awava okugulu wadda omuggo* ("Friendship is irreplaceable, you lose a leg and get a crutch"); *Omukwano gutta bingi* ("Friendship wipes out a lot"; it covers a lot of problems). *Ekitali kyogere kizimba ku mwoyo* ("The unspoken pain swells on the heart"; one should tell one's troubles or needs to others, and not suffer alone) (Nason, "Proverbs of the Baganda," 249–51, 259); *Ogubula ey'eguya teguwangala* ("Any friendship in which there is no tolerant party cannot last"; friendship should be based on mutual tolerance (*kugumiikiriza*); *Musajja gy'agenda, gyasanga banne* ("Wherever a man goes, he ought to be welcome") (Bakabulindi, "Wisdom of the Baganda," 12, 17).

236. Nsimbi, *Amaanya Amaganda n'Ennono zaago*, 27–35; Sempebwa, *Reality of a Bantu*, 149–51.

237. The Ganda proverbial wisdom illustrates: *Akulunza embuzi, omulunza nte*: If one herds your goats (*mbuzi*), which is an act of goodwill, you are not expected to return the same act of herding his goats; but if you reciprocate his act of goodwill, then herd his cows (*nte*), and *not* his goats, which is a new act of goodwill. Sempebwa, *Reality of a Bantu*, 152; Roscoe, *Baganda*, 6.

238. Bujo, *African Ethic*, 137.

act of hospitality. Moreover, owing to Nyerere's socialist *Ujamaa* aspect of hospitality in Tanzania, practically, one could hardly distinguish between weak and strong members given the solidary and generous structure of the social community.[239] Such positive attitudes or moral ideals of well-being (like generosity, solidarity) are easily cultivated and translated from childhood to adulthood. It is with the same positive attitude that Africans freely greet each other on the streets of Vienna as family members even when they had never seen each other as we mentioned (sis, bro, mama). Contrary to principlism, it is morally reasonable to presume that all morally committed persons share an admiration of and endorsement of at least the moral ideal of mutual solidarity, and more so valuing of friendly relationships, as well as holding them firmly as part of their shared substantive moral beliefs. The Baganda counsel, "*Egindi wala nga tekuli mumanyi (gw'oyagala)*," literally reveals to us that a place is a long way from here only when no friend of yours lives there. The real meaning is that friendship shortens time and distance. It is likewise necessarily universally praiseworthy and gives us all reasons to embrace it not only as an obligatory ethical imperative of every particular morality, but more also of the universal morality.[240]

The *second instance of justice* is corrective justice. Among Ganda, the central purpose of corrective justice is to mutually prevent a recurrence in the relational community, but not to solely reform the agent:

> Although the Ganda believe that punishment can deter the recurrence of wrongs, they do not believe that punishment can necessarily reform the agent. For them, the purpose of corrective justice is to prevent a recurrence and rectify the situation rather than the agent. By committing a wrong the agent has so to speak upset the social equilibrium. This social equilibrium must not only be restored but steps have to be taken to prevent the recurrence of the wrong done. Thus the Ganda have two methods of correction: the remedial method, and the deterrent

239. Nyerere notes this care: "Both the 'rich' and the 'poor' individuals were completely secure in African society. Natural causes brought famine, but they brought famine to everybody—'poor' or 'rich.' No one starved, either for food or for human dignity, because he lacked personal wealth; he could depend on the wealth possessed by the community of which he was a member. That was socialism. That is socialism." Nyerere, *Ujamaa*, 3–4.

240. See Metz, "African Moral Theory," 342; in contrast to *PBE*, 6. About counsels to do with the necessity of friendliness, see Bakabulindi, "Wisdom of the Baganda," 12 and 17.

method—where deterrent refers to the situation and not to the agent.[241]

In remedying a situation with an implication of the relational community, the Ganda like other African peoples always have a conception of compensation for the wrong suffered. Here compensation, fine, damages, or recompense is *ngassi*. *Ngassi* is derived from the verb stem *-gatta*, which means "add," "join," "combine," "put together," "unite," "unify," or "compensate."[242] Time and again, both remedial and deterrent corrective methods are applied through a family-clan communicative palaver to sort for example disputes, some amongst families, others amongst clan communities respectively. *First, a* family may herd livestock on ancestral graves of another family. Contempt of graves of a *muntu's* ancestors attenuates prestige and self-respect while upsetting the social equilibrium. A goat as compensation may be one of the remedial requirements.[243] Since the purpose of corrective justice is to intercept a recurrence and correct the circumstances rather than the moral agent, the value of the goat as compensation (*ngassi*) is to restore the prestigious unity lost with the injured family, besides undoing the offence by the injuring family. The portrayal of *ngassi* is intended not only to be equal to the value of the damage suffered which is prestige lost, but also equal to the value of the damage caused which is the offence.[244] For both the injured family and the injuring family in question, a corrective remedial method mutually offers assistance in preventing the relational recurrence of the wrong done and

241. Sempebwa, *Reality of a Bantu*, 212–13; Haydon, *Law and Justice in Buganda*, 259.

242. Haydon, *Law and Justice in Buganda*, 259; Murphy, *Luganda-English Dictionary*, 90, 413.

243. Haydon, *Law and Justice in Buganda*, 266. Goats are used as *ngassi* because their meat is very good, they are most preferred, easily kept by each peasant to pay future fines, and they have no taboos connected to them. So, goats are regularly used as *ngassi* to remedy wrongs done—so as to restore the upset social equilibrium. If the *ngassi* is not paid (situation remaining upset), one family may bear a grudge, great anger, wrath, rancor or fury (*kiruyi*) on another family, and this could lead to *kuwoolera ggwanga* (to avenge, take vengeance on). The verb *kuwoolera*, comes from the verb stem *-woola* which means putting things in order. *Kuwoolera gwanga* indicates putting things in order or take them apart, but in an unpleasant manner. Carrying out vengeance is always avoided because solidarity is most valued. Roscoe, *Baganda*, 422–23; Haydon, *Law and Justice in Buganda*, 279; Sempebwa, *Reality of a Bantu*, 216–17; Murphy, *Luganda-English Dictionary*, 203, 600.

244. See Sempebwa, *Reality of a Bantu*, 214–15

reconciles the damaged state of affairs without necessarily changing the injuring family (agent).

Second, concerning disputes amongst clan communities, according to the Baganda negative social injunctions (offenses against the community) like, "You ought not to steal" (stealing, *bubbi*), a community member is obliged to take immediate action on seeing a thief,[245] which is believed as an inevitable way to maintain the deep solidarity of the corporate life and clan kinship. If a person steals a goat, relatedness of life is involved because the goat belongs to a member of the corporate community, perhaps to the cousin or the niece of the thief.[246] This offense does not only extend to the thief but equally also to the whole corporate community. If a royal was offended (as the owner of the goat) or one member had a sexual affair with the royal's wife, and given that royals are higher in rank, the awarded *ngassi* has to commensurate with the offended hierarchization, e.g., a death penalty.[247] Here death comes in as a deterrent method in one way. In deterring the situation, the Baganda moral wisdom stretches beyond reforming the individual, because payment of *ngassi* by the thief may not reform the individual: *Mboneredde tatta muze* (I am sorry does not kill the bad habit).[248] Pessimistically, this moral wisdom rests on the substantive belief that, notwithstanding the *ngassi* payment, offenses can be committed again. An alternative terminology for committing an offense or crime is *kuzza musango*, which almost means a repetition of crime. This is because the verb *kuzza* is derived from the stem *-zza*, which means to cause to return or bring back.[249] As far as the "upset social equilibrium," the deterring method of justice is effective when it rectifies the situation and prevents the recurrence of the offense than deterring the moral doer. The moral notion of deterring the recurrence of an offense can be reconstructed to mean a form of punishment (*kibonerezo*) that will make it very difficult if not impossible for one to commit the same offense again, since the nature of *kibonerezo*

245. See Sempebwa, *Reality of a Bantu*, 153.

246. See Mbiti, *African Religions and Philosophy*, 205.

247. See Haydon, *Law and Justice in Buganda*, 276.

248. A couple of proverbs prove Ganda's pessimistic view—that the main purpose of punishment isn't to reform the individual: *Nkumanyi muze takuganya kw'etonda* (Once a person knows you commit wrongs, he does not believe that your being sorry will prevent you from committing those wrongs again). Nason, "Proverbs of the Baganda," 258.

249. See Sempebwa, *Reality of a Bantu*, 217. *Kuzza musango*, to commit crime. Murphy, *Luganda-English Dictionary*, 378.

is suffering.[250] Like Haydon and Roscoe put it, in the village the *kibonerezo* to food theft or robbery has always been death with impunity. Even though minor stealing is punished with preventative considerations, in many contemporary communities, the appropriate punishment to the moral gravity of stolen crops is death.[251] It is not to mean that Africans are ratifying lynch justice. The focus, though with some exceptions,[252] is to simply institute a deterring *kibonerezo* in the administration of justice.

Third and last, there are also *instances of distributive justice* among Africans, though these go beyond the limited communal tenets of fairness, equality, and proportionality amongst humans that are steadily emphasized in principlism, to accommodate the good as conceived within an integral ecological-cosmic nature of the *muntu's* relational ontological reality. Indeed, the cosmic moral nature of sanctions or taboos can be highlightened, for instance, in cases of suicide that disharmoniously decrease life for the whole. According to Tangwa, "*suicide* is strictly taboo to the Nso' and is considered an offence against the earth *(nsag).*" This is because "a person who commits suicide shows contempt and disregard for the family and the community . . . S/he dies 'whole' *(wu kpu i fur)* without indicating a successor, without indicating his/her unpaid debts, unsettled quarrels and disputes, unfulfilled promises etc., all matters of

250. However, the Baganda do not believe in the idea of retaliation: "an eye for an eye, a tooth for a tooth." Sempebwa, *Reality of a Bantu*, 218. The Ganda understanding of the nature of punishment (notion of suffering), is derived from *bonaabona* (*–bonyeebonye*) (suffer, be in misery or pain, experience trouble); *bonyaabonya* (*–bonyaabonyeezza*) (cause to suffer, torture); *bonereza* (*–bonerezza*) (punish). Murphy, *Luganda-English Dictionary*, 28.

251. Haydon, *Law and Justice in Buganda*, 281; Sempebwa, *Reality of a Bantu*, 223; Roscoe, *Baganda*, 15.

252. Some tolerable exceptions may include: "Only the chief's food might be stolen by his peasant with impunity," as these Baganda proverbs prove: *Omwami takayanira mmere* ("The chief doesn't quarrel about food"); *Embuga ye biribwa* ("The chief's headquarters means things to eat"); *Alya ekya mukamawe nga tasenguse taba mubbi* ("He who takes his master's food without his permission while he is still his subject is no thief"). Haydon, *Law and Justice in Buganda*, 281. "Do not steal, except in war." In warfare, Buganda acquired wealth. The acquired wealth included goods and people from the enemy during a war (*kunyanga*, to plunder, rob, steal, seize by force, carry off), which was a praiseworthy action. When one gave up at the point of success, it was referred to as *kukoowa nga banyaga* (to give up or despair when one is on the point of success, i.e., become tired when (others) are enjoying the spoils). Mair, *African People*, 66; Haydon, *Law and Justice in Buganda*, 182; Murphy, *Luganda-English Dictionary*, 456. It is why some proverbs support courage in war: *Bamutta akira eyafa* ("It is better to be killed in battle than to die at home"). *Ekirya atabaala ky'ekirya n'asigadde eka* ("A warrior expects death so he does not stay at home"). Sempebwa, *Reality of a Bantu*, 282.

great importance on which the health, well-being and prosperity of the family and the community depend."[253] Here distributive justice cuts across the three-dimensional community relationally. It caters for the wellness of the forthcoming generation that depends on the ancestor who departs from the living community without settling disputes mutually. If such a mutual cost would be carried down generations without being settled, it would be unethical for Bantu and it would require extensive reconciliatory rites. Assessing the understanding of distributive justice among Bantu as connoting mutual cost, mutual assistance, or mutual aid in relation to the *muntu*'s neighbor and the community seems not parallel to tenets of equality, proportionality, and fairness as espoused in bioethical principlism, even when we focus on communitarian theories.

In principlism, communitarian theories articulate distributive justice as resting on principles of fair, equal, and proportionate distribution derived from conceptions of the good as understood in each moral community. Rightly, these theories uphold relationships between individual roles and their social communal embeddedness, in which individualism and individual rights have no place.[254] One of the adherents is Charles Taylor, a communitarian, who rejects the priority of the individual (rights) as an isolated atom in preference to the conception of the common human good that evolves along varied community frameworks (structures).[255] These structures are seen as pluralistic principles of justice, deriving from as many different conceptions of the good as there are diverse moral communities. It is why, according to Daniel Callahan's avowedly communitarian account, we should enact public policies from a shared consensus about the good rather than based on individual rights. Not to emphasize individual rights, it is here that some recall theories associated with equality. Among these, are other adherents of communitarian justice, where some are egalitarian theorists. These theories are historically as old as religious traditions. They hold that all human persons must be treated as equals because they are created as equals and have equal moral status. In their articulation of distributive justice, egalitarian

---

253. Tangwa, "Bioethics: An African Perspective," 196.

254. Communitarian theories lay claim to traditions traceable to Aristotle, Georg Wilhelm Hegel, and David Hume. *PBE*, 270–1, 275–76.

255. The term "atomism" loosely characterizes the doctrines of social contract theory, which arose in the seventeenth century, and successor doctrines, which are constituted by individuals for the fulfillment of ends that are primarily individual. Taylor, "Atomism," 39–62.

theorists focus on creating an equal measure of liberty and equal access to the goods in life that every rational person values. Despite that, principlists affirm that there is no prominent egalitarian theory that has contained a distributive principle that requires equal sharing of all social benefits and burdens to all persons.[256] Even if we were to consider Rawls's account of justice as equal respect for persons and fairness,[257] he relies on the society's affirmative responsibility to regulate fairness equally, so that each and every member of each particular society has equal access to at least some sufficient—though not maximal—level of good medical care.[258] The question related to distributive justice remains open whether there is an equal interactive sharing of all social benefits and burdens among community members (relationally).

To handle injustices of unequal sharing of social goods, the need of determining one's welfare comes in. Under communitarian justice, the capability theories pioneered by Amartya Sen try to address problems of justice, welfare, and human rights.[259] For Sen, one's welfare is not determined by traditions that look to intrinsic goods—think of utilitarianism and Rawls's account of social primary goods; rather, they are concerned with actual opportunities for living well, i.e., the value of health that is essential for a flourishing life. So social institutions can and should protect and promote means necessary for everyone to exercise capabilities essential for a flourishing life or do things one has reason to value.[260] Likewise for Martha Nussbaum, in her capability approach, to secure a right to a citizen's flourishing is to put them in a position of capability that effectively and affirmatively functions with institutional support.[261] Even though Nussbaum claims each of the ten capabilities she identifies[262] is equally important, she places special emphasis on two of them: practical reason and affiliation. Her emphasis, according to principlists, is rightly

256. See PBE, 270–71, 274, 276.

257. See Rawls, Justice as Fairness, 58–60.

258. See Rawls, Justice as Fairness, 42–44.

259. The capability account represents the various functionings one is able to do or be. Sen, "Capability and Well-Being."; Gaertner, "Amartya Sen: Capability and Well-Being,"

260. See PBE, 270–1, 277.

261. Nussbaum, Frontiers of Justice, 155–23, 281–91, especially 287.

262. Human capabilities of every world citizen, in each and every nation, include life; bodily health; bodily integrity; senses, imagination, and thought; emotions; practical reason; affiliation; living with other species like animals, plants, and the world of nature; play; and control over one's environment. Nussbaum, "Frontiers of Justice," 76–78.

justifiable since she explicitly finds the core behind the intuition of human functioning in a dignified free person who constructs her way of life in reciprocity with others, and not merely following or being shaped by others.[263] For her, welfarism is to put citizens in a position of capability. Here welfarism is primarily hinged on a free autonomous individual who is promised theoretical positions of capabilities without indicating how she can *actually* achieve these capabilities. But there are also well-being theories of social justice that try to fill up this gap in Nussbaum's welfarism. Originally, the theory was conceived by Madison Powers and Ruth Faden and explicitly directed at public health and health policy, emphasizing core dimensions of human welfare.[264] Power and Faden emphatically reject Nussbaum's language of capabilities. For them, justice is not the capability or freedom to achieve well-being, but rather the actual achievement of well-being. Justice is to be actually healthy; it is to be actually secure, not the mere capability to be healthy. Under "Capabilities, Functioning, and Well-Being," Powers and Baden confirm:

> We want to form attachments to others for our own sake and for reasons of sustaining just institutions and practices. What we want is the success of these attachments, not simply that some can form them if they so choose. Justice requires not just the capability for attachment if individuals wish to exercise it; rather it is essential for the capability to be developed and exercised for the success of other-regarding morality . . . *The central concern of justice, then, is with achievement of well-being, not the freedom or capability to achieve wellbeing* . . . We cannot achieve many things for others, but we can do things that support or enable others to achieve those outcomes themselves . . . We can design institutions that enable persons to be more likely to have and exercise their capabilities for affiliation and respect, and this is the outcome that justice aims for. At bottom, we care centrally about well-being, but we recognize also that many societal duties related to bringing it about involve a supportive role.[265]

We can conclude with two points. One, justice for welfarists is an actual achievement of well-being with support of others and supportive institutions, not an individual's capability or freedom to achieve well-being.

263. See *PBE*, 278.

264. Madison Powers and Faden Ruth characterize the dimensions of well-being as the adequate theory of justice in six core dimensions: health, personal security, reasoning, respect, attachment, and self-determination. *Social Justice*, 15–29, especially 16.

265. Powers and Faden, *Social Justice*, 37–41, here specifically 40–41. Emphasis added.

We can locate some similarities with the African approach to justice. This is basically why corrective justice as mutual aid in the African sense frustrates a recurrence and corrects the actual community circumstances rather than the individual. African logic finds it plausible to rectify not only the upset social equilibrium/well-being but also to deter the recurrence of the wrong done.

Second, Nussbaum's capability approach extends justice to the life of other species like animals, plants, and the world of nature, which notion to an extent connects with African moral logic. We can quickly note that the African approach to justice crosses relationally from human communities to the natural world. Take an instance of distribution of acquired goods ,which may be regulated under the *muntu*'s achieved personhood, valuable skills, or services, but not limited to her. There are practical examples in nature intersubsistence activities amongst Ganda like hunting, where the one who spots the animal (*muzizi*) renders a more valuable service and skill than the tracker who locates the source for others to follow (*abagoberezi*).[266] The reader shouldn't be surprised with the numerous examples related to nature such as animal life or savannah or hunting. It is because nature intersubsistence is wholly supported by the volcanic forest vegetation, and it is why Africans south of the Sahara are ordinarily farming oriented given their remarkably fertile volcanic soils, which scientifically give birth to a tropical rainy climate throughout the year.[267] So with the example of the *muntu*'s hunting in the natural fertile world, two issues of justice come up. One, the *muzizi* receives a better share since spotting an animal to be hunted is a more valuable service than hunting. Two, hunting is not a violation (unjustified action) of nature rights but rather an infringement (justified action). For the virgin natural savannahs of Sub-Saharan Africans, hunting legitimately overrides nature rights because it culls an overpopulated habitat to sustain nature intersubsistence; i.e., hunting is mutual

266. For more on Ganda notion of the merits of valuable services and skills, see Sempebwa, *Reality of a Bantu*, 204–8; Haydon, *Law and Justice in Buganda*, 181–82; Nsimbi, *Amaanya Amaganda n'Ennono zaago*, 29. Mugoberezi, follower, adherent. *Abagoberezi b'ensolo*, those who follow/pursue game (in a hunt). Murphy, *Luganda-English Dictionary*, 354, 397.

267. There are scientific reasons why some African communities turned to agriculture than animal husbandry. The Baganda became agriculturalists rather than pastoralists, just because, the "luxurious forest vegetation rendered Buganda unsuitable for cattle grazing." That favored them becoming primarily agriculturalists and remotely pastoralists. Kiwanuka, *History of Buganda*, 26.

aid exercised between Bantu communities and the natural world. This is against any therapeutic hunting that would be purely human predation and for anthropocentric reasons. Distributive justice as mutual assistance (aid) is a harmless simplification backing a relational ethics of an African thought, given that it isn't limited to interhuman spheres of life only, which could even be majorly physical; rather it encompasses the ecological-cosmic nature of life.

6

Conclusion

Résumé of Part II and Conclusive Indications of a Comprehensive and Inclusive Approach to Bioethics

FROM CONTEXT TO GLOBAL

As we come to the end of the previous chapter and the book as a whole, Tangwa's studies—as one of the discussed Africanists—can still guide us. He succinctly compares and contrasts bioethical principlism that is based on the industrialized Western culture with his interpretation of African moral thought as based on an *eco-bio-communitarian* culture of "live and let live":

> Western culture is predominantly an anthropocentric, literate culture, technologically advanced, epistemologically and morally driven by an obsession with certainty and a Manichean syndrome that strictly distinguishes good from evil. It embodies an attitude that leaves the impression of knowing all that is knowable, and shaped in many ways by free market forces and profit motives, it is outward-looking, dominant, and domineering of other cultures which it tries to assimilate or at least proselytize into its ever expanding universal vision and operations. By contrast . . . African culture is a "live and let live" culture,

predominantly oral rather than literate, and marked by great
variety and diversity . . . African culture is, in essence, eco-bio-
communitarian . . . , tolerant, cautious, non-aggressive, non-
proselytizing, and inward-looking.[1]

Relying on other scholarship, in addition to Western principlism
being couched in economic terms, along a culture that knows it all with
an ethical key to the universalizable common morality, Tangwa emphasizes that "the Western worldview can be described as predominantly
anthropocentric and individualistic, and contrasted with its African
counterpart, which *he describes* as eco-bio-communitarian."[2] We have
proposed appropriating bioethical principlism to an integrative normative approach. Therefore, if consequentialistic bioethical principlism is
to sit well in an African setting, an eco-biocentric attitude of "live and
let live" is proffered by Tangwa as a contributory factor in strategizing a
united entire system: "a more humble motivation for the pursuit of science and technology based on the eco-bio-centric attitude of *live and let
live* can be substituted for the aggressive motivation of domination to
the immeasurable advantage of the whole of humankind."[3] Tendering in
an eco-biocentric attitude is to counteract even benevolent undertakings
that may yield unpredicted immoral consequences that cannot simply
be attributed to our originally intended benevolence. Tangwa gives an
example: "If I rescue a man from drowning and he heads home and immediately murders his pregnant wife and child, it cannot, by any stretch
of the rational or moral imagination, be the case that my rescue action
was morally wrong because its bad consequences by far outweigh the
good consequences." It is here that Tangwa appeals:

> The prospect of "evidence-based ethics" is no less appealing
> than that of "evidence-based medicine," but neither of them,
> as a holistic approach, it seems to me, can be satisfactory. As
> a general moral theory of human action, consequentialism is
> quite unconvincing and rationally indefensible; for the simple
> reason that no amount of supposedly good consequences can

---

1. Tangwa, "African Thought," 104.

2. Tangwa, "Biomedical and Environmental Ethics," 392. Emphasis is mine to actu-
alize grammar. The original is "I have described."

3. Tangwa, "Biomedical and Environmental Ethics," 394; see also Tangwa, *African
Bioethics in a Western Frame*, 157.

ever justify the intentional, free and voluntary performance of an action known or even only believed to be morally wrong.[4]

The unconvincing principlistic consequentialism visible in modern recolonization tendencies can be subverted through a decolonization of ethics, and this project can be completed partly by integrating the ex-colonies' world of treasures and voices within agreed global regulatory documents or codifications. This clarion call begs bioethical principlism to distance itself, as it has been the standard norm, from *offering Global North cultural particulars as universals for the whole globe*:

> Since the Nuremberg Code, and until the very recent revisions of the Helsinki and CIOMS guidelines, what had taken place in the formulation of "international" ethical guidelines were basically attempts to universalize and globalize a particularly powerful paradigm, a Western paradigm; even if it has much to recommend it and has also admittedly attempted to borrow and incorporate a few foreign ideas into its framework. The trend to involve other voices, perspectives, and cultures in the formulation of international ethical guidelines is one that needs encouraging and enhancing, in view of the fact that much of the so-called developing world are also ex colonies of much of the developed world.[5]

Though consequentialism is plausibly viable if we are to pursue a partial moral theory, its fatal general limitation along bioethical principlism, for instance, is to increasingly emphasize empirical bioethics ("evidence-based ethics" or "evidence-based medicine"), a universalized common morality, for sorting global ethical problems, which approach seems very attractive to people with a background in the empirical disciplines. Such bioethics accords rather well with consequentialist utilitarianism, which is inept in Black African contexts as we have extensively presented exploitive and corruptive cases in biomedical interventions in Africa. It is clear that African moral thought prioritizes context. Equally, just like other African scholars, Tangwa's proposal is that bioethics should be promoted, first and foremost, by the authorities of each locality, country, or region and, secondarily, by international or global agencies and institutions.[6] More or less, this is what Bujo proffers as context-sensitive universalism,

4. Tangwa, "Moral Status of Embryonic Stem Cells," 452.

5. Tangwa, "Universalism and relativism," 67.

6. Tangwa, "Leaders in Ethics Education," 104; Tangwa, *African Bioethics in a Western Frame*, 7.

where the starting point of self-articulation lies not in abstract universal principles but in practical concrete communities. In dialogue with Bujo and Tangwa, I challenge not only principlistic claims of absoluteness of universal common morality,[7] the principle-based structures that reject *particular cultural frameworks*[8] (customary moralities), but also any project of *world ethos*,[9] such as of Hans Küng, that purely grounds bioethics on anthropocentric standards. Tangwa's concern "with Western ethical thinking (and Western thinking in general) is its obsession with details and the incorrigible conviction that there is one correct principle from which everything is deducible and to which everything is relatable, as well as the perennial search for such a principle."[10] It would be egocentric for some "individuals to perceive their own culture as *the* culture,"[11] since "all human cultures, like all human beings themselves, are *morally* equal."[12] If we approach every culture with an open and unprejudiced mindset, we cannot fail enormously reaping something positive. Therefore, "no single human culture is privileged with *a holistic* and comprehensive view of reality . . . There is no culture that possesses *monopoly* of disinterested objective thinking, while the others are left with, at best, only *the ability to narrate their cultural opinions*, prejudices, biases and their folkloric tribal myths and legends . . . no particular culture, in itself, is superior or inferior to any other."[13] We necessarily need a combined contribution of approaches from each cultural context, even when all cultures are limited

7. More claims of the universal common morality in the PBE, see Walter, *Das Absolute in der Ethik*; Küng, *Das Projekt Weltethos*.

8. Tangwa, "African Thought," 105.

9. Hans Küng tries to answer the question why we need a global ethos in his book *Projekt Weltethos* (1990). His reason is because an ethos is needed for the entire humanity (*die Notwendigkeit eines Ethos für die Gesamtmenschheit*) (p. 14). Basically, he answers this question anthropocentrically in three parts: The first part is "no survival without a word ethos" ("kein Überleben ohne ein Weltethos") (pp. 19–96). The second part is "no world peace without peace of religions" ("kein Weltfriede ohne Religionsfriede") (pp. 97–135). The third part is "no peace of religions without dialogue between religions" ("kein Religionsfriede ohne Religionsdialog") (pp. 138–71). In Küng's effort for a global ethos, he seeks for a principle-based way of life to yield a harmonious coexistence. He emphasizes moral principles that basically remain focussed on anthropocentrism.

10. Tangwa, "Bioethics: An African Perspective," 193; see Tangwa, *African Bioethics in a Western Frame*, 175.

11. Tangwa, *African Bioethics in a Western Frame*, 85. Emphasis original.

12. Tangwa, *African Bioethics in a Western Frame*, 88. Emphasis original.

13. Tangwa, *African Bioethics in a Western Frame*, 98. Emphasis added.

and imperfect, towards achieving a more comprehensive and inclusive functional approach to bioethics.

This is why even when Western culture is scientifically-cum-technologically more successful, the imperialistic nature of its principlistic culture should not exploitatively dominate non-Western worlds or further loudly assert Western culture as an infallible oracle along globalization, while propagating a high degree of universal immunity and imperviousness against non-Western cultural influences. *On the one hand*, such immunities are promoted by old colonial doctrines like "might is right," "knowledge is power," "economic determinism," Europeanization (Westernization) of other cultures disguised as humanization. All these lead to a pervasive desire to know everything, coordinate everything, harmonize, control, impose a highly exploitative Western economic system, commercialize, and justifiably monopolize everything (unholy alliances between African governments and biotechnologists and big multinational corporations/pharmaceutical industries), while condescending listening to ("a big mouth but small ears") and benefiting from other universal cultures. We can never claim that one culture can fix the whole world, since other cultures too have enriching perspectives on life and need to be appreciatively listened to.[14] So the four biomedical principles, along their theories, have a Global North cultural orientation, terms, language, and idioms, and remain "very much a paradigm of the Western industrialized world, where their relevance and urgent applicability have been made abundantly manifest by various activities that violated or that run

14. See Tangwa, "Bioethics: An African Perspective," 185; Tangwa, *African Bioethics in a Western Frame*, 4–6, 40–3, 60–1, 73–75, 96, 104, 124; Tangwa, "African Thought," 104, 159–61, 191–92. Even though Tangwa equally argues for the equality of cultures *qua* cultures and of the importance of different cultural perspectives, given the limitations of each and every particular culture, he insists that issues of cross-cultural relevance should not promote epistemological superiority. "The obsession with certainty, and the illusion that may be induced of having achieved it in many domains of human concern, is what has given the Western world its spirit of epistemological over-confidence, an *over-sabi* bordering on arrogance, its evangelical and proselytising impulse, its high sense of self-righteousness, that could easily result in heedless recklessness at the level of practice." Tangwa, "Bioethics, Biotechnology and Culture," 125–38. Tangwa argues that even-though cultures are limited with imperfections, they are all morally equal without any being any superior to another. Cultures remain different lenses through which we observe reality. Tangwa is of the view that morality and cultures are like dancing masquerades where one must change one's position to have an "adequate but necessarily partial of view" of the masquerade. It is why principle-based cultures should develop the habit of also standing up and moving around a bit, to view the dancing masquerade from different perspectives, lest they exclude some other enriching wisdom that may lead to their complete demise. Tangwa, "Morality and Culture," 20.

the risk of violating them."[15] *On the other hand*, Africans have positively benefited from the indisputably historical dominant Western cultures by honestly endeavoring to enrich their indigenous approaches to bioethics and the globalized Western-based principlistic approaches. But unfortunately, in so doing, African indigenous cultural vital aspects, in turn, have scarcely neither humanized nor enriched (basically no genuine conversational ethics) the so-called universalized Western cultural approaches to bioethics.[16] In different contexts, we witness more "subtle forms of colonization and exploitation such as biological colonization."[17] Examples are evident in African efforts to improve native practices of medicine that are unfortunately condemned through principlistic oriented bioethical interventions. In order to *bring together* rather than *take apart* African interventions in bioethics and universally oriented principlistic interventions, local efforts ought to be positively received and amalgamated in a broader bioethical discourse. In addition to Prof. Victor Anomah Ngu's efforts in securing an HIV vaccine as presented before, similar efforts in which African traditional practices of medicines have added value to the medical world include a pharmaceutical company in South Africa that developed a local herb called *Sunderlandia* into standardized tablets as well as two traditional medicines standardized in Nigeria—dopravil and conavil—that are effective for the management of HIV/AIDS and also progressed in clinical trial phases.[18] We should, however, note how these efforts struggled to find a global reception, even when many would have understood it as an easy venture of collaboration, given a contemporary scientific-cum-technological advancement era that is backed by the WHO, and in which era we preach absence of moral superiority and a context-sensitive bioethics.

Here context sensitivity is not to tolerate moral sensitivity in the sense where "a drug trial which cannot be carried out in the Western/first/developed world on *ethical grounds* can be carried out in Africa or any other part of the so-called third/underdeveloped world, on the grounds that those on whom the tests are carried out are those facing epidemic scale of infections, are too poor to buy expensive drugs and

15. Tangwa, "Ethical Principles," s4.

16. See Tangwa, "Bioethics: An African Perspective," 199.

17. Tangwa, *African Bioethics in a Western Frame*, 81.

18. See Nyika, "African Traditional Medical Practice," S36; Yauri, and Awaisu, "Natural Medicinal Products in Nigeria," 120–30; African Advisory Committee for Health Research and Development, "Report of the Nineteenth Meeting."

stand to benefit the most from the result of such trials."[19] Though such practices as ethics dumping are rampant in Africa, these trails remain disreputable, and it would be indeed an unethical reading of what context sensitivity ought to entail. Before substantiating further on context sensitivity, besides Tangwa, Bujo counteracts not only "reiterative universalism," which allows universal principles to take on validity in particular instances, but also "covering law universalism," in which the principles for good life are established in general terms without any real possibility of self-articulation in particular contexts. It becomes compelling to suggest an African relational community ethic because it not only involves a contextualistic or context-sensitive universalism that is neither reduced to a "covering-law universalism" nor "reiterative universalism," but rather favors both the cultural group and individuals since they can-articulate themselves within the sociality of the community. Besides, context-sensitive self-articulation doesn't claim any absolute validity but remains open to dialogue and homeomorphically receptive of other global partner contexts. Then universal claims must be brought down to earth by local practices that concretize the universal, the particular which must allow itself to be questioned and justified in formal universal terms.[20] Africanists like Tangwa, in the first instance, think that we need to evolve from recognition of local identities and the fact that "the dominant Western identity is also a local identity imposing itself globally."[21] It is why it is critical "for every people, every culture and every part of the world to reflect on the applicability of these principles in their own particular context and situation."[22] Put differently, bioethical principles (contrary to rules of thumb) must have a "particular relevance and importance . . . in relation to a particular context and situation in all their peculiar circumstances" without claiming any absoluteness "in the abstract or in advance of the particular concrete situation to which they apply," given that, any of the principles "could justifiably be bent or even broken, if the particular situation and context so warrants."[23] This moral guidance would be more meaningful if we are to embrace a genuine global discourse, as any global agreed framework can be helpful only as a constitution guiding precise contexts—such as favoring construal of local proverbs, sanctions

19. Tangwa, *African Bioethics in a Western Frame*, 100. Emphasis original.
20. See Bujo, *An African Ethic*, 8, 23.
21. Tangwa, *African Bioethics in a Western Frame*, 5.
22. Tangwa, "Ethical Principles," s4–5.
23. Tangwa, "Ethical Principles," s6.

and imperatives, idiomatic expressions, and way of community life—but not with comprehensively prepared rules of thumb, rules with all the exceptions pertaining to varied cultural normative differentiations. Here Tangwa identifies not the claimed principlistic universalized common morality, but local moral contexts as having procedural contents:

> Biomedical research rules of thumb are best elaborated at the local rather than international level. To the extent that any of the guidelines expresses a genuine ethical imperative, to that same extent would it be universally relevant and applicable. But to apply it in a particular concrete situation, it must necessarily be shaped and colored, like water in a container, by all the data furnished by particular context and perspective. To attempt *determining such details for one milieu from another milieu* is to run the high risk of serious error. In any case no such details, determined a priori, can provide advance justification for violation of ethical imperatives without prior in situ contextual appreciation of all the particularizing, constraining, or compelling concrete circumstances.[24]

Argued differently, even if we were to base a genuine global discourse on bioethical principlism, it doesn't have detailed procedural rules of thumb that can be applied *with certainty, equally, and uniformly* everywhere as Tangwa maintains. Bioethical principlism is not absolute and so must apply equally to global communities without necessarily attempting to uniformize customary moral rules. For bioethical principlism, even when it has different levels of standards, to hold good universally and timelessly, its concrete application in African settings must be markedly interpreted and adapted in culture-specific contexts.[25] In dialogue, Orobator firmly concludes that like all other sciences, for example, even "theology does not float above culture and context. Doing theology is not an exercise in conceptual weightlessness. It develops within the particular culture and context of the community that attempts to utter a word or two on the reality of God and demands of faith for daily living."[26] For example, an African approach to bioethics has comprehensively illustrated this through means of a practical communicative palaver, a palaver that relationally embodies the three-dimensional community in both its contextual and global forms.

---

24. Tangwa, "Universalism and Relativism," 66. Emphasis added.

25. See Tangwa, "Medical Research in Developing Countries," 46–47.

26. Orobator, *Theology Brewed in an African Pot*, 139.

Conclusive Moral Prescriptions Arising from African Moral Thought and an Indication of a Possible Contribution

In Part II, I have established what is essential for an African interpretation of bioethics. While taking the criticism of various Africanists and African scholars seriously, I examined how an approach to bioethics within African moral thought can be unfolded so as to *bring together* rather than *take apart* African approaches to bioethics and a universally oriented principlism. Even though every region around the globe needs to fashion its own approach to bioethics to address those pertinent bioethical problems impinging on its day-to-day living, a global dimension to bioethics is inevitable as most of the bioethical issues have a global scope while each culture has a different approach.

In order to tame globalized exploitations and vernacularize the efficacy of bioethical principlism,[27] the global nature of principlism bidirectionally necessitates not only a universally inclusive restructuring but also a locally strategized approach. *On the one hand,* African scholarships have criticized the fact that dominant Western paradigms in bioethics have been unresponsive to local needs and values in Africa, neglecting much-needed population-based familiar approaches, advocacy, and health promotion that could better address the local social determinants of health by globally embracing the bioethics from below (especially from those that are marginalized).[28] In revising the two major controversial principles of principlism, autonomy is replaced by *relational respect for persons* or what Tangwa has called respect for other human beings as moral equals, while justice is replaced by *relational mutual harmony.*[29] In reconstructing principlistic ethical codes by harmonizing and balancing individual and community moral needs, as well as lending a stronger moral authority to the kinds of ethical guidelines being localized in Africa, an African approach to bioethics strongly suggests a richer and deeper concept of respect for persons that appreciates *respect for the creativity of each one's autonomous decisions without isolating oneself from the broader relational communal life.* Respect for the *relational person* is

27. See Tangwa, "Ethics in Occupational Health," 4. Appadurai, contributes a new framework for the cultural study of globalization in *Modernity at Large.*

28. See Tangwa, "Ethics in Occupational Health," 9; Azétsop, "African Bioethics," 14–15; Behrens, "Indigenous African Bioethics," 34; Maura, "Beyond a Western Bioethics?," 158–77.

29. Behrens and Wareham, "Africanized Bioethics," 111–12; Tangwa, "Universalism and Relativism," 64; Tangwa, *African Bioethics in a Western Frame,* 162–63.

the key to consent, human dignity, or status as opposed to individual "autonomy." What is operationalized here is the moral theory of *personhood* as "a person is a person through other persons," as well as "mutual aid" or "mutual assistance" or "mutual cost" operationalized as justice. Here justice as mutual aid is *relational harmony*, not simply fairness. The individual *muntu* is inclusively embedded in the broader caring clan family. In interpreting the African moral theory of personhood, I further suggested *respectful sympathy* beyond strong anthropocentrism to include other entities hierarchically (hierarchy of life forces as one characteristic of African moral thought). It means not excluding inanimate life in the moral world by embracing the notion of *community sympathy* (sympathetic community dignity), a practical normative perspective that overrides all moral controversies, inadequacies, and conflicts across cosmic entities to include all entities and very creation. *On the other hand*, in the pursuit of a more inclusive and comprehensive efficacious biomedical ethics, I go beyond Tangwa's emphasis of the Africanness of bioethics. Despite his persuasive input, I depart from his preoccupation on disputes about which ethical principles deserve prioritization between African and Western approaches and appreciate what can enrich bioethics without setting Western bioethical principlism and an African approach apart. For a global biomedical ethics to succeed, we need some respectful humility and flexibility that appreciates conversion towards a broader perspective that unwinds both global and local problems, an approach that appreciates the sacredness of the interconnectedness of all members of the triadic community (more so beyond one's own land[30]). The sought approach could be added as the fifth principle to the four traditional bioethical principles or the hunted approach could permeate all the other principles.

Narrowing down, the distinctive moral prescription of Africans in their approach to bioethics interconnectedly relies on the substantive belief that every *muntu* ought to promote life of the corporate community, i.e., the living, the living dead, those not yet born, and all living and non-living entities. In order not to simply promote life but to promote it fully, the African ethical system appreciates *promoting an integral community life* by continuously promoting the norm of *solidarity, continued vitality*, the *hierarchization of life* of every entity within the *corporate community*. The social reality of the community palaver indeed supplies the relevant

30. Tangwa et al., "Global Health Inequalities," 242; Bujo, *Ethical Dimension of Community*, 21–22.

normative content to the life principle, to which the individual must not blindly follow, but rather responsibly interact with, without her identity, freedom, or conscience being crushed. This normative structure presupposes the ontological social reality that all material reality in the physical corporate community has spiritual life, corroborating that there is no creation that is lifeless in Bantu moral thought. Therefore, in reconstructing Tangwa's eco-biocentric attitude of "live and let live," we cannot erect a Black African approach to bioethics outside an integral social interaction of *all* existents of the three-dimensional community (including the visible and invisible worlds, the victims and victimized of the history). Besides, their continuity and vitality, solidarity and hierarchical order must be harmoniously and vitally maintained in a relational social balance if biomedical principlism is to sit well in an African milieu. And if this mutuality of life is abundantly promoted, then we can talk of a *holistic* prescription for an African approach to bioethics, while if this holistic solidarity of life is obstructed, an African remains trapped in a difficult position to promote life fully and holistically, both locally and globally.

Approvingly, the African ontological social reality of the interconnectedness of all creation translates into an inclusive and more comprehensive holistic normative sense of solidarity (none can be better than us all) that equally extends ethically to systems of life, institutions, and nations, biodiversity and ecology, the created living and non-living entities—in which sense I condense everything in the *Other*. It is this holistic milieu that offers a platform in which the morally challenging consequentialistic exploited relationships in Africa as discussed in the second chapter are handled: 1) exploited cosmic-ecological relationships, 2) exploited interpersonal and humanistic collaborations, and 3) exploited structural-universal collaborations. It is to these challenges and other similar ones, which I summarize as the *Other*, that I am proffering an approach of *holistic bioethics*, conceivable as the *normative approach of solidarity* hinged on *respectful sympathy*, a contribution to a comprehensive and inclusive approach to bioethics. This holistic bioethics will be explored in my next scholarly work.

Concluding Résumé

In six chapters, this work has investigated how bioethical principlism can be complemented, both locally and globally, so that we can unite rather than separate African approaches to bioethics and a universally oriented principlism. Western principlism as majorly presented by Thomas L. Beauchamp and James F. Childress can hardly be incarnated in the African context unless it is complemented by a normative understanding of solidarity, on the one hand, while on the other hand the local context as presented by African scholars—like Bujo Bénézet, Godfrey B. Tangwa and many others—needs to simultaneously be enhanced in order to enrich bioethical principlism. Without claiming to offer the last word on intercultural bioethics or getting preoccupied with disputes about which bioethical approaches deserve prioritization, this work has interactively relied on ordinary moral experiences practically visible in the social realities surrounding Black Africans south of the Sahara and principle-based studies.

In particular, the African social-ethical system promotes a relational *community life* by continuously promoting the norms of *solidarity, continued vitality,* and the *hierarchization of life* of every entity. Within this social reality, all material reality in the physical corporate community has spiritual life, corroborating that there is no creation that is lifeless in Bantu moral thought. Therefore, in reconstructing African regional moralities like Tangwa's eco-biocentric attitude of "live and let live," we cannot erect a Black African approach to bioethics outside a relational social interaction of all existents. So, African moral thought proposes a more integral approach to solidarity, suggesting how every entity and force can promote or weaken the life of the relational corporate community, i.e., the living, the living-dead (including the victimized ancestors of history), those not yet born, and all living and non-living entities.

Bibliography

Achebe, Chinua. *Things Fall Apart*. London: Heinemann, 1958.

African Advisory Committee for Health Research and Development Meeting. "African Advisory Committee for Health Research and Development (AACHRD): Report of the Nineteenth Meeting." Harare: WHO Regional Office for Africa, 27 March–31 March 2000.

Aggleton, Peter. "'Just a Snip'? A Social History of Male Circumcision." *International Journal on Sexual and Reproductive Health and Rights* 15 (2007) 15–21.

Aine, Bob. "Museveni Castigates Circumcision as Away [*sic*] to Prevent HIV/AIDS." *PML Daily*, October 20, 2018. https://www.pmldaily.com/news/2018/10/museveni-castigates-circumcision-as-away-to-prevent-hiv-aids.html.

Allen, Anita. *Unpopular Privacy: What Must We Hide?* New York: Oxford University Press, 2011.

American Medical Association. "Confidentiality." *Code of Medical Ethics*, Opinion 3.2.1. https://code-medical-ethics.ama-assn.org/ethics-opinions/confidentiality#:~:text=Physicians%20in%20turn%20have%20an,personal%20health%20information%20is%20disclosed.

Andoh, Cletus T. "Bioethics and the Challenges to Its Growth in Africa." *Open Journal of Philosophy* 1 (2011) 67–75.

Andorno, R. "Global Bioethics at UNESCO: In Defense of the Universal Declaration on Bioethics and Human Rights." *Journal of Medical Ethics* 33 (2007) 150–54.

Anstötz, Christoph. "Should a Brain-Dead Pregnant Woman Carry Her Child to Full Term? The Case of the 'Erlanger Baby.'" *Bioethics* 7 (1993) 340–50.

Appadurai, Arjun. *Modernity at Large: Cultural Dimensions of Globalization*. Minneapolis: University of Minnesota Press, 1996.

Aquinas, Thomas. *Summa Theologiae*. Translated by Laurence Shapcote and edited by John Mortensen and Enrique Alarcón. Lander, Wyoming: Aquinas Institute for the Study of Sacred Doctrine, 2012.

———. *The Summa Contra Gentiles*. Translated by Vernon J. Bourke. Notre Dame, IN: University of Notre Dame Press, 1975.

Archer, Alfred. "Are Acts of Supererogation Always Praiseworthy?" *Theoria* 82 (2016) 238–55.

Areen, Judith. "The Legal Status of Consent Obtained from Families of Adult Patients to Withhold or Withdraw Treatment." *Journal of the American Medical Association* 258 (1987) 229–35.

Arendt, Hannah. *The Life of the Mind*. London: Harcourt Trade, 1978.

Aristotle. *The Complete Works of Aristotle*. Edited by Jonathan Barnes. Princeton, NJ: Princeton University Press, 1984.

———. *Ethica Nicomachea*. Greek version. Edited by J. Bywater. Oxford: Clarendon, 1894.

———. *Eudemian Ethics*. Greek version. Edited by F. Susemihl. Leipzig: Teubner, 1884.

———. *Eudemian Ethics*. English version. Translated by H. Rackham. Cambridge, MA: Harvard University Press, 1981.

———. *Nicomachean Ethics*. Translated by H. Rackham. Cambridge, MA: Harvard University Press, 1934.

Arras, John D., and Fenton M. Elizabeth. "Bioethics Human Rights: Access to Health-Related Goods." *The Hastings Center Report* 39 (2009) 27–38.

Avila, D. "Assisted Suicide and the Inalienable Right to Life." *Issues in Law and Medicine* 16 (2000) 111–41.

Azetsop, Jacquineau. "Theological Creativity, Christian Imagination, and Ecclesial Practices in a Time of AIDS." *HIV and AIDS in Africa: Christian Reflection, Public Health, Social Transformation*, edited by Jacquineau Azetsop, xv–xxiii. Maryknoll, NY: Orbis, 2016.

Bagheri, Alireza, et al. "Experts' Attitudes towards Medical Futility: An Empirical Survey from Japan." *BMC Medical Ethics* 7 (2006). doi:10.1186/1472-6939-7-8.

Bailey, Robert C., et al. "Male Circumcision for HIV Prevention in Young Men in Kisumu, Kenya: A Randomized Controlled Trial." *Lancet* 369 (2007) 643–56.

Bakabulindi, J. "The Traditional Wisdom of the Baganda Concerning Moral Behavior." *Occasional Research Papers in African Traditional Religions and Philosophy* 17 (1974) 12–17.

Bartlett, Robert H. "Clinical Research in Acute Fatal Illness." *Journal of Intensive Care Medicine* 31 (2016) 456–65.

Battin, Margaret P. "Development of the AAS Statement on 'Suicide' and 'Physician Aid in Dying.'" *Suicide and Life-Threatening Behavior* 49 (2019) 774–76.

Beauchamp, Tom L. "The Medical Ethics of Physician-Assisted Suicide." *Journal of Medical Ethics* 25 (1999) 437–39.

———. "Rights Theory and Animal Rights with Animals." In *Oxford Handbook of Animal Ethics*, edited by Tom L. Beauchamp and R. G. Frey, 198–227. New York: Oxford University Press, 2011.

Beauchamp, Tom L., and James F. Childress. *Principles of Biomedical Ethics*. 8th ed. New York: Oxford University Press, 2019.

Beets, Peter A. D. "Strengthening Morality and Ethics in Educational Assessment through Ubuntu in South Africa." *Educational Philosophy and Theory* 44 (2012) 68–83.

Behrens, Kevin G. "African Philosophy, Thought and Practice and Their Contribution to Environmental Ethics." PhD diss., University of Johannesburg, 2011.

———. "Towards an Indigenous African Bioethics." *South African Journal of Bioethics and Law* 6 (2013) 32–35.

———. "Two 'Normative' Conceptions of Personhood." *Quest: An African Journal of Philosophy* 25 (2013) 103–19.

Behrens, Kevin G., and C. S. Wareham. "Bioethics beyond Borders: Toward an Africanized Bioethics Curriculum." *Cambridge Quarterly of Healthcare Ethics* 30 (2021) 103–13.

Bellieni, Carlo V. "The Pain Principle: An Ethical Approach to End-of-Life Decisions." *Ethics and Medicine* 36 (2020) 41–49.

Berge, Erling, et al. "Lineage and Land Reforms in Malawi: Do Matrilineal and Patrilineal Landholding Systems Represent a Problem for Land Reforms in Malawi?" *Land Use Policy* 41 (2014) 61–69.

Bermúdez, José Luis. "Mindreading and Moral Significance in Nonhuman Animals." In *Oxford Handbook of Animal Ethics*, edited by Tom L. Beauchamp and R. G. Frey, 407–40. New York: Oxford University Press, 2011.

Bernat, James L. "Medical Futility: Definition, Determination, and Disputes in Critical Care." *Neurocritical Care* 2 (2005) 198–205.

Bernstein, Ira M., et al. "Maternal Brain Death and Prolonged Fetal Survival." *Obstetrics and Gynecology* 74 (1989) 434–37.

Biko, Steve. *I Write What I Like: A Selection of His Writings*. Johannesburg: Picador Africa, 2004. Originally published 1978.

Bimwenyi-Kweshi, Oscar. *Alle Dinge erzählen von Gott. Grundlegung afrikanischer Theologie*. Freiburg/Basel/Wien: Herder, 1982. Originally published as *Discours théologique négro-africain. Problèmes des fondements* (Paris, 1981).

Binsbergen, Wim van. "The Roman Catholic Church, and the Hermeneutics of Race, as Two Contexts for African Philosophy." *Quest: An African Journal of Philosophy* 19 (2005) 2–30.

Blair, Irene V., et al. "Unconscious (Implicit) Bias and Health Disparities: Where Do We Go from Here?" *Permanente Journal* 15 (2011) 71–78.

Borysowski, Jan, et al. "Ethics Review in Compassionate Use." *BMC Medicine* 15 (2017) 136–36.

Bostick, Nathan A., et al. "Report of the American Medical Association Council on Ethical and Judicial Affairs: Withholding Information from Patients: Rethinking the Propriety of 'Therapeutic Privilege.'" *The Journal of Clinical Ethics* 17 (2006) 302–6.

Bradie, Michael. "The Moral Life of Animals." In *The Oxford Handbook of Animal Ethics*, edited by Tom L. Beauchamp and R. G. Frey, 547–73. New York: Oxford University Press, 2011.

Bradley, Andrew. "Positive Rights, Negative Rights and Health Care." *Journal of Medical Ethics* 36 (2010) 838–41.

Bridges, John F. P., et al. "Engaging Families in the Choice of Social Marketing Strategies for Male Circumcision Services in Johannesburg, South Africa." *Social Marketing Quarterly* 16 (2010) 60–76.

The British Museum. "Egyptian Death and Afterlife: Mummies. About 2686 BC–AD 39." https://www.britishmuseum.org/collection/galleries/egyptian-death-and-afterlife-mummies.

Brody, Baruch A., and Amir Halevy. "Is Futility a Futile Concept?" *The Journal of Medicine and Philosophy* 20 (1995) 123–44.

Brown, Rebecca C. H. "Moral Responsibility for (Un)healthy Behavior." *Journal of Medical Ethics* 39 (2013) 695–98.

Buhles, William C. "Compassionate Use: A Story of Ethics and Science in the Development of a New Drug." *Perspectives in Biology and Medicine* 54 (2011) 304–15.

Bujo, Bénézet. "Differentiations in African Ethics." In *The Blackwell Companion to Religious Ethics*, edited by William Schweiker, 423–37. Oxford: Blackwell, 2005.

———. *The Ethical Dimension of Community: The African Model and the Dialogue between North and South.* Nairobi: Paulines Publications Africa, 1998. Originally published in German as *Die ethische Dimension der Gemeinschaft: Das afrikanische Modell im Nord-Süd-Dialog* (Universitätsverlag Freiburg, Schweiz Paulusdrukerei, 1993).

———. *Foundations of an African Ethic: Beyond the Universal Claims of Western Morality.* Translated by Brian McNeil. New York: Crossroad, 2001. Originally published in German a *Wider den Universalanspruch westlicher Moral* (Freiburg im Breisgau: Herder, 2000).

Cadman, J. "Ganciclovir Implants: One Year Later." *GMHC (Gay Men's Health Crisis) Treat Issues* 11(1997) 3–6.

Cain, Joanna M. "Is Deception for Reimbursement in Obstetrics and Gynecology Justified?" *Obstetrics and Gynecology* 82 (1993) 475–78.

Callahan, Daniel. "Individual Good and Common Good: A Communitarian Approach to Bioethics." *Perspectives in Biology and Medicine* 46 (2003) 496–507.

———. "Principlism and Communitarianism." *Journal of Medical Ethics* 29 (2003) 287–91.

Camilla, Yusuf, and Fessha Yonatan. "Female Genital Mutilation as a Human Rights Issue: Examining the Effectiveness of the Law against Female Genital Mutilation in Tanzania." *Africa Human Rights Law Journal* 13(2013) 356–82.

Caplan, Arthur, and Folkers Kelly McBride. "Charlie Gard and the Limits of Parental Authority." *The Hastings Center Report* 47 (2017) 15–16.

Chatfield, Kate, and David Morton. "The Use of Non-Human Primates in Research." In *Ethics Dumping*, edited by D. Schroeder et al., 81–89. SpringerBriefs in Research and Innovation Governance. Berlin: Springer, 2018. https://doi.org/10.1007/978-3-319-64731-9-10.

Chattopadhyay, Subrata, and Raymond De Vrie. "Bioethical Concerns Are Global, Bioethics is Western." *Eubios Journal of Asian and International Bioethics* 18 (2008) 106–9.

Chemhuru, Munamato. "Using the African Teleological View of Existence to Interpret Environmental Ethics." *Philosophia Africana* 18 (2016) 41–51.

Childress, James, and Tom L. Beauchamp. "Common Morality Principles in Biomedical Ethics: Responses to Critics." *Cambridge Quarterly of Healthcare Ethics* 31 (2022) 164–76.

———. *Principles of Biomedical Ethics.* 5th ed. Oxford University Press, 2001.

Chuwa, Leonard Tumaini. *African Indigenous Ethics in Global Bioethics: Interpreting Ubuntu.* Advancing Global Bioethics. New York: Springer Science and Business Media, 2014.

Chwang, Eric. "Futility Clarified." *The Journal of Law, Medicine, and Ethics* 37 (2009) 487–95.

"Circumcision for the Correction of Sexual Crimes among the Negro Race." *Maryland Medical Journal* 30(1894) 345–46.

Close, Elian, et al. "Doctors' Perceptions of How Resource Limitations Relate to Futility in End-of-Life Decision Making: A Qualitative Analysis." *Journal of Medical Ethics* 45 (2019) 373–79.

Clouser, Danner K., and Gert Bernard. "A Critique of Principlism." *Journal of Medicine and Philosophy* 15 (1990) 219–36.

Collopy, Bart J. "Autonomy in Long-Term Care: Some Crucial Distinctions." *The Gerontologist* 28 (1988) 10–17.

Coulson-Smith, Peta, et al. "In Defense of Best Interests: When Parents and Clinicians Disagree." *American Journal of Bioethics* 18 (2018) 67–69.

Council for International Organizations Medical Sciences (CIOMS). *International Ethical Guidelines for Biomedical Research Involving Human Subjects.* Geneva: CIOMS, 2002.

Damman, Ernest. *Die Religionen Afrikas.* Stuttgart: Kohlhammer, 1963.

D'Angelo, Abby, et al. "Assessing Genetic Counselors' Experiences with Physician Aid-in-Dying and Practice Implications." *Journal of Genetic Counseling* 28 (2019) 164–73.

Daskal, Steven. "Support for Voluntary Euthanasia with No Logical Slippery Slope to Non-Voluntary Euthanasia." *Kennedy Institute of Ethics Journal* 28 (2018) 23–48.

Dave, S. S., et al. "Male Circumcision in Britain: Findings from a National Probability Sample Survey." *Sexually Transmitted Infections* 79 (2003) 499–500.

DeGrazia, David. "Common Morality, Coherence, and the PBE." *Kennedy Institute of Ethics Journal* 13 (2003) 219–30.

———. "Moral Status as a Matter of Degree?" *Southern Journal of Philosophy* 46 (2008) 181–98.

———. *Taking Animals Seriously.* New York: Cambridge University Press, 1996.

Dehon, Erin, et al. "A Systematic Review of the Impact of Physician Implicit Racial Bias on Clinical Decision Making." *Academic Emergency Medicine* 24 (2017) 895–904.

Dempwold, O. "Sprachforschung und Mission." *Das Buch der deutschen Weltmission,* Herausgeber D. J. Richter. Gotha, 1935.

Deutsche Gesellschaft für Internationale Zusammenarbeit (GIZ). "A Guide to Peaceful Coexistence on Private Mailo Land." February 28, 2018. https://landportal.org/fr/node/87490.

Dheensa, Sandi, et al. "'Is This Knowledge Mine and Nobody Else's? I don't Feel That.' Patient Views about Consent, Confidentiality, and Information-Sharing in Genetic Medicine." *Journal of Medical Ethics* 42 (2016) 174–79.

Dhooper, Surjit Singh. "Organ Transplantation—Who Decides?" *Social Work* (New York) 35 (1990) 322–27.

Dickert, Neal W., et al. "Reframing Consent for Clinical Research: A Function-Based Approach." *The American Journal of Bioethics* 17 (2017) 3–11.

Dworkin, Gerald, et al. *Euthanasia and Physician-Assisted Suicide.* New York: Cambridge University Press, 1998.

Edwardes, Allen. *Death Rides a Camel: A Biography of Sir Richard Burton.* London: Julian, 1983.

Éla, Jean-Marc. *Mein Glaube als Afrikaner. Das Evangelium in Schwarz-afrikanischer Lebenswirklichkeit.* Theologie der Dritten Welt 10. Freiburg im Breisgau: Herder, 1987.

———. *My Faith as an African.* Translated by John Pairman Brown and Susan Perry. Nairobi: Action, 2001.

———. "Symbolique Africaine et Mystère Chrétien." *Les Quatre fleuves* 10 (1979) 91–109.

Elwyn, Todd S., et al. "Cancer Disclosure in Japan: Historical Comparisons, Current Practices." *Social Science and Medicine* 46 (1998) 1151–63.

Emerson, Claudia I., et al. "Access and Use of Human Tissues from the Developing World: Ethical Challenges and a Way Forward Using a Tissue Trust." *BMC Medical Ethics* 12 (2011). doi:10.1186/1472-6939-12-2.

ENACT (Enhancing Africa's Response to Transnational Organized Crime). "Trafficking of Human Beings for the Purpose of Organ Removal in North and West Africa." Interpol, July 2021.

Engelhardt, Tristram H. *The Foundations of Bioethics*. New York: Oxford University Press, 1996.

Etieyibo, Edwin E. "African Philosophy and Nonhuman Nature." In *Debating African Philosophy: Perspectives on Identity, Decolonial Ethics and Comparative Philosophy*, edited by George Hull, 164–81. London: Routledge, 2019.

———. "Ubuntu, Cosmopolitanism and Distribution of Natural Resources." *Philosophical Papers* 46 (2017) 139–62.

Etzioni, Amitai. "On a Communitarian Approach to Bioethics." *Theoretical Medicine and Bioethics* 32 (2011) 363–74.

European Union. "Directive 2010/63/EU of the European Parliament and of the Council of 22 September 2010 on the Protection of Animals Used for Scientific Purposes." Text with EEA relevance. OJ L 276/33. https://eur-lex.europa.eu/legal-content/EN/TXT/?uri=CELEX:32010L0063.

Eze, Michael Onyebuchi. "Menkiti, Gyekye, and Beyond: Towards a Decolonisation of African Political Philosophy." *Filosofia Theoretica* 7 (2018) 1–17.

———. "What Is African Communitarianism? Against Consensus as a Regulative Ideal." *South African Journal of Philosophy* 27 (2008/9) 386–99.

Faik-Nzuji, Clémentine Madiya. *Die Macht des Sakralen. Menschen, Natur und Kunst in Afrika. Eine Reise nach innen*. Solothurn/Düsseldorf: Walter, 1993. Originally published as *La puissance du sacré*.

Fayemi, Kazeem Ademola. "African Bioethics vs. Healthcare Ethics in Africa: A Critique of Godfrey Tangwa." *Developing World Bioethics* 16 (2016) 98–106.

Fenton, Elizabeth M., and John D. Arras. "Bioethics and Human Rights: Curb Your Enthusiasm." *Cambridge Quarterly of Healthcare Ethics* 19 (2010) 127–33.

Fox, Marie, and Michael Thomson. "HIV/AIDS and Male Circumcision: Discourses of Race and Masculinity." In *Exploring Masculinities: Feminist Legal Theory Reflections*, edited by M. A. Fineman and M. Thomson, 97–113. London: Routledge, 2016.

Frader, Joel, and John D. Lantos. "ECMO and the Ethics of Clinical Research in Pediatrics." *The New England Journal of Medicine* 323 (1990) 409–13.

Frederick, Danny. "Pro-Tanto versus Absolute Rights." *The Philosophical Forum* 45 (2014) 375–94.

Freeman, Victor G., et al. "Lying for Patients: Physician Deception of Third-Party Payers." *Archives of Internal Medicine* 159 (1999) 2263–70.

Freund, Karen M., et al. "Disparities by Race, Ethnicity, and Sex in Treating Acute Coronary Syndromes." *Journal of Women's Health* 21(2012) 126–32.

Frey, R. C. "Utilitarianism and Animals." In *The Oxford Handbook of Animal Ethics*, edited by Tom L. Beauchamp and R. G. Frey, 172–97. New York: Oxford University Press, 2011.

Friedman, Marilyn. "Autonomy and Social Relationships: Rethinking the Feminist Critique." In *Autonomy, Gender, Politics: Studies in Feminist Philosophy*, 81–97. New York: Oxford University Press, 2003.

Gaertner, Wulf. "Amartya Sen: Capability and Well-Being." In *The Quality of Life*, edited by Martha Nussbaum and Amartya Sen, 62–66. Oxford: Oxford University Press, 1993. https://doi.org/10.1093/0198287976.003.0005.

Gaylin, Willard. "The Competence of Children: No Longer All or None." *The Hastings Center Report* 12 (1982) 33–38.

German Medical Association (Die Bundesaerztekammer). "Confidentiality" §9 *Schweigepflicht*, Abs. 1, of the Professional Code of the German Medical Association. https://.www.bundesaerztekammer.de/fileadmin/user_upload/downloads/pdf-Ordner/Recht/MBO-AE_Beschluesse_124._DAET_2021_engl._Fassung.pdf.

Gert, Bernard, et al. "Distinguishing between Patient's Refusals and Requests." *The Hastings Center Report* 24 (1994) 13–15.

———. "Principlism." Ch. 5 in *Bioethics: A Systematic Approach*. 2nd ed. Oxford University Press, 2006. https://doi.org/10.1093/0195159063.003.0005.

Gillon, Raanan. "Autonomy and the Principle of Respect for Autonomy." *British Medical Journal* (Clinical Research Edition) 290 (1985) 1806–8.

———. "Bioethics, Overview." In *The Concise Encyclopedia of the Ethics of New Technologies*, edited by Ruth Chadwick, 1–12. San Diego: Academic Press, 2001.

Glass, Michael. "Forced Circumcision of Men" (abridged). *Journal of Medical Ethics* 40 (2014) 567–71.

Goldberg, Aryeh. "A Futile Use of Futility." *The Hastings Center Report* 50 (2020) 4–5.

Gómez-Torres, María José, et al. "Effectiveness of Human Spermatozoa Biomarkers as Indicators of Structural Damage during Cryopreservation." *Cryobiology* 78 (2017) 90–94.

Gómez-Vírseda, Carlos, and Rafael Amo Usanos. "Relational Autonomy: Lessons from COVID-19 and twentieth-century philosophy." *Medicine, Health Care, and Philosophy* 24 (2021) 493–505.

Greely, Henry T. "Human/Nonhuman Chimeras: Assessing the Issues." In *Oxford Handbook of Animal Ethics*, edited by Tom L. Beauchamp and R. G. Frey, 671–98. New York: Oxford University Press, 2011.

Green, Alexander R., et al. "Implicit Bias among Physicians and Its Prediction of Thrombolysis Decisions for Black and White Patients." *Journal of General Internal Medicine* 22 (2007), 1231–38.

Greene, Mark, et al. "Ethics: Moral Issues of Human-Non-Numan Primate Neural Grafting." *Science* 309 (2005) 385–86.

Gudmundur, Alfredsson, and Eide Asbjorn. *The Universal Declaration of Human Rights: A Common Standard of Achievement*. The Hague: Martinus Nijhoff, 1992.

Gwandure, Calvin. "The ethical concerns of using medical male circumcision in HIV prevention in sub-Saharan Africa." *South African Journal of Bioethics and Law* 4 (2011) 89–94.

Gyekye, Kwame, and Kwasi Wiredu, eds. *Person and Community: Ghanaian Philosophical Studies*. Washington, DC: Council for Research in Values and Philosophy, 1992.

H3Africa Consortium. "Research Capacity: Enabling the Genomic Revolution in Africa." *Science* 344 (2014) 1346–48.

Hallich, Oliver. "Embryo Donation or Embryo Adoption? Conceptual and Normative Issues." *Bioethics* 33 (2019) 653–60.

Hannan, Edward L., et al. "Access to Coronary Artery Bypass Surgery by Race/Ethnicity and Gender among Patients Who Are Appropriate for Surgery." *Medical Care* 37 (1999) 68–77.

Harris, John. *Bioethics: Oxford Reading in Philosophy*. Oxford: Oxford University Press, 2001.

Hawkins, Jennifer S., and Emanuel J. Ezekiel. *Exploitation and Developing Countries*. Princeton, MJ: Princeton University Press, 2008.

Haydon, Edwin Scott. *Law and Justice in Buganda*. London: Butterworths, 1960.

Hazra, Nisha C., et al. "'Fair innings' in the face of ageing and demographic change." *Health Economics Policy Law* 13 (2018) 209–17.

Hegel, George W. F. *The Philosophy of History*. Translated by J. Sibree. New York: Wiley, 1990.

Heidegger, Martin. *Being and Time*. Translated by J. Macquarrie and E. Robinson. New York: Harper and Row, 1962.

Hein, Irma M. "Key Factors in Children's Competence to Consent to Clinical Research." *BMC Medical Ethics* 16 (2015) 74–74.

Hein, Irma M., et al. "Why Is It Hard to Make Progress in Assessing Children's Decision-Making Competence?" *BMC Medical Ethics* 16 (2015) 1–1.

Henk, Ten Have. *Global Bioethics: An Introduction*. London: Routledge, 2016.

Herissone-Kelly, Peter. "Determining the common morality's norms in the sixth ed. of *Principles of Biomedical Ethics*." *Journal of Medical Ethics* 37 (2011) 584–87.

Herring, Andrew A., et al. "Insurance Status of U.S Organ Donors and Transplant Recipients: The Uninsured Give, but Rarely Receive." *International Journal of Health Services* 38 (2008) 641–52.

Herzberg, Stephan. "Die thomanische Deutung des Gewissens." In *Gewissen. Dimensionen eines Grundbegriffs medizinischer Ethik*, edited by Franz-Josef Bormann and Verena Wetzstein, 189–208. Berlin/Boston: De Gruyter, 2014.

Hess, Paul L. "Sex and Race/Ethnicity Differences in Implantable Cardioverter-Defibrillator Counseling and Use among Patients Hospitalized with Heart Failure: Findings from the Get With The Guidelines–Heart Failure Program." *Circulation* 134 (2016) 517–26.

Heyland, Daren K., et al. "Understanding Cardiopulmonary Resuscitation Decision Making: Perspectives of Seriously Ill Hospitalized Patients and Family Members." *Chest* 130 (2006) 419–28.

Hezayehei, Maryam, et al. "Sperm Cryopreservation: A Review on Current Molecular Cryobiology and Advanced Approaches." *Reproductive biomedicine* 37 (2018) 327–39.

Ho, Anita. "Relational Autonomy or Undue Pressure? Family's Role in Medical Decision-Making." *Scandinavian Journal of Caring Sciences* 22 (2008) 128–35.

Hoffman, Kelly M. "Racial Bias in Pain Assessment and Treatment Recommendations, and False Beliefs about Biological Differences between Blacks and Whites." *Proceedings of the National Academy of Sciences* 113 (2016) 4296–301.

Hoffmann, Tobias. "Conscience and Synderesis." In *The Oxford Handbook of Aquinas*, edited by Brian Davies and Eleonore Stump, 255–64. Oxford/New York: Oxford University Press, 2012.

Holm, Søren. "Principles of Biomedical Ethics." 5th ed. *Journal of Medical Ethics* 28 (2002) 329–34.

Horikawa, Naoshi, et al. "Changes in Disclosure of Information to Cancer Patients in a General Hospital in Japan." *General Hospital Psychiatry* 22 (2000) 37–42.

Horn, Lyn, and Gabriel Mwaluko. "Public Health Research Ethics." In *Research Ethics in Africa: A Resource for Research Ethics Committees*, edited by Paul Ndebele et al., 99–107. Stellenbosch: Sun Media, 2014.

Hyder, A. A., et al. "Ethical Review of Health Research: A Perspective from Developing Country Researchers." *Journal of Medical Ethics* 30 (2004) 68–72.

Hyun, Su Kim, and Diane Kjervik. "Deferred Decision Making: Patients' Reliance on Family and Physicians for CPR Decisions in Critical Care." *Nursing Ethics* 12 (2005) 493–506.

Ikuenobe, Polycarp A. "The Communal Basis for Moral Dignity: An African Perspective." *Philosophical Papers* 45 (2017) 437–69.

———. "Good and Beautiful: A Moral-Aesthetic View of Personhood in African Communal Traditions." *Essays in Philosophy* 17 (2016) 124–63.

———. "Human Rights, Personhood, Dignity, and African Communalism." *Journal of Human Rights* 17 (2018) 589–604.

———. *Philosophical Perspectives on Communalism and Morality in African Traditions.* Lanham, MD: Lexington, 2006.

Ilogu, Edmund Christopher Onyedum. *Christian Ethics in an African Background: A Study of the Interaction of Christianity and Ibo Culture.* Leiden: Brill, 1974.

Jahn, Janheinz. *Muntu. Umrisse der neoafrikanischen Kultur.* Düsseldorf: E. Diederichs, 1958.

———. "Value Conceptions in Sub-Saharan Africa." In *Cross-Cultural Understanding: Epistemology in Anthropology*, edited by F. S. C. Northrop and H. L. Livingston, 55–69. New York: Harper and Row, 1964.

Jason, Hanna. *In Our Best Interest: A Defense of Paternalism.* Oxford: Oxford University Press, 2018.

Jeffrey, Ian David. "Relational Ethical Approaches to the COVID-19 Pandemic." *Journal of Medical Ethics* 46 (2020) 495–98.

Jegede, Samuel. "African Ethics, Health Care Research and Community and Individual Participation." *Journal of Asian and African Studies* 44 (2009) 239–53.

Jha, Aisha K., et al. "Racial Trends in the Use of Major Procedures among the Elderly." *New England Journal of Medicine* 2005 (353) 683–91.

Jones, David Albert. "Is There a Logical Slippery Slope from Voluntary to Nonvoluntary Euthanasia?" *Kennedy Institute of Ethics Journal* 21 (2011) 379–404.

Joynt, M. G. "The Critical Care Society of Southern Africa Consensus Statement on ICU Triage and Rationing (ConICTri)." *South African Medical Journal* 109 (2019) 613–29.

Kagabo, Liboire. "Alexis Kagame (1912–1981): Life and Thought." In *A Companion to African Philosophy*, edited by Kwasi Wiredu, 231–42. Oxford: Blackwell, 2004.

Kagame, Alexis. "The Empirical Apperception of Time and the Conception of History in Bantu Thought." In *African Philosophy: A Classical Approach*, edited by Parker English and Kibujjo M. Kalumba, 81–90. Upper Saddle River, NJ: Prentice Hall, 1996.

———. *La Philosophie bântu-rwandaise de l'Être.* Bruxelles: Académie Royale des Sciences Coloniales, 1956.

———. *Sprache und Sein: die Ontologie der Bantu Zentralafrikas.* Heidelberg: Kivouvou Verlag, 1985.

Kamm, Frances. *Intricate Ethics: Rights, Responsibilities, and Permissible Harm*. New York: Oxford University Press, 2006.

Karlström, Mikael. "Imagining Democracy: Political Culture and Democratization in Buganda." *Africa: Journal of the International African Institute* 66 (1996) 485–505.

Kasozi, Ferdinand Mutaawe. *Introduction to an African Philosophy: The Ntu'ology of the Baganda*. München/Freiburg: Karl Alber, 2011.

Katongole, J. C. "Ethos Transmission through African-Bantu Proverbs: Proverbs as a Means for Transmitting Values and Beliefs among Africans with the Example of Bantu-Baganda." PhD diss., Würzburg, 1997.

Keating, Gregory C. "Principles of Risk Imposition and the Priority of Avoiding HHarm." *Revus: Journal for Constitutional Theory and Philosophy of Law* 36 (2018) 7–39.

Kelbessa, Workineh. "Environmental Philosophy in African Traditions of Thought." *Environmental Ethics* 40 (2018) 309–23.

Kilama, Wen. "From Research to Control: Translating Research Findings into Health Policies, Operational Guidelines and Health Products." *Acta Tropica* 112 (2009) S91–101.

King, Patricia. "The Authority of Families to Make Medical Decisions for Incompetent Patients after the Cruzan Decision." *Law, Medicine, and Health Care* 19 (1991) 76–79.

Kirungi, W. L., et al. "Trends in Antenatal HIV Prevalence in Urban Uganda Associated with Uptake of Preventive Sexual Behavior." *Sexually Transmitted Infections* 82 (2006) i36–41.

Kiwanuka, Semakula, and Matia Mulumba. *A History of Buganda: From the Foundation of the Kingdom to 1900*. London: Longman, 1971.

Kleinlugtenbelt, Ydo V., and Kim Madden. "Cochrane in CORR. sup. (R): Decision Aids for People Facing Health Treatment or Screening Decisions." *Clinical Orthopaedics and Related Research* 475 (2017) 1298. https://doi.org/10.1007/s11999-017-5254-4.

Knox, Peter. *Aids, Ancestors and Salvation: Local Beliefs in Christian Ministry to the Sick*. Nairobi: Paulines Publications Africa, 2008.

Kopelman, Loretta M. "Conceptual and Moral Disputes about Futile and Useful Treatments." *The Journal of medicine and philosophy* 20 (1995) 109–21.

Kopelman, Loretta M., and Anton A. van Niekerk. "AIDS and Africa." *Journal of Medicine and Philosophy* 27 (2002) 139–42.

Kristinsson, Sigurdur. "Autonomy and Informed Consent: A Mistaken Association?" *Medicine, Health Care, and Philosophy* 10 (2007) 253–64.

Küng, Hans. *Das Projekt Weltethos*. München/Zürich: Piper, 1990.

Kyewalyanga, Francis-Xavier Serufusa. *Traditional Religion, Customs and Christianity in Uganda: As Illustrated by the Ganda with Some References to Other African Cultures and Islam*. Freiburg im Breisgau: Krause Renner, 1976.

Lamb, Rae M. "Hospital Disclosure Practices: Results of a National Survey." *Health Affairs* 22 (2003) 73–83.

———. "Open disclosure: The Only Approach to Medical Error." *Quality and Safety in Health Care* 13 (2004) 3–5.

Lamont, Mark. "Forced Male Circumcision and the Politics of Foreskin in Kenya." *African Studies* 77(2018) 293–311.

Larmore, Charles. *The Morals of Modernity*. Cambridge: Cambridge University Press, 1996.

Laumann, Edward O., et al. "Circumcision in the United States: Prevalence, Prophylactic Effects, and Sexual Practice." *Journal of the American Medical Association* 277 (1997) 1052–57.

Lavrysen, Laurens. *Human Rights in a Positive State: Rethinking the Relationship between Positive and Negative Obligations under the European Convention on Human Rights*. Cambridge: Intersentia, 2016.

LenkaBula, Puleng. "Beyond Anthropocentricity-Botho/Ubuntu and the Quest for Economic and Ecological Justice in Africa." *Religion and Theology* 15 (2008) 375–94.

Levi-Strauss, Claude. *L'Analyse structurale en Linguistique et en Anthropologie*. Paris: Pon, 1958.

Lewis, Penny. "The Empirical Slippery Slope from Voluntary to Non-Voluntary Euthanasia." *The Journal of Law, Medicine, and Ethics* 35 (2007) 197–210.

Liddell, Henry George, and Robert Scott. *A Greek-English Lexicon*. Rev. by Sir Henry Stuart Jones with Roderick McKenzie. Oxford: Clarendon, 1940.

———. *An Intermediate Greek-English Lexicon*. Oxford: Clarendon, 1889.

Lillehammer, Hallvard. "Voluntary Euthanasia and the Logical Slippery Slope Argument." *Cambridge Law Journal* 61 (2002) 545–50.

Lindemann, Hilda Nelson. "The Architect and the Bee: Some Reflections on Postmortem Pregnancy." *Bioethics* 8 (1994) 247–67.

Locke, John. *An Essay Concerning Human Understanding*. London: Oxford University Press, 1964.

Loewy, Rich H. "In Defense of Paternalism." *Theoretical Medicine and Bioethics* 26 (2005) 445–68.

Low-Beer, Daniel, and Musoke Sempala. "Social Capital and Effective HIV Prevention: Community Responses." *Global Health Governance* 4 (2010) 1–18.

Low-Beer, Daniel, and Rand L. Stoneburner. "AIDS Communications through Social Networks: Catalyst for Behavior Changes in Uganda." *African Journal of AIDS Research* 3 (2004) 1–13.

———. "Behaviour and Communication Change in Reducing HIV: Is Uganda Unique?" *African Journal of AIDS Research* 2 (2003) 9–21.

———. "Population level HIV declines and behavioral risk avoidance in Uganda." *Science* 304 (2004) 714–18.

Lurie, Nicole. "Health Disparities-Less Talk, More Action." *The New England Journal of Medicine* 353 (2005) 727–29.

Lutz, David W. "African Ubuntu Philosophy and Global Management." *Journal of Business Ethics* 84 (2009) 313–28.

Macbeath, Alexander. *Experiments in Living*. New York: Macmillan, 1952.

MacIntrye, Alasdair. *After Virtue: A Study in Moral Theory*. 2nd ed. Notre Dame, IN: University of Notre Dame Press, 1981.

Maffioli, Elisa M. "Is Traditional Male Circumcision Effective as an HIV Prevention Strategy? Evidence from Lesotho." *PLoS One* 12 (2017). doi:10.1371/journal.pone.0177076.

Magesa, Laurenti. *African Religion: The Moral Traditions of Abundant Life*. Nairobi: Paulines Publications Africa, 1998.

———. "Contextualizing HIV and AIDS in the African Reality." In *HIV and AIDS in Africa: Christian Reflection, Public Health, Social Transformation*, edited by Jacquineau Azetsop, 15–26. Maryknoll, NY: Orbis, 2016.

———. *What Is Not Sacred? African Spirituality*. Maryknoll, NY: Orbis, 2013.

Mair, Lucy Philip. *An African People in the Twentieth Century*. London: Routledge, 1934.

Mallinson, R. K., and B. T. Sibandze. "HIV and Male Circumcision in Swaziland." In *Case Studies in Global Health Policy Nursing*, edited by G. E. Dorman and M. de Chesnay, 121–45. New York: Springer, 2018.

Manninen, A. B. "A Case for Justified Non-Voluntary Active Euthanasia: Exploring the Ethics of the Groningen Protocol." *Journal of Medical Ethics* 32 (2006) 643–51.

Maquet, Jacques. *Africanity: The Cultural Unity of Black Africa*. Translated by Joan R. Rayfield. Oxford University Press, 1972.

Marouf, Fatma E., and Bryn S. Esplin. "Setting a Minimum Standard of Care in Clinical Trials: Human Rights and Bioethics as Complementary Frameworks." *Health and Human Rights* 17 (2015). https://www.hhrjournal.org/2015/06/setting-a-minimum-standard-of-care-in-clinical-trials-human-rights-and-bioethics-as-complementary-frameworks/.

Masolo, Dismas A. *Self and Community in a Changing World*. Bloomington: Indiana University Press, 2010.

———. "Western and African Communitarianism: A Comparison." In *A Companion to African Philosophy*, edited by Kwasi Wiredu, 483–98. London: Oxford, 2003.

Matevia, Marilyn. "Justice for All: Revisiting the Prospects for a Biocommunitarian Theory of Interspecies Justice." *Journal of International Wildlife Law & Policy* 19 (2016) 189–202.

Matolino, Bernard, and Wenceslaus Kwindingwi. "The End of Ubuntu." *South African Journal of Philosophy* 32 (2013) 197–205.

Maura, Ryan A. "Beyond a Western Bioethics?" *Theological Studies* 65 (2004) 158–77.

Mbih, Jerome Tosam. "African Perspectives in Global Bioethics." *Developing World Bioethics* 18 (2018) 208–11.

———. "The Philosophical Foundation of Kom Proverbs." *Journal on African Philosophy* 9 (2014) 1–27.

———. "The Relevance of Kom Ethics to African Development." *International Journal of Philosophy* 2 (2014) 36–47.

Mbiti, John S. *African Religions and Philosophy*. London/Nairobi: Heinemann, 1969.

———. *African Religions and Philosophy*. 2nd ed. Oxford: Heinemann Educational, 1989.

———. *Afrikanische Religion und Weltanschauungen*. Berlin/New York: De Gruyter, 1974.

———. *Concepts of God in Africa*. Southampton: SPCK, 1970.

McCune, M. J. "The SCID-Hu Mouse: A Small Animal Model for the Analysis of Human Hematolymphoid Differentiation and Function." *Bone Marrow Transplant* 9 (1992) 74–76.

McElwee, Brian. "Supererogation across Normative Domains." *Australasian Journal of Philosophy* 95 (2017) 505–16.

Meinhof, Carl. *Grundzüge einer vergleichen Grammatik der Bantu-sprachen*. Hamburg: Eckardt and Messtorf, 1948.

Mendy, Maimuna, et al. "Infrastructure and Facilities for Human Biobanking in Low-and Middle-Income Countries: A Situation Analysis." *Pathobiology* 81(2014) 252–60.

Menkiti, Ifeanyi. "Person and Community in African Traditional Thought." *African Philosophy: An Introduction*, edited by R. Wright, 171–81. Lanham, MD: University Press of America, 1984.

———. "Person and Community—A Retrospective Statement." *Filosofia Theoretica* 7 (2018) 162–67.

Merkies, Hubertus Cornelis Gerardus. *Ganda Classification: An Ethno-Semantic Survey.* Nijmegen: De Katholieke Universiteit te Nijmegen, 1980.

Metz, Thaddeus. "African and Western Moral Theories in a Bioethical Context." *Developing World Bioethics* 10 (2010) 49–58.

———. "African Moral Theory and Public Governance: Nepotism, Preferential Hiring, and Other Partiality." In *African Ethics: An Anthology of Comparative and Applied Ethics*, edited by Munyaradzi Felix Murove, 335–56. Durban: KwaZulu-Natal, 2009.

———. "An African Theory of Moral Status: A Relational Alternative to Individualism and Holism." *Ethical Theory and Moral Practice: An International Forum* 14 (2012) 387–402.

———. "Ethics in Africa and in Aristotle: Some Points of Contrast." *Phronimon* 13 (2012) 99–117.

———. "How the West Was One: The Western as Individualist, the African as Communitarian." *Educational Philosophy and Theory* 47 (2015) 1175–84.

———. "Human Dignity, Capital Punishment, and an African Moral Theory: Toward a New Philosophy of Human Rights." *Journal of Human Rights* 9 (2010) 81–99.

———. "Introduction: Engaging with the Philosophy of D. A. Masolo." *An African Journal of Philosophy* 25 (2013) 7–16.

———. "Toward an African Moral Theory." *Journal of Political Philosophy* 15 (2007) 321–41.

———. "Two Conceptions of African Ethics." *An African Journal of Philosophy* 25 (2013) 141–62.

———. "Ubuntu as a Moral Theory and Human Rights in South Africa." *African Human Rights Law Journal* 11 (2011) 532–59.

Mike, Valerie, et al. "ECMO: Clinical trials and the ethics of evidence." *Journal of Medical Ethics* 19 (1993) 212–18.

Mill, John Stuart. *Utilitarianism.* First published 1864. Cambridge: Cambridge University Press, 2014. https://doi.org/10.1017/CBO9781139923927.

Mitchell, Gready R. "Medical futility, treatment withdrawal and the persistent vegetative state." *Journal of Medical Ethics* 19 (1993) 71–76.

Mkhize, Nhlanhla. "Ubuntu-Botho Approach to Ethics: An Invitation to Dialogue." In *African Perspectives on Ethics for Healthcare Professionals*, edited by Nico Nortjé et al., 25–48. Advancing Global Bioethics 13. Berlin: Springer, 2018.

Mokgoro, Yvonne. "Ubuntu and the Law in South Africa." *Potchefstroom Electronic Law Journal* 1(1998) 15–26.

Molefe, Motsamai. An African Perspective on the Partiality and Impartiality Debate: Insights from Kwasi Wiredu's Moral Philosophy." *South African Journal of Philosophy* 36 (2017) 470–82.

———. *An African Philosophy of Personhood, Morality and Politics.* New York: Palgrave Macmillan, 2019. doi:10.1007/978-3-030-15561-2.

———. "A Rejection of Humanism in African Moral Tradition." *Theoria* 62 (2015) 59–77.

———. "Solving the Conundrum of African Philosophy Through Personhood: The Individual or Community?" *Journal of Value Inquiry* 54 (March 2019) 41–57. https://doi.org/10.1007/s10790-019-09683-8.

Moosa, R. M., and M. Kidd. "The Dangers of Rationing Dialysis Treatment: The Dilemma Facing a Developing Country." *Kidney International* 70 (2006) 1107–14.

Morata, Lauren. "An Evolutionary Concept Analysis of Futility in Health Care." *Journal of Advanced Nursing* 74 (2018) 1289–300.

Mudimbe, Valentin-Yves. *Tales of Faith: Religion as Political Performance in Central Africa.* London: Athlone, 1997.

Mulago, Vincent. "Die lebensnotwendige Teilhabe: Bantu Strukturprinzipien der Gemeinschaft." In *Theologie und Kirche in Afrika*, edited by H. Bürkle, 66–72. Stuttgart: n.p., 1968.

———. "L'union vitale chez les Bashi, les Banyarwanda et les Barundi face à l'unité vitale ecclésiale." Thèse de doctorat en Théologie. Rome: Universalité de la Propagande, 1955.

Munyaka, Mluleki, and Mokgethi Motlhabi. "Ubuntu and Its Socio-Moral Significance." In *African Ethics: An Anthology of Comparative and Applied Ethics*, edited by F. M. Murove, 324–31. Pietermaritzburg: University of KwaZulu-Natal Press, 2009.

Muraya, MarthaW. "Mau Mau War, Female Circumcision and Social-Cultural Identity among the Agikuyu of Kiambu, Kenya." *International Journal of Culture and History* 2 (2015) 26–45.

Murphy, John D. *Luganda-English Dictionary.* Washington, DC: The Catholic University of American Press, 1972.

Museveni, Yoweri K. "AIDS and Its Impact on the Health, Social and Economic Infrastructure in Developing Countries." In *Science Challenging AIDS*, edited by G. B. Rossi et al., xi–xvi. Naples: Karger, 1993.

Nagel, Thomas. "The Problem of Global Justice." *Philosophy and Public Affairs* 33 (2005) 113–47.

Nair-Collins, Michael. "Laying Futility to Rest." *The Journal of Medicine and Philosophy* 40 (2015) 554–83.

Namara-Lugolobi, Emily, et al. "Twenty Years of Prevention of Mother to Child HIV Transmission: Research to Implementation at a National Referral Hospital in Uganda." *African Health Sciences* 22 (2022) 22–33.

Namuunda, Mutombo, et al. "Male Circumcision and HIV Infection among Sexually Active Men in Malawi." *BMC Public Health* 15 (2015). doi: 10.1186/s12889-015-2384-z.

Nason, C. S. "Proverbs of the Baganda." *Uganda Journal* 3 (1936) 247–58.

National Institutes of Health. "National Institutes of Health Guidelines for Human Stem Cell Research." *The Federal Register / Federal Information and News Dispatch* 74 (2009) 32170–75.

Nchangwi, Munung S., et al. "Genomics Governance: Advancing Justice, Fairness and Equity through the Lens of the African Communitarian Ethic of Ubuntu." *Medicine, Health Care, and Philosophy* 24 (2021) 377–88.

Ndebele, Paul, et al. "History of Research Ethics in Africa." *Research Ethics in Africa: A Resource for Research Ethics Committees*, edited by Paul Ndebele et al., 3–10. Stellenbosch: Sun Media, 2014.

———. "HIV/AIDS Reduces the Relevance of the Principle of Individual Medical Confidentiality among the Bantu People of Southern Africa." *Theoretical Medicine and Bioethics* 29 (2008) 331–40.

Ndombe, Cesar Mawanzi. *Das symbolische Denken als Schlüssel zum Verständnis der negro-afrikanischen (Bantu-) Weltanschauung: Eine religionsphilosophische Deutung im Anschluss an die Kulturphilosophie Ernst Cassirers*. Frankfurt am Main: Peter Lang, 2008.

New American Bible (Revised Edition). Confraternity of Christian Doctrine, 2010.

Ngu, Victor Anomah. "Vaccines for the HIV: Past Efforts and Future Prospects." Paper presented to the Cameroon Academy of Sciences, February 12, 1999.

Ngu, Victor Anomah, and F. A. Ambe. "Effective Vaccines against and Immunotherapy of the HIV: A Preliminary Report." *Journal of the Cameroon Academy of Sciences* 1 (2001) 2–8.

Ngu, Victor Anomah, and Tangwa B. Godfrey. "Effective Vaccine against and Immunotherapy of the HIV: Scientific Report and Ethical Considerations from Cameroon." *Journal of the Cameroon Academy of Sciences* 12 (2015) 77–84.

Nicolaides, Angelo. "Gender Equity, Ethics and Feminism: Assumptions of an African Ubuntu Oriented Society." *Journal of Social Sciences* 42 (2015) 191–210.

Niekerk, Van Jaci, and Rachel Wynberg. "Human Food Trial of a Transgenic Fruit." In *Ethics Dumping*, edited by D. Schroeder et al., 91–98. SpringerBriefs in Research and Innovation Governance. Berlin: Springer, 2018. https://doi.org/10.1007/978-3-319-64731-9_11.

Nili, Shmue. "Our Problem of Global Justice." *Social Theory and Practice* 37 (2011) 629–53.

Nkemnkia, Martin Nkafu. *African Vitalogy: A Step Forward in African Thinking*. Nairobi: Paulines Publications Africa, 1999.

Nnabagereka, Development Foundation, and Jimmy Spire Ssentongo. *Obuntubulamu: The Moral Concept and Way of Life*. Kampala: Uganda Martyrs University Press, 2022.

Nsimbi, Michael B. *Amaanya Amaganda n'Ennono zaago*. Nairobi: East African Literature Bureau, 1956.

———. *Olulimi Oluganda* (Ganda language). Kampala: Eagle, 1955.

Ntabaza, Kagaragu. *Emigani Bali Bantu: Proverbs et maximes des Bashi*. 4th ed. Bukavu: Libreza, 1984.

Nussbaum, Marth C. *Frontiers of Justice: Disability, Nationality, Species Membership*. Cambridge, MA: Harvard University Press, 2006.

Nwankwo, C. K., et al. "Attitudes of Cancer Patients in a University Teaching Hospital in Southeast Nigeria on Disclosure of Cancer Information." *Psycho-Oncology* 22/8 (2013) 1829–33.

Nyamiti, Charles. "The Incarnation Viewed from the African Understanding of Person." *African Christian Studies* 6 (1990) 3–27.

Nyamnjoh, Anye-Nkwenti, et al. "The Contribution of Theories of Personhood in the Revaluation of Children in African Societies." *Current Sociology* 70 (2021) doi:10.1177/00113921209852869.

Nyerere, Julius Kambarage. *Freedom and Unity/Uhuru na Umoja: A Selection from Writings and Speeches 1952–65*. Dar es Salaamm: Oxford University Press, 1966.

————. *Ujamaa: Essays on Socialism*. Oxford University Press, 1968.

Nyika, Aceme. "The Ethics of Improving African Traditional Medical Practice: Scientific or African Traditional Research Methods?" *Acta Tropica* 112S (2009) S32–36.

Odozor, Paulinus Ikechukwu. "An African Moral Theology of Inculturation: Methodological Considerations." *Theological Studies* 69 (2008) 583–609.

Ohm, Thomas. "Stammesreligionen im sündlichen Tanganyika-Territorium." *Reihe: Arbeitsgemeinschaft für Forschung des Landes Nordrhein-Westfalen* 5/5, edited by L. Brandt. Köln, 1953.

O'Neill, Onora. "Some Limits of Informed Consent." *Journal of Medical Ethics* 29 (2003) 4–7.

Onwubiko, Alozie Oliver. "Re-Encountering African Culture in Living Christianity in My Father's Home." *An African Journal of Philosophy/ Revue Africaine de Philosophie* 19 (2005) 91–108.

Orobator, Agbonkhianmeghe Emmanuel. *From Crisis to Kairos: The Mission of the Church in the Time of HIV/AIDS-Refugees and Poverty*. Nairobi: Pauline Publications Africa, 2005.

————. "Ethics of HIV/AIDS Prevention: Paradigms of a New Discourse from an African Perspective." In *Applied Ethics in a World Church*, edited by Linda Hogan, 147–54. Maryknoll, NY: Orbis, 2008.

————. *Religion and Faith in Africa: Confessions of an Animist*. Maryknoll, NY: Orbis, 2018.

————. *Theology Brewed in an African Pot: An Introduction to Christian Doctrine from an African Perspective*. Nairobi: Paulines Publications Africa, 2008.

Oruka, Henry O, and Jump Calestous. "Ecophilosophy and Parental Earth Ethics (On the Complex Web of Being)." In *Philosophy, Humanity and Ecology*, edited by Henry O. Oruka, 115–29. Nairobi: ACTS Press, 1994.

Oyowe, Oritsegbubemi A. "Personhood and Strong Normative Constraints." *Philosophy East and West* 68/3 (2018) 783–801.

Oyowe, Oritsegbubemi A., and Olga Yurkivska. "Can a Communitarian Concept of African Personhood Be both Relational and Gender-Neutral?" *South African Journal of Philosophy* 33/1 (2014) 85–99.

Pagden, Anthony. "Human Rights, Natural Rights, and Europe's Imperial Legacy." *Political Theory* 31/2 (2003) 171–99.

Panikkar, Raimon. "Śatapathaprajñâ: Should We speak of Philosophy in Classical India? A Case of Homeomorphic Equivalents." In *Contemporary philosophy: A new survey*, edited by Guttorm Fløistad, 11–67. La philosophie contemporaine, Chroniques nouvelles. Netherlands: Kluwer Academic, 1993.

Pereira, J. "Legalizing Euthanasia or Assisted Suicide: The Illusion of Safeguards and Controls." *Current Oncology* 18 (2011) e38–45.

Perrett, W. Roy. "Killing, Letting Die and the Bare Difference Argument." *Bioethics* 10 (1996) 131–39.

Persad, Govind, et al. "Principles for Allocation of Scarce Medical Interventions." *Lancet* 373 (2009) 423–31.

Peterson, Trudy Huskamp. "The Universal Declaration of Human Rights: An Archival Commentary." *Comma* 2020/1–2 (August 2021) 33–86. https://doi.org/10.3828/comma.2020.4.

Phillip, Rieder, et al. "The End of Medical confidentiality? Patients, Physicians, and the State in History." *Medical Humanities* 42 (2016) 149–54.

Pimm, Stuart L., and Robert Leo Smith. "Ecology." *Encyclopedia Britannica*, 2019. https://www.britannica.com/science/ecology.

Powers, Madison, and Faden Ruth. *Social Justice: The Moral Foundation of Public Health and Health Policy*. Oxford: Oxford University Press, 2006. See especially chapters 1 and 4–7.

Powner, David J., and Ira M. Bernstein. "Extended Somatic Support for Pregnant Women after Brain Death." *Critical care medicine* 31 (2003) 1241–49.

Praeg, Leonhard. *A Report on Ubuntu*. Pietermaritzburg: University of KwaZulu-Natal Press, 2014.

Presbey, Gail M. "Maasai Concepts of Personhood: The Roles of Recognition, Community, and Individuality." *International Studies in Philosophy* 34 (2002) 57–82.

Quill, Timothy E., and Cristine K. Cassel. "Nonabandonment: A Central Obligation for Physicians." *Annals of Internal Medicine* 122 (1995) 368–74.

Ramose, Magobe. *African Philosophy through Ubuntu*. Harare: Mond, 1999.

———. "Towards Emancipative Politics in Africa." In *African Ethics: An Anthology of Comparative and Applied Ethics*, edited by F. Murove, 107–29. Pietermaritzburg: University of KwaZulu-Natal Press, 2009.

Rasita, Vinay, et al. "Ethics of ICU Triage during COVID–19." *British Medical Bulletin* 138 (2021) 5–15.

Rawls, John. *Justice as Fairness: A Restatement*. Edited by Kelly Erin. Cambridge, MA: Belknap Press of Harvard University Press, 2001.

Reiss, Thomas H., et al. "'When I Was Circumcised, I Was Taught Certain Things': Risk Compensation and Protective Sexual Behavior among Circumcised Men in Kisumu, Kenya." *PLoS One* 5 (2010). doi: 10.1371/journal.pone.0012366.

Republic of Uganda. The Anti-Homosexuality Act. 2023.

Rivlin, Michael M. "Why the Fair Innings Argument Is Not Persuasive." *BMC Medical Ethics* (2000). doi: 10.1186/1472–6939-1–1.

Rodriguez-Osorio, Carlos A., and Guillermo Dominguez-Cherit. "Medical Decision Making: Paternalism versus Patient-Centered (Autonomous) Care." *Current Opinion in Critical Care* 4 (2008) 708–13.

Roscoe, John. *The Baganda: An Account of Their Native Customs and Beliefs*. 2nd ed. London: Frank Cass, 1965.

Ruch, Ernest Albert, and K. C. Anyanwu. *African Philosophy: An Introduction to the Main Philosophical Trends in Contemporary Africa*. Rome: Catholic Book Agency, 1981. Published online by Cambridge University Press, December 2011.

Russell, Margaret L., et al. "Paying Research Subjects: Participants' Perspectives." *Journal of Medical Ethics* 26 (2000) 126–30.

Rwiza, Richard N. *Environmental Ethics in the African Context*. Nairobi: CUEA Press, 2021.

Safurdeen, Salim Abdool Karim, and Cheryl Baxter. "New Prevention Strategies under Development and Investigation." In *HIV/AIDS in South Africa*, edited by Abdool Karim Salim Safurdeen and Abdool Karim Quarraisha, 268–82. Cape Town: Cambridge University Press, 2010.

Savulescu, Julian. "Genetically Modified Animals: Should There Be Limits to Engineering the Animal Kingdom?" In *The Oxford Handbook of Animal Ethics*, edited by Tom L. Beauchamp and R. G. Frey, 641–70. Oxford University Press, 2011.

———. "Is it in Charlie Gard's Best Interest to Die?" *The Lancet* (British ed.) 389 (2017) 1868–69.

Schneiderman, Lawrence J. "Defining Medical Futility and Improving Medical Care." *Journal of Bioethical Inquiry* 8 (2011) 123–31.

———. "Medical futility: Its Meaning and Ethical Implications." *Annals of Internal Medicine* 112 (1990) 949–54.

———. "Medical futility: Response to Critiques." *Annals of Internal Medicine* 125 (1996) 669–74.

Schneiderman, Lawrence J., et al. The Abuse of Futility." *Perspectives in Biology and Medicine* 60 (2017) 295–313.

Schoeman, Marelize I. "Research Ethics Governance—An African Perspective." In *Social Science Research Ethics in Africa*, edited by N. Nortjé et al., 1–15. Research Ethics Forum 7. Switzerland: Springer Nature, 2019.

Schroeder, Doris, et al. "Ethics Dumping: Introduction." In *Ethics Dumping*, edited by D. Schroeder et al., 1–8. SpringerBriefs in Research and Innovation Governance. Berlin: Springer, 2018. https://doi.org/10.1007/978-3-319-64731-9_1.

Sebunnya, Gerald M. "Beyond the Sterility of a Distinct African Bioethics: Addressing the Conceptual Bioethics Lag in Africa." *Developing World Bioethics* 17 (2017) 22–31.

Segalo, Puleng, and Lien Molobela. "Considering Africanist Research Ethics Practices in Social Science Research in Africa." *Social Science Research Ethics in Africa*, edited by N. Nortjé et al., 35–46. Research Ethics Forum 7. Cham: Springer, 2019.

Sekadde, Ssaalongo Y., and Yosamu Semugoma. *Ndimugezi (I Am Wise): Kitabo kya Ngero za Luganda* (Ganda proverbs). London: Macmillan, 1952.

Sempebwa, Joshau Wantate. *The Ontological and Normative Structure in the Social Reality of a Bantu Society: A Systematic Study of Ganda Ontology and Ethics*. Heidelberg: Ruprecht-Karl-Universität, 1978.

Sen, Amartya. "Capability and Well-Being. In *The Quality of Life*, edited by Martha Nussbaum and Amartya Sen. Oxford: Oxford University Press, 2003. doi: 10.1093/0198287976.001.0001.

Senghor, Léopold Sédar. "Der Geist der negro-afrikanischen Kultur." In *Dunkle Stimmen: Schwarzer Orpheus, Schwarze Ballade*, edited by Jahn Janheinz, 343–63. Köln/Düsseldorf: Eugen Diederichs, 1957. From "Présence Africaine" VIII–X, Paris, 1956.

Sharalaya, Zarina. "Racial Disparities in Cardiovascular Care: A Review of Culprits and Potential Solutions." *Journal of Racial and Ethnic Health Disparities* 1 (2014) 171–80.

Sindima, Harvey. *Religious and Political Ethics in Africa: A Moral Inquiry*. Westport, CT: Greenwood, 1998.

Singer, Peter. "Famine, Affluence, and Morality." *Philosophy and Public Affairs* 1 (1972) 229–43.

Singer, Peter, and Jim Mason. *The Ethics of What We Eat: Why Our Food Choices Matter*. Emmaus, PA: Rodale, 2006.

Slutkin, Gary, et al. "How Uganda Reversed Its HIV Epidemic." *AIDS and Behavior* 10 (2006) 351–60.

Smedira, G. Nicholas. "Allocating hearts." *The Journal of Thoracic and Cardiovascular Surgery* 131 (2006) 775–76.

Smith, Edwin William. *African Ideas of God*. London: Edinburgh House, 1950.

Smith, Stephen W. "Evidence for the Practical Slippery Slope in the Debate on Physician-Assisted Suicide and Euthanasia." *Medical Law Review* 13 (2005) 17–44.

Sørensen, Georg. "Globalization, Values and Global Governance." In *Universal Ethics-Perspectives and Proposals from Scandinavian Scholars*, edited by Göran Bexell Dan-Erik Andersson, 143–48. The Hague: Martinus Nijhoff, 2002.

Soubrier, Jean-Pierre. "Self-Crash Murder-Suicide." *Crisis: The Journal of Crisis Intervention and Suicide Prevention* 37 (2016) 399–401.

Sriwilaijaroen, Nongluk, and Yasuo Suzuki. "Sialoglycovirology of Lectins: Sialyl Glycan Binding of Enveloped and Non-Enveloped Viruses." *Methods in Molecular Biology* 2132 (2020) 483–45.

Stacey, Dawn, et al. "Decision Aids for People Facing Health Treatment or Screening Decisions." *Cochrane Database of Systematic Reviews* 4 (2017). https://doi.org/10.1002/14651858.CD001431.pub5.

Statista. "Number of Living Languages in Africa as of 2022, by Bountry." July 18, 2023. https://www.statista.com/statistics/1280625/number-of-living-languages-in-africa-by-country/.

Staunton, Ciara, and Keymanthri Moodley. "Challenges in Biobank Governance in Sub-Saharan Africa." *BMC Medical Ethics* 14 (2013). http://www.biomedcentral.com/1472-6939/14/35).

Steinberg, Jacob, and Caroline Davies. "Kurt Zouma Fined £250,000 and Cats Taken by RSPCA after Video Emerges." *The Guardian*, February 9, 2022.

Stoljar, Natalie. "Informed Consent and Relational Conceptions of Autonomy." *The Journal of Medicine and Philosophy* 36 (2011) 375–84.

Sulmasy, D. P. "Futility and the Varieties of Medical Judgment." *Theoretical Medicine* 18 (1997) 63–78.

Sunstein, Cass R., and Thaler H. Richard. "Libertarian Paternalism Is Not an Oxymoron." *The University of Chicago Law Review* 70 (2003) 1159–202.

Taber, David J. "Twenty Years of Evolving Trends in Racial Disparities for adult Kidney Transplant Recipients." *Kidney International* 90 (2016) 878–87.

Táíwò, Olúfémi. *How Colonialism Preempted Modernity in Africa*. Bloomington: Indiana University Press, 2010.

Tangwa, Godfrey B. *African Perspectives on some Contemporary Bioethics Problems*. Cambridge, MA: Scholars, 2019.

———. "ART and African Sociocultural Practices: Worldview, Belief and Value Systems with Particular Reference to Francophone Africa." *Current Practices and Controversies in Assisted Reproduction Report of WHO Meeting on "Medical, Ethical and Social Aspects of Assisted Reproduction" Held at WHO Headquarters in Geneva, Switzerland 17–21 September 2001*, edited by Effy Vayena et al., 55–59. Geneva: World Health Organization, 2002.

———. "Between Universalism and Relativism: A Conceptual Exploration of Problems in Formulating and Applying International Biomedical Ethical Guidelines." *Journal of Medical Ethics* 30 (2004) 63–67.

———. "Bioethics: An African Perspective." *Bioethics* 10 (1996) 183–200.

———. "Bioethics and Ubuntu: The Transformative Global Potential of an African Concept." *The Tenacity of Truthfulness: Philosophical Essays in Honor of Mogobe Bernard Ramose*, edited by H. Lauer and H. Yitah, 239–49. Pretoria: EARS, 2019.

———. "Bioethics, Biotechnology and Culture: A Voice from the Margins." *Developing World Bioethics* 4 (2004) 125–38.

———. "Cameroon." In *Handbook of Global Bioethics*, edited by H. Ten Have and B. Gordijn, 941–58. Dordrecht: Springer Science and Business, 2014.

———. "Capacity Building in Health Research Ethics in Central Africa: Key Players, Current Situation and Recommendations." *Bioethica Forum* 6 (2013) 4–11.

———. "Colonialism and Linguistic Dilemmas in Africa: Cameroon as Paradigm." *Quest* 13 (1999) 3–17.

———. *Elements of African Bioethics in a Western Frame*. Mankon, Bamenda: Langaa Research and Publishing, 2010.

———. "Ethical Issues Concerning GM Crops, Foods and Feeds." *Journal of the Cameroon Academy of Sciences* 6 (2006) 69–73.

———. "Ethical Principles in Health Research and Review Process." *Acta Tropica* 112S (2009) S2–7.

———. "Ethics Committees: Moral Agency, Moral Worth and the Question of Double Standards in Medical Research in Developing Countries." *Developing World Bioethics* 1 (2001) 156–62.

———. "Giving Voice to African Thought in Medical Research Ethics." *Theoretical Medicine and Bioethics* 38 (2017) 101–110.

———. "Globalization or Westernization? Ethical Concerns in the Whole Bio-Business," *Bioethics* 13 (1999) 218–26.

———. "How Not to Compare Western Scientific Medicine with African Traditional Medicine." *Developing World Bioethics* 7 (2007) 41–44.

———. "International Regulations and Medical Research in Developing Countries: Double Standards or Differing Standards?" *Notizie di Politeia* 18 (2002) 46–50.

———. "Leaders in Ethics Education: Godfrey B. Tangwa." *International Journal of Ethics Education* 1 (2016) 91–105.

———. "Moral Status of Embryonic Stem Cells: Perspective of an African Villager." *Bioethics* 21 (2007) 449–57.

———. "Morality and Culture: Are Ethics Culture-Dependent?" *Bioethics in a Small World*, edited by F. Thiele and R. E. Ashcroft, 17–21. Heidelberg: Springer Berlin, 2005.

———. "Some African Reflections on Biomedical and Environmental Ethics." In *A Companion to African Philosophy*, edited by Kwasi Wiredu, 387–95. Oxford: Blackwell, 2004.

———. "The Structure and Function of Research Ethics Committees in Africa: A Case Study." *PLoS Medicine* 4 (2007) 0026–31.

———. "Third Party Assisted Conception: An African Perspective." *Theoretical Medicine and Bioethics* 29 (2008) 297–306.

———. "The Traditional African Perception of a Person: Some Implications for Bioethics." *The Hastings Center Report* 30 (2000) 39–43.

———. "Vulnerable Human Beings." *Acta Tropica* 112S (2009) S16–20.

Tangwa, Godfrey B., et al. "Addressing Ethical Issues in H3Africa research—the Views of Research Ethics Committee Members." *The HUGO Journal* 9 (2015) 3. doi:10.1186/s11568-015-0006-6.

———. "Capacity Building of Ethics Review Committees across Africa Based on the Results of a Comprehensive Needs Assessment Survey." *Developing World Bioethics* 9 (2009) 149–56.

———. "Composition, Training Needs and Independence of Ethics Review Committees across Africa: Are the Gatekeepers Rising to the Emerging Challenges?" *Journal of Medical Ethics* 35 (2009) 189–93.

———. "Ebola Vaccine Trials." *Ethics Dumping*, edited by Schroeder et al., 49–59. SpringerBriefs in Research and Innovation Governance. Switzerland: Springer Nature, 2018. https://doi.org/10.1007/978-3-319-64731-9_6.

———. "Ethics in Occupational Health: Deliberations of an International Workgroup Addressing Challenges in an African context." *BMC Medical Ethics* 15 (2014) 9. doi:10.1186/1472-6939-15-48.

———. "A Framework for Tiered Informed Consent for Health Genomic Research in Africa." *Nature Genetics* (2019). https://doi.org/10.1038/s41588-019-0520-x (accessed 11 November 2022).

———. "Global Health Inequalities and the Need for Solidarity: A View from the Global South." *Developing World Bioethics* 18 (2018) 241–49.

———. "A New Tuskegee? Unethical Muman Experimentation and Western Neocolonialism in the Mass Circumcision of African Men." *Developing World Bioethics* (2020) 1–16.

———. "Perspectives of Different Stakeholders on Data Use and Management in Public Health Emergencies in Sub-Saharan Africa: A Meeting Report." *Wellcome Open Research* 6 (2021). https://doi.org/10.12688/wellcomeopenres.16494.1.

———. "Small Is Beautiful: Demystifying and Simplifying Standard Operating Procedures: A Model from the Ethics Review and Consultancy Committee of the Cameroon Bioethics Initiative." *BMC Medical Ethics* 17 (2016). doi:10.1186/s12910-016-0110-8.

Tangwa, Godfrey B., and Nchangwi S. Munung. "COVID–19: Africa's Relation with Epidemics and Some Imperative Ethics Considerations of the Moment." *Research Ethics and COVID–19* 16 (2020) 1–11.

———. "Sprinting Research and Spot Jogging Regulation: The State of Bioethics in Cameroon." *Cambridge Quarterly of Healthcare Ethics* 20 (2011) 356–66.

Tangwa, Godfrey B., and Henk J. Van Rinsum. "Colony of Genes, Genes of the Colony: Diversity, Difference and Divide." *Third World Quarterly* 25 (2004) 1031–43.

Tasioulas, John, et al. "The place of human rights and the common good in global health policy." *Theoretical Medicine and Bioethics* 37 (2016) 365–82.

Taylor, Charles W. "Atomism." In *Powers, Possessions and Freedom: Essays in Honour of C.B. Macpherson,* edited by C. B. Macpherson and Alkis Kontos, 39–62. Toronto: University of Toronto Press, Scholarly Publishing Division, 1979.

Tempels, Frans Placid. *Bantu-Philosophie: Ontologie und Ethik.* Heidelberg: Rothe, 1956. (French ed.: *La philosophie bantoue.* Paris: Présence Africaine, 1945; English: *Bantu Philosophy.* Paris: Présence Africaine, 1959).

———. *Bantu Philosophie.* Heidelberg: Wolfgang Rothe, 1956.

Thiagarajan, Ravi R., et al. "ECMO to Aid Cardiopulmonary Resuscitation in Infants and Children." *Circulation* 116 (2007) 1693–1700.

Thornton, V. "Who Gets the Liver Transplant? The Use of Responsibility as the Tie Breaker." *Journal of Medical Ethics* 35 (2009) 739–42.

Tindana, Paulina, et al. "Engaging Research Ethics Committees to Develop an Ethics and Governance Framework for Best Practices in Genomic Research and Biobanking in Africa: The H3Africa Model." *BMC Medical Ethics* 20 (2019). doi:10.1186/s12910-019-0398-2.

Tönnies, Ferdinand. *Gemeinschaft und Gesellschaft: Abhandlung des Communismus und des Sozialismus als empirischer Kulturformen.* Leipzig: Fues's, 1887.

Traphagan, W. John. *Rethinking Autonomy: A Critique of Principlism in Biomedical Ethics.* Albany: State University of New York Press, 2013.

Treadwel, Marsha J., et al. "Stakeholder Perspectives on Public Health Genomics Applications for Sickle Cell Disease: A Methodology for a Human Heredity and Health in Africa (H3Africa) Qualitative Research Study." *OMICS: A Journal of Integrative Biology* 21 (2017) 323–32.

Trials of War Criminals before the Nuremberg Military Tribunals under Control Council Law No. 10. Nuremberg, October 1946–April 1949. Washington, DC: US Government Printing Office, 1949–1953.

Trianosky, Gregory W. "Supererogation, Wrongdoing, and Vice." *The Journal of Philosophy* 83(1986) 26–40.

Truog, Robert D. "Is 'Best Interests' the Right Standard in Cases Like That of Charlie Gard?" *Journal of Medical Ethics* 46 (2020) 16–17.

———. "Triage in the ICU." *The Hastings Center Report* 22 (1992) 13–17.

Tutu, Desmond. *No Future Without Forgiveness: A Personal Overview of South Africa's Truth and Reconciliation Commission.* New ed. London: Rider, 2000.

Ubel, Peter A., et al. "Allocation of Transplantable Organs: Do People Want to Punish Patients for Causing Their Illness?" *Liver Transplantation* 7 (2001) 600–607.

UNAIDS. "Good Participatory Practices: Guidelines for Biomedical HIV Prevention Trails." 2nd ed. JC1853E. Geneva: UNAIDS, June 2011.

UNESCO. "Bioethics Core Curriculum, Section 1: Syllabus Ethics Education Programme." 2008. http://unesdoc.unesco.org/images/0016/001636/163613e.pdf.

———. "Records of the General Conference, 33rd Session, Paris, 3–21 October 2005." http://unesdoc.unesco.org/images/0014/001428/142825e.pdf.

———. "Universal Declaration on Bioethics and Human Rights." 2005. http://unesdoc.unesco.org/images/0014/001461/146180e.pdf.

UNESCO Chair in Bioethics. "Casebook on Bioethics for Judges." Israel National Commission for UNESCO. Haifa, 2016. http://www.unesco-chair-bioethics.org/?mbt_book-casebook-on-bioethics-for-judges.

United Nations General Assembly. "International Day of Human Fraternity." UN Resolution 75/200, adopted December 21, 2020. https://documents.un.org/doc/undoc/gen/n20/377/94/pdf/n2037794.pdf?token=tUQmYsopOl6CSFFc63&fe=true.

———. "Universal Declaration of Human Rights." UN Resolution 3/217, adopted December 10, 1948. https://www.ohchr.org/en/human-rights/universal-declaration/translations/english.

Upton, Rebecca L. "Illness and Healing: Africanist Anthropology." In *A Companion to the Anthropology of Africa*, edited by R. R. Grinker et al., 97–117. New York: Wiley, 2019.

Visagie, Retha, et al. "Informed Consent in Africa-Integrating Individual and Collective Autonomy." In *Social Science Research Ethics in Africa*, edited by N. Nortjé et al., 165–79. Research Ethics Forum 7. Cham: Springer, 2019.

Vreeman, Rachel, et al. "A Qualitative Study Using Traditional Community Assemblies to Investigate Community Perspectives on Informed Consent and Research Participation in Western Kenya." *BMC Medical Ethics* 13 (2012). doi:10.1186/1472-6939-13-23.

Walter, Jennifer K., and Lainie Friedman Ross. "Relational Autonomy: Moving beyond the Limits of Isolated Individualism." *Pediatrics* 133 (2014) S16–23.

Walter, Kerber, ed. *Das Absolute in der Ethik.* Munich: Kindt, 1991.

Warner, Teddy D., and Gluck John Paul. "What Do We Really Know about Conflicts of Interest in Biomedical Research?" *Psychopharmacology* 171 (2003) 36–46.

Washington, Sylvia Bâ. *The Concept of Negritude in the Poetry of Leopold Sedar Senghor.* Princeton, NJ: Princeton University Press, 1973.

Wasunna, Angela Amondi, et al. "The Discourses of Bioethics in Sub-Saharan Africa." In *The Cambridge World History of Medical Ethics*, edited by Robert Baker and Laurence B. McCullough, 525–30. Cambridge: Cambridge University Press, 2008.

Wasunna, Christine, et al. "Informed Consent in an African Context." *Research Ethics in Africa: A Resource for Research Ethics Committees*, edited by Paul Ndebele et al., 57–62. Stellenbosch: Sun Media, 2014.

Wathuta, Jane. "A Look at Uganda's Early HIV Prevention Strategies through a Moderate 'African' Communitarian Lens." *Developing World Bioethics* 18 (2018) 109–18.

Wendler, David, and Annette Rid. "Systematic Review: The Effect on Surrogates of Making Treatment Decisions for Others." *Annals of Internal Medicine* 54 (2011) 336–46.

White, Ben, et al. "What Does "Futility" Mean? An Empirical Study of Doctors' Perceptions." *The Medical journal of Australia* 204 (2016) 318. doi:10.5694/mja15.01103.

WHO. "Traditional Male Circumcision among Young People: A Public Health Perspective in the Context of HIV Prevention." Geneva: WHO, 2009.

Wilkinson, Dominic J. C., and Julian Savulescu. "Knowing When to Stop: Futility in the ICU." *Current Opinion in Anaesthesiology* 24 (2011) 160–65.

Wilkinson, Martin, and Andrew Moore. "Inducement in Research." *Bioethics* 11(1997) 373–89.

Williams, Alan. "Intergenerational Equity: An Exploration of the 'Fair Innings' Argument." *Health Economics* 6 (1997) 117–32.

Willoughby, W. C. *The Soul of Bantu: A Sympathetic Study of the Magico-Religious Practices and Beliefs of the Bantu Tribes of Africa.* London: SCM, 1928.

Wilton, David. *Word Myths: Debunking Linguistic Urban Legends.* Oxford: Oxford University Press, 2004.

Wiredu, Kwasi. *Cultural Universals and Particulars: An African Perspective.* Indianapolis: Indiana University Press, 1996.

———. "Introduction: African Philosophy in Our Time." In *Companion to African Philosophy*, edited by Kwasi Wiredu, 1–27. Oxford: Blackwell, 2004.

———. "Moral Foundations of an African Culture." In *Person and Community: Ghanaian Philosophical Studies*, edited by Wiredu and K. Gyekye, 192–206. Washington, DC: Council for Research in Values and Philosophy, 1992.

———. "An Oral Philosophy of Personhood: Comments on Philosophy and Orality." *Research in African Literatures* 40/1 (2009) 8–18.

———. "Social Philosophy in Postcolonial Africa: Some Preliminaries Concerning Communalism and Communitarianism." *South African Journal of Philosophy* 27 (2008) 332–39.

Wittgenstein, Ludwig. *Philosophical Investigations.* Translated by G. E. M. Anscombe. Oxford: Blackwell, 1953.

WMA. "Declaration of Helsinki." *Bulletin of the World Health Organisation* 79 (2001) 373–74.

———. 64th General Assembly. Fortaleza, Brazil, October 2013. https://www.wma.net/policies-post/wma-declaration-of-helsinki-ethical-principles-for-medical-research-involving-human-subjects/. Emphasis is on Art. 23.

Wolinsky, Howard. "Bioethics for the World." *The European Molecular Biology Organization Reports* 7 (2006) 354–58.

Wynberg, Rachel, Doris Schroeder, and Roger Chennells, eds. *Indigenous Peoples, Consent and Benefit Sharing: Lessons from the San-Hoodia Case.* Berlin: Springer, 2009. doi:10.1007/978-90-481-3123-5.

Yauri, Bala Muhammad, and Ahmed Awaisu. "The Need for Enhancement of Research, Development, and Commercialization of Natural Medicinal Products in Nigeria: Lessons from the Malaysian Experience." *African Journal of Traditional, Complementary, and Alternative Medicines* 5 (2008) 120–30.

Yotsukura, Sayo. "Ethnolinguistic Introduction to Japanese Literature." In *Language and Thought: Anthropological Issues,* edited by W. McCormack and Stephen Wurm, 261–70. The Hague/Paris: Mouton, 1977.

www.ingramcontent.com/pod-product-compliance
Lightning Source LLC
Chambersburg PA
CBHW060329100426

42812CB00003B/925